AF356603

Digital Workflows and Material Sciences in Dental Medicine

Digital Workflows and Material Sciences in Dental Medicine

Editor

Tim Joda

MDPI • Basel • Beijing • Wuhan • Barcelona • Belgrade • Manchester • Tokyo • Cluj • Tianjin

Editor
Tim Joda
Reconstructive Dentistry
University Center for Dental
Medicine Basel (UZB),
University of Basel
Basel
Switzerland

Editorial Office
MDPI
St. Alban-Anlage 66
4052 Basel, Switzerland

This is a reprint of articles from the Special Issue published online in the open access journal *Journal of Clinical Medicine* (ISSN 2077-0383) (available at: www.mdpi.com/journal/jcm/special_issues/Digital_Workflows).

For citation purposes, cite each article independently as indicated on the article page online and as indicated below:

LastName, A.A.; LastName, B.B.; LastName, C.C. Article Title. *Journal Name* **Year**, *Volume Number*, Page Range.

ISBN 978-3-0365-2587-7 (Hbk)
ISBN 978-3-0365-2586-0 (PDF)

Contents

About the Editor

Tim Joda

Professor Dr. Joda is Vice Chair, Head of Dental Technology & Digital Dental Solutions, and Director of the Postgraduate Education Program at the Department of Reconstructive Dentistry, University Center for Dental Medicine Basel (UZB), Switzerland. He is Principle Investigator and actively participates in clinical and translational research related to implant workflows and prosthetic treatment concepts in the field of digital dental technologies and e-health data management using AI and ML.

Preface to "Digital Workflows and Material Sciences in Dental Medicine"

The trend of digitalization is an omnipresent phenomenon nowadays –in social life and in the dental community. Advancement in digital technology has fostered research into new dental materials for the use of these workflows, particularly in the field of prosthodontics and oral implantology.

CAD/CAM-technology has been the game changer for the production of tooth-borne and implant-supported (monolithic) reconstructions: from optical scanning, to on-screen designing, and rapid prototyping using milling or 3D-printing. In this context, the continuous development and speedy progress in digital workflows and dental materials ensure new opportunities in dentistry.

The objective of this Special Issue is to provide an update on the current knowledge with state-of-the-art theory and practical information on digital workflows to determine the uptake of technological innovations in dental materials science. In addition, emphasis is placed on identifying future research needs to manage the continuous increase in digitalization in combination with dental materials and to accomplish their clinical translation.

This Special Issue welcomes all types of studies and reviews considering the perspectives of the various stakeholders with regard to digital dentistry and dental materials.

Tim Joda
Editor

Journal of
Clinical Medicine

Review

Dental Caries Diagnosis and Detection Using Neural Networks: A Systematic Review

María Prados-Privado [1,2,3,*], Javier García Villalón [1], Carlos Hugo Martínez-Martínez [4], Carlos Ivorra [1] and Juan Carlos Prados-Frutos [3,5]

[1] Asisa Dental, Research Department, C/José Abascal, 32, 28003 Madrid, Spain; javier.villalon@asisadental.com (J.G.V.); carlos.ivorra@asisadental.com (C.I.)

[2] Department of Signal Theory and Communications, Higher Polytechnic School, Universidad de Alcala de Henares, Ctra, Madrid-Barcelona, Km. 33,600, 28805 Alcala de Henares, Spain

[3] IDIBO GROUP (Group of High-Performance Research, Development and Innovation in Dental Biomaterials of Rey Juan Carlos University), Avenida de Atenas s/n, 28922 Alcorcon, Spain; juancarlos.prados@urjc.es

[4] Faculty of Medicine, Universidad Complutense de Madrid, Plaza de Ramón y Cajal, s/n, 28040 Madrid, Spain; carlos.martinez@asisa.es

[5] Department of Medical Specialties and Public Health, Faculty of Health Sciences, Universidad Rey Juan Carlos, Avenida de Atenas, 28922 Alcorcon, Spain

* Correspondence: maria.prados@uah.es

Received: 29 September 2020; Accepted: 3 November 2020; Published: 6 November 2020

Abstract: Dental caries is the most prevalent dental disease worldwide, and neural networks and artificial intelligence are increasingly being used in the field of dentistry. This systematic review aims to identify the state of the art of neural networks in caries detection and diagnosis. A search was conducted in PubMed, Institute of Electrical and Electronics Engineers (IEEE) Xplore, and ScienceDirect. Data extraction was performed independently by two reviewers. The quality of the selected studies was assessed using the Cochrane Handbook tool. Thirteen studies were included. Most of the included studies employed periapical, near-infrared light transillumination, and bitewing radiography. The image databases ranged from 87 to 3000 images, with a mean of 669 images. Seven of the included studies labeled the dental caries in each image by experienced dentists. Not all of the studies detailed how caries was defined, and not all detailed the type of carious lesion detected. Each study included in this review used a different neural network and different outcome metrics. All this variability complicates the conclusions that can be made about the reliability or not of a neural network to detect and diagnose caries. A comparison between neural network and dentist results is also necessary.

Keywords: artificial intelligence; caries; images; detection

---

1. Introduction

Machine learning is an application of artificial intelligence (AI) that provides systems the ability to automatically learn and improve from experience without being explicitly programmed [1,2]. Machine learning needs input data, such as images or text, to obtain an output through a model.

Neural networks can be classified according to their typology or network structure or according to their learning algorithm. According to its topology, we can distinguish, as a characteristic of a network, the number of layers; the type of layers, which can be hidden or visible; input or output; and the directionality of the neuron connections. Depending on the typology, we can distinguish monolayer or multilayer networks.

According to its learning algorithm or how the network learns the patterns, we can distinguish as characteristics if it is supervised, unsupervised, competitive, or by reinforcement [3]. The model in supervised learning is trained, employing a labeled database. By contrast, the expected output is unknown in unsupervised learning [1,4]. Reinforcement learning is a model that falls between supervised and unsupervised learning.

The most common form of machine learning is supervised learning. To work with images, it is necessary, first, to collect a large data set of images, and second, to label each category in each image, in this case, with caries detected by a dentist. Then the training process begins. During the training process, the user/modeler feeds the data to network, it passes through the network, and an output is computed based on the current set of model weights. To obtain the best score of all categories, an objective function to measure the error is computed. Then the algorithm modifies its internal parameters to have the highest score of all categories. Finally, after training, the performance of the system is measured on a different set of images called a test dataset. The validation test serves to test the ability of the model to obtain good answers on new images (inputs) that it has never seen during the training process [5].

Convolutional neural network (CNN) is a type of deep and feedforward network. CNNs are designed to process data that come in the form of multiple arrays as images, and their architecture is composed of several stages.

Artificial intelligence is used in dentistry to identify and detect different variables from images, such as teeth, caries, and implants. Deep learning has been demonstrated to be a good collection of techniques to assist medical practitioners in medical fields such as radiology [6,7].

One of the most frequent activities in dental practice is to detect early caries lesions or to provide treatment preventing more invasive therapies [8]. The International Caries Detection and Assessment System (ICDAS) was developed by an international team of caries researchers to integrate several new criteria systems into one standard system for caries detection and assessment [9]. A workshop was organized to discuss and reach consensus about definitions of the most common terms in cariology [10]. Full agreement was obtained of the definition of dental caries:

> "Dental caries is a biofilm-mediated, diet-modulated, multifactorial, non-communicable, dynamic disease resulting in net mineral loss of dental hard tissues. It is determined by biological, behavioral, psychosocial, and environmental factors. As a consequence of this process, a caries lesion develops". [10]

The definition of initial caries lesions was also obtained with full agreement and was defined as a frequently used term for noncavited caries lesions that refer to the stage of severity. Sound enamel/dentin was defined, with a 100% agreement, as a tooth structure without clinically detectable alterations of the natural translucency, color, or texture. However, other terms, such as secondary caries/recurrent caries, residual caries, or "hidden" caries, did not obtain full agreement [10]. During the first two sessions, ICDAS was formed and, afterwards, those criteria were revised, modified, and called ICDAS II [11].

The visual-tactile detection method is generally used in dental practice, followed by radiographic caries detection [8]. Bitewing radiography is the most frequent technique in carious lesion detection. Several studies have compared the performance of CBCT to conventional or digital intraoral radiography, histology, or micro CT for enamel and dentin caries detection [12,13]. The conclusion was that CBCT did not improve the accuracy of caries detection [14].

Dental caries is the most prevalent dental disease worldwide, and neural networks and artificial intelligence are increasingly being used in the field of dentistry. Many studies have contributed to the field of dental caries detection using neural networks with different dental images. This review aims to evaluate studies investigating caries detection with artificial intelligence and neural networks. This literature review analyzed in each study the type of image, the total image database and its characteristics, the neural network employed to detect caries, the exclusion criterion of images, and whether the database had been modified before the training process. Then, it was analyzed as to how caries were defined, what type of caries were detected, and the outcome metrics and values.

2. Materials and Methods

2.1. Review Questions

(1)   What are the neural networks used to detect and diagnosis dental caries?

(2)   How is the database used in the construction of these networks?

(3)   How are caries lesions defined, and in which teeth are they detected?

(4)   What are the outcome metrics and the values obtained by those neural networks?

2.2. Search Strategy

The research questions were elaborated considering each of the components of the PICO(S) [15] strategy research questions, which are explained as follows: (P) neural networks and caries detection; (I) caries definition and which teeth are detected; (C) studies with neural network are used to detect and diagnosis dental caries; (O) outcome metrics and values; (S) neural networks.

An electronic search was performed in the following databases up until 15 August 2020: MEDLINE/PubMed, Institute of Electrical and Electronics Engineers (IEEE) Xplore, and ScienceDirect.

The search strategy used is detailed in Table 1.

Table 1. Search strategy.

Database	Search Strategy	Search Data
MEDLINE/PubMed	(deep learning OR artificial intelligence OR neural network *) AND caries NOT review	15 August 2020
IEEE Xplore	(deep learning OR artificial intelligence OR neural network) AND caries AND (detect OR detection OR diagnosis)	15 August 2020
ScienceDirect	(deep learning OR artificial intelligence OR neural network) AND caries AND (detect OR detection OR diagnosis)	15 August 2020

2.3. Study Selection and Items Collected

M.P.-P. and J.G.-V. performed the bibliographic search and selected the articles that fulfilled the inclusion criteria. Both authors collected all the data from the selected articles in duplicate and independently of each other. Disagreements between the two authors were reviewed using full text by a third author (J.C.P.-F.) to make the final decision. The references of the articles included in this study were manually reviewed.

The following items were collected: study (journal and year), type of image, total image database, database characteristics (pixels and examiners), neural network, image exclusion criteria, database modification (resized pixel), caries definition, caries type detected, teeth in which caries lesions were detected, outcome metrics (accuracy, sensitivity, specificity), and outcome metrics values.

2.4. Inclusion and Exclusion Criteria

The inclusion criteria were full manuscripts, including conference proceedings, that reported the use of neural networks for the detection and diagnosis of caries. There were no restrictions on the language or date of publication. Exclusion criteria were reviews, no dental caries application, no images, and no neural network employed.

2.5. Study Quality Assessment

The risk of bias from neural networks studies was evaluated by two of the authors (C.M.-M. and C.I.). To this end, the guidelines presented in the Cochrane Handbook [7] were followed, which incorporates seven domains: random sequence generation (selection bias); allocation concealment (selection bias); masking of participants and personnel (performance bias); masking of outcome assessment (detection bias); incomplete outcome data (attrition bias); selective reporting (reporting bias); and other biases.

The studies were classified into the following categories: low risk of bias—low risk of bias for all key domains; unclear risk of bias—unclear risk of bias for one or more key domains; high risk of bias—high risk of bias for one or more key domains.

2.6. Statistical Analysis

The mean, standard deviation (SD), median, and percentage were calculated for several variables. Statistical calculations were performed with IBM SPSS Statistics (SAS Institute Inc., Cary, NC, USA).

3. Results

3.1. Study Selection

Figure 1 details a flowchart of the study selection. All of the electronic search strategies resulted in 187 potential manuscripts. A total of 178 studies were excluded because they did not meet the inclusion criteria. Additionally, a manual search was carried out to analyze the references cited in ten of the articles that were included in this work. Finally, three more articles were incorporated from the manual search. In the end, a total of thirteen studies were analyzed.

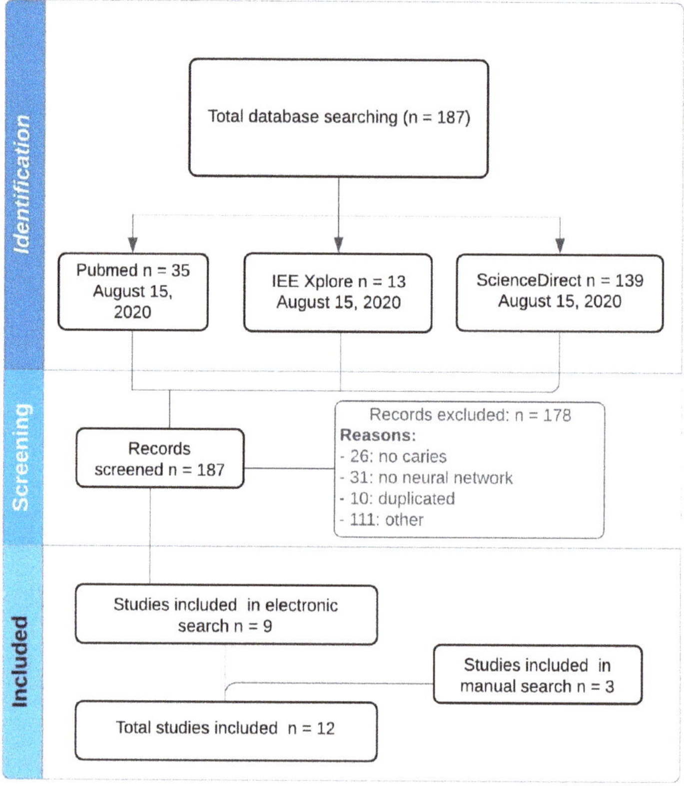

Figure 1. Flowchart.

3.2. Relevant Data about the Image Database and Neural Network of the Included Studies

Table 2 details the main characteristic of the studies included in the manuscript. Included studies were conducted between 2008 and 2020. All studies were published in English. Regarding the types of images, the most used were the periapical, the near-infrared light transilluminations, and the bitewings, each one appearing twice in each of the studies (16.66% in each study). The rest of the images used by a single study were panoramic radiographs, radiovisiography, intra-oral, in vivo with an intraoral camera and, and X-ray images (8.33%). Only two studies did not detail the type of image employed.

Image databases also varied from 87 to 3000 images, with a mean of 669.27 images, a standard deviation of 1153.76, and a median of 160 images. Seven (58.33%) of the included studies labeled the dental caries in each image by experienced dentists. Of those seven articles that used experts to indicate caries, five (71.42%) indicated that they use two experts ($n = 2$), one (14.28%) employed one expert, one employed four experts, and one employed 25 examiners. Three studies (25%) did not indicate the number of examiners.

Three studies detailed the exclusion criteria of the images. Six of the included studies detailed how images were standardized by resizing the number of pixels.

3.3. Relevant Data about Caries of the Included Studies

Table 3 details the main characteristic of carious lesion detection and outcome metrics of the included studies. Three studies (23.07%) detailed how caries are defined in their studies. One study explained that a caries lesion was considered where a radiolucent area appears on the structure. The other two considered the caries definition that follows the ICDAS II classification system. Seven (53.84%) of the included studies detailed the type of caries detected: three studies detected occlusal caries, while the other four studies detected proximal, enamel, and dentinal lesions; pre-cavitated lesions; and initial caries. Six of the included studies did not detail the type of caries detected by their neural network or define what they considered as caries.

Regarding the teeth where caries lesions are detected, seven studies (53.84%) did not detail in which teeth caries were detected, four studies (30.7%) employed molar and premolar teeth, and two (15.38%) of the included studies used posterior extracted teeth.

Eight of the included studies analyzed accuracy, obtaining the following outcomes: a range from 68.57 to 99% (mean ± SD of 90 ± 7%, median of 89%), a precision range from 0.615 to 0.987 (mean ± SD of 0.801 ± 0.263), and an AUC from 0.74 to 0.971 (mean ± SD of 0.815 ± 0.1).

Table 2. Main characteristics of image database and neural network.

Authors	Neural Network Task	Image	Total Image Database	Database Characteristics (Pixels and Examiners)	Neural Network	Image Exclusion Criterion	Database Modification (Resized and Other)	Journal	Year
Schwendicke et al. [16]	Classification	Near-infrared light transillumination	226	Pixel: $435 \times 407 \times 3$. Examiners: two (clinical experience, 8–11 years)	Resnet18, Resnext50	-	Resized pixel: 224×224	Journal of Dentistry	2020
Geetha et al. [17]	Classification	Intra-oral digital radiography	105	Pixel: Examiners: a dentist	ANN with 10-fold cross validation	-	Resized pixel: 256×256	Health Information Science and Systems	2020
Casalengo et al. [18]	Segmentation	Near-infrared transillumination	217	Pixel: Examiners: by experts	CNN trained on a semantic segmentation task	-	Resized pixel: 256×320	Journal of Dental Research	2019
Moutselos et al. [19]	Segmentation and classification	In vivo with an intraoral camera	87	-	DNN Mask R-CNN, which extends Faster R-CNN by adding an FCN for predicting object masks.	1. Teeth with hypoplastic and/or hypomineralized. 2. Teeth with sealants on the occlusal surfaces.	-	Conf Proc IEEE Eng Med Biol Soc	2019
Lee et al. [20]	Classification	Periapical	3000	Pixel: Examiners: four calibrated board-certified dentists	CNN	1. Moderate-to-severe noise, haziness, distortion, and shadows. 2. Full crown or large partial inlay restoration. 3. Deciduous teeth.	Resized pixel: 299×299 Other: standardized contrast between gray/white matter and lesions.	Journal of Dentistry	2018
Sornam et al. [21]	Classification	Periapical	120	-	Feedforward Neural Network	-	-	IEEE International Conference on Power, Control, Signals, and Instrumentation Engineering (ICPCSI-2017)	2017

Table 2. *Cont.*

Authors	Neural Network Task	Image	Total Image Database	Database Characteristics (Pixels and Examiners)	Neural Network	Image Exclusion Criterion	Database Modification (Resized and Other)	Journal	Year
Singh et al. [22]	Detection	Panoramic radiographs	93	-	Radon Transformation (RT) and Discrete Cosine Transformation (DCT).	-	Resized pixel: 500 × 500	2017 8th International Conference on Computing, Communication and Networking Technologies (ICCCNT)	2017
Srivastava et al. [23]	Segmentation	Bitewing	3000	Pixel: Examiners: by certified dentists	FCNN (deep fully convolutional neural network)	-	-	NIPS 2017 workshop on Machine Learning for Health (NIPS 2017 ML4H)	2017
Prajapati et al. [24]	Classification	Radiovisiography	251	-	CNN	-	Resized pixel: 500 × 748	5th International Symposium on Computational and Business Intelligence	2017
Berdouses et al. [25]	Detection and classification	-	103	Pixel: Examiners: two	-	-	-	Computers in Biology and Medicine	2015
Devito et al. [26]	Detection	Bitewing	160	Pixel: Examiners: 25	Multilayer perceptron neural	-	-	Oral Med Oral Pathol Oral Radiol Endod	2008
Kuang et al. [27]	Segmentation	X-ray images	-	Pixel: 1000 × 800 Examiners: -	Back propagation Neural Network	-	-	Second International Symposium on Intelligent Information Technology Application	2008

CNN: Convolutional neural network.

Table 3. Main data about caries of the included studies.

Authors	Type of Study	Caries Definition	Caries Type Detected	Teeth	Outcome Metrics	Outcome Metrics Values
Schwendicke et al. [16]	in vitro	-	Occlusal and/or proximal caries	Premolar and molar	AUC, sensitivity, specificity, and positive/negative predictive values	0.74, 0.59, 0.76, 0.63, and 0.73
Geetha et al. [17]	in vitro	Loss of mineralization of these structures (radiolucent)	-	-	Accuracy, false positive rate, ROC, and precision	0.971, 0.028, 0.987
Casalengo et al. [18]	clinical	-	-	Upper and lower molars and premolars	IOU/AUC	72.7/83.6 and 85.6%
Moutselos et al. [19]		Classified from 1 to 6 using the ICDAS II classification system.	Caries on occlusal surfaces	-	Accuracy	0.889
Lee et al. [20]	in vitro	-	Dental caries, including enamel and dentinal carious lesions	Premolar, molar, and both premolar and molar	Accuracy, sensitivity, specificity, PPV, NPV, ROC curve, and AUC	82, 81, 83, 82.7, 81.4
Sornam et al. [21]	in vitro	-	-	-	Accuracy	99%
Singh et al. [22]	in vitro	-	-	-	Accuracy	86%
Srivastava et al. [23]	in vitro	-	-	-	Recall/Precision/ F1-Score	0.805/0.615/0.7
Prajapati et al. [24]	in vitro	-	-	-	Accuracy	0.875
Berdouses et al. [25]	in vitro	ICDAS II	Pre-cavitated lesion and cavitated occlusal lesion	Posterior extracted human teeth	Accuracy	80%
Devito et al. [26]	in vitro	-	sound, enamel caries, enamel-dentine junction caries and, dentinal caries	Premolar and molar	ROC	0.717
Kuang et al. [27]	in vitro	-	Initial caries	-	Accuracy	68.57%

ICDAS: The International Caries Detection and Assessment System.

3.4. Study Quality Assessment

Evaluation of selection bias: All studies blinded image data.

Evaluation of performance bias: None of the studies indicated a blinding of staff or assessors.

Assessment of detection bias: All study results were blinded.

Evaluation of attrition bias: Not all of the studies reported complete results. Berdouses et al. [25] detailed all the results analyzed in the present review.

Evaluation of notification bias: Not all of the studies provided detailed information about the neural network parameters. Lee et al. [20] provided all the information.

Figure 2 shows a detailed description of the risk assessment of bias in the included studies.

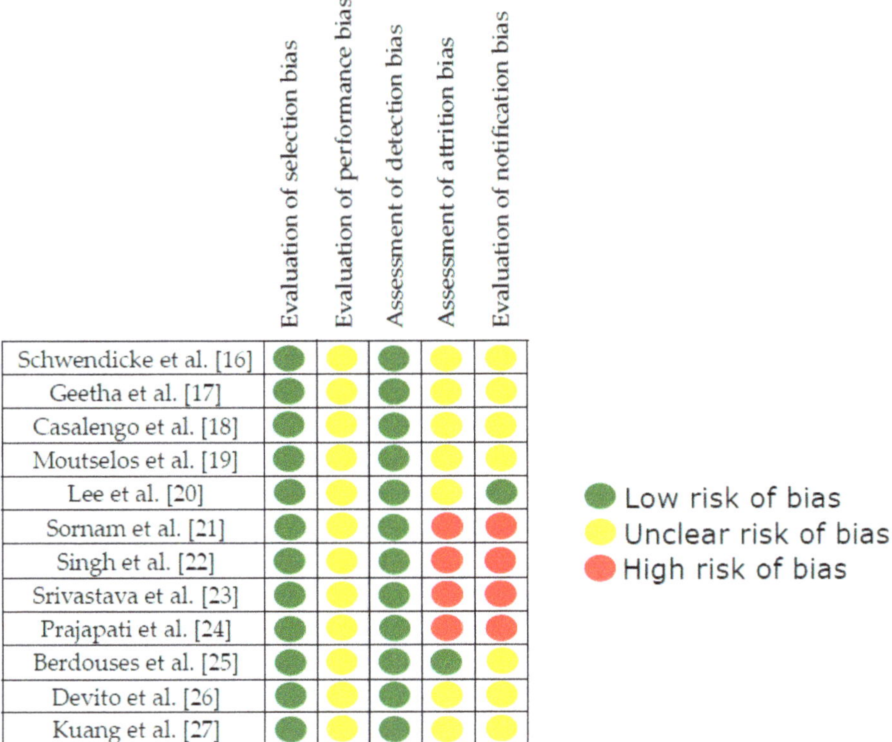

Figure 2. Assessment of risk of bias of included studies.

4. Discussion

The goal of this review is to visualize the state of the art of neural networks in detecting and diagnosing dental caries. The way in which each of the studies analyzes caries (definition, type, tooth), as well as the parameters of each neural network (type of network, characteristics of the database, and results), were studied.

A good definition of what is meant by caries and the type of caries lesions to be analyzed is essential to compare and analyze the results obtained in each study. Studies included in this review that detailed the use of ICDAS II obtained an accuracy between 80 and 88.9% (mean ± SD of 85.45 ± 6.29%), while the study that defined caries as a loss of mineralization of these structures (radiolucent) obtained an accuracy of 97.1%. However, 76% of the studies included in the present review did not detail how a caries lesion is defined.

Another bias factor is related to the training dataset. Images employed during the training process must be labeled by experts. Seven (58.33%) of the included studies indicated that examiners were used to label the images, although the experience and the number of those examiners varied from one study to another. Some studies analyzed the relation between caries detection and dentist experience. Bussaneli et al., concluded in their study that the experience of the examiner is not determinant to occlusal lesions in primary teeth but influenced the treatment decision of initial lesions [28]. However, when an artificial intelligence is trained with human observer's scores, the system can never exceed the trainer and, therefore, the performance depends on the quality of the input.

An important fact for artificial intelligence technology is the overfitting. Burnham and Anderson describes "the essence of overfitting is to have unknowingly extracted some of the residual variation as if that variation represented underlying model structure" [29]. A model is overfitted when it is so specific to the original data that trying to apply it to data collected in the future would result in problematic or erroneous outcomes and therefore less-than-optimal decisions [30].

The included studies in this review that detailed the use of examiners to obtain an accuracy ranged from 80 to 97% (mean ± SD of 88.7 ± 8.55%). The best result was obtained in the study that used only one examiner, and, therefore, the same criteria in caries detection was always used, followed by the study where four experts analyzed the images. Finally, the worst result in terms of accuracy was obtained by the study with two examiners. Regarding the experience of the examiners, only one study detailed the number of years of experience. However, these results were not completely related to the number of examiners; other factors such as neural network, dataset, and caries definition must be kept in mind. In this sense, the results detailed in Table 3 must be analyzed with caution, since each of the networks used in the studies has a different purpose, which means that the results are not comparable between them. The data analyzed in a general way helps us to get an idea about what percentages of accuracy, on average, are obtained in caries detection and diagnosis studies using neural networks.

One of the limitations of this review is that studies using artificial intelligence with different tasks have been taken into account. This means that, although the studies obtain the same metrics, they cannot be compared with each other. The reason is that each artificial intelligence is designed for one thing that makes comparison of results impossible. From each to future reviews, it is recommended to include studies whose artificial intelligence has the same purpose. The use of a large dataset is crucial for the performance of the deep learning model. It is possible to improve the technical capability employing a technique called data augmentation. This technique artificially inflates the training database by oversampling or data warping. Oversampling creates synthetic instances and adds them to the training dataset. However, data warping transforms the existing images [31].

The study from Geetha et al. [17] presents the highest accuracy of all the studies included in the present review. However, this study is a special case because the authors built their own feature extractor, which is rare nowadays, and used a very shallow neural network with only one hidden layer. Authors of that study used only 105 images and did 10-fold cross-validation, and, therefore, their model was not evaluated on a hold-out test set.

A great variety of architectures has been found in the studies included in this literature review. ResNets are residual networks that are CNNs designed to allow thousands of convolutional layers. Mask R-CNN is an extension of Faster R-CNN by adding a branch for predicting segmentation masks on each Region of Interest (ROI) [5]. Semantic image segmentation is the task of classifying each pixel in an image from a predefined set of classes, which has several applications in medical images. Six (50%) of the included studies detailed that, before starting the training process, they homogenized the size of the images (Table 2).

Shokri et al., analyzed the effect of filters on detecting proximal and occlusal caries employing intraoral images and concluded that the lowest accuracy in caries diagnosis was noted for the detection of enamel lesions on original radiographs (52%). However, this in vitro study induced caries by a demineralizing solution, and therefore induced carious lesions were more regular than those that developed naturally [32]. Belém et al. [33] analyzed the accuracy of detection of subsurface demineralization by different imaging modalities and concluded that original images had an accuracy of 73% and a sensitivity of 62%. Kositbowornchai et al. compared the accuracy of detecting occlusal caries lesions on original images and obtained a mean Receiver Operating Characteristic (ROC) curve of 0.75 [34]. Here, two of the studies analyzed enamel lesions with an accuracy of 82% and a ROC curve of 0.717. The studies that analyzed occlusal lesions in this review obtained an accuracy of 80 and 88.9%, and a precision of 45.3%.

Several studies analyzed the precision in the detection of caries depending on the type of image used. Schwendicke et al., concluded in their systematic review that fluorescence-based images showed a significantly higher accuracy, sensitivity, and specificity in detecting initial lesions than conventional radiographic images, and generally that radiographic caries detection is especially suitable for detecting dentine lesions and cavitated proximal lesions [8]. Here, two of the studies employed near-infrared transillumination images and obtained similar outcome metrics to the other studies with different image types.

Supervised learning is one where the learning process of the algorithm from the training dataset can be considered to be a process supervised by a teacher. The correct answer is previously known, and the algorithm iteratively makes its predictions at the same time as it is corrected by the teacher. Seven (58.33%) of the included studies labeled the dental caries in each image by experienced dentists. However, in addition to knowing the number of examiners and their experience, it is very important to know what the intra-examiner agreement is, that is, to know if the examiner's answers are the same if the categorization of the images is repeated a second time. It is also very important to know the inter-examiner agreement, that is, for the same image, how many examiners provide the same answer. None of the included studies mentioned the inter- and intra-examiner agreement. Intra- and inter-examiner agreement is evaluated by calculating Cohen's Kappa. According to Bulman and Osborn [35], values of Cohen's Kappa between 0.81 and 1.00 indicate almost perfect agreement.

Other graphical methods such as ROC (Receiver Operating Characteristic) curve or Bland-Altman plot can be employed to obtain information on those samples in which there is less agreement.

It is important to emphasize that manual labeling by experts provides a reference that is necessary for training and evaluating the model but does not necessarily represent ground truth [18]. The use of a histologic gold standard method is indispensable for the validation of a caries diagnostic method. None of the studies included in the present review mentioned the reference standard employed.

A quality analysis of the included studies was done using the Cochrane Handbook tool, which was employed to assess the risk of bias, concluding that in most domains, no data were given related to the transparency of the studies. This ensures that the data collected and analyzed have been managed in a controlled manner, avoiding all possible methodological errors. The criteria for allocation masking and randomization were not detailed in all of the studies, which is considered to be an unclear risk of bias. The present systematic review focused on the use of artificial intelligence in carious lesion diagnostic and detection; the bias is located in the lack of a reference standard and the inclusion of studies with different algorithms. However, the data presented above cannot be analyzed in isolation. Inter- and intra-examiner agreement must be taken into account in studies involving multiple examiners in order to obtain comparable and reliable results. This is a fundamental parameter to correctly define the variables that the neural network has to learn. That is, good agreement between examiners is essential to obtain good results once the image passes through the neural network. None of the studies using multiple examiners and included in this review detailed the concordance mentioned above in their respective studies. Neither was there a single parameter to compare the results obtained by the neural network, nor common parameters for the database. All these factors complicate the conclusions that can be made about the reliability or not of a neural network to detect and diagnose caries.

To be able to know if the neural networks give certain results, it is necessary to make a comparison with the results provided by the dentists, who should also have similar training and experience in order for comparisons to be made between them.

The diagnostic performance of artificial intelligence models varies between the different algorithms used and is still necessary to verify the generalizability and reliability of these models. For this, it would be necessary to use the ability to compare the results of the tasks of each algorithm before transferring and implement these models in clinical practice.

Author Contributions: Conceptualization, M.P.-P. and J.C.P.-F.; methodology, M.P.-P. and J.G.V.; data curation, M.P.-P. and J.G.V.; writing—original draft preparation, M.P.-P.; writing—review and editing, C.H.M.-M. and J.C.P.-F.; visualization, M.P.-P., J.G.V., C.H.M.-M., J.C.P.-F. and C.I.; supervision, C.I.; funding acquisition, C.H.M.-M. and C.I. All authors have read and agreed to the published version of the manuscript.

Funding: This research was funded by Asisa Dental S.A.U.

Conflicts of Interest: The authors declare no conflict of interest.

References

1. Pauwels, R. A brief introduction to concepts and applications of artificial intelligence in dental imaging. *Oral Radiol.* **2020**. [CrossRef]
2. Chen, Y.W.; Stanley, K.; Att, W. Artificial intelligence in dentistry: Current applications and future perspectives. *Quintessence Int.* **2020**, *51*, 248–257. [CrossRef]
3. Kohli, M.; Prevedello, L.M.; Filice, R.W.; Geis, J.R. Implementing Machine Learning in Radiology Practice and Research. *Am. J. Roentgenol.* **2017**, *208*, 754–760. [CrossRef]
4. Clarke, A.M.; Friedrich, J.; Tartaglia, E.M.; Marchesotti, S.; Senn, W.; Herzog, M.H. Human and Machine Learning in Non-Markovian Decision Making. *PLoS ONE* **2015**, *10*, e0123105. [CrossRef]
5. LeCun, Y.; Bengio, Y.; Hinton, G. Deep learning. *Nature* **2015**, *521*, 436–444. [CrossRef]
6. Schwendicke, F.; Golla, T.; Dreher, M.; Krois, J. Convolutional neural networks for dental image diagnostics: A scoping review. *J. Dent.* **2019**, *91*, 103226. [CrossRef]
7. Mazurowski, M.A.; Buda, M.; Saha, A.; Bashir, M.R. Deep learning in radiology: An overview of the concepts and a survey of the state of the art with focus on MRI. *J. Magn. Reson. Imaging* **2019**, *49*, 939–954. [CrossRef] [PubMed]
8. Schwendicke, F.; Tzschoppe, M.; Paris, S. Radiographic caries detection: A systematic review and meta-analysis. *J. Dent.* **2015**, *43*, 924–933. [CrossRef]
9. Gupta, M.; Srivastava, N.; Sharma, M.; Gugnani, N.; Pandit, I. International Caries Detection and Assessment System (ICDAS): A New Concept. *Int. J. Clin. Pediatr. Dent.* **2011**, *4*, 93–100. [CrossRef]
10. Machiulskiene, V.; Campus, G.; Carvalho, J.C.; Dige, I.; Ekstrand, K.R.; Jablonski-Momeni, A.; Maltz, M.; Manton, D.J.; Martignon, S.; Martinez-Mier, E.A.; et al. Terminology of Dental Caries and Dental Caries Management: Consensus Report of a Workshop Organized by ORCA and Cariology Research Group of IADR. *Caries Res.* **2020**, *54*, 7–14. [CrossRef]
11. Dikmen, B. Icdas II Criteria (International Caries Detection and Assessment System). *J. Istanb. Univ. Fac. Dent.* **2015**, *49*, 63. [CrossRef]
12. Valizadeh, S.; Tavakkoli, M.A.; Vasigh, H.K.; Azizi, Z.; Zarrabian, T. Evaluation of Cone Beam Computed Tomography (CBCT) System: Comparison with Intraoral Periapical Radiography in Proximal Caries Detection. *J. Dent. Res. Dent. Clin. Dent. Prospect.* **2012**, *6*, 1–5. [CrossRef]
13. Abogazalah, N.; Ando, M. Alternative methods to visual and radiographic examinations for approximal caries detection. *J. Oral Sci.* **2017**, *59*, 315–322. [CrossRef]
14. Zhang, Z.; Qu, X.; Li, G.; Zhang, Z.; Ma, X. The detection accuracies for proximal caries by cone-beam computerized tomography, film, and phosphor plates. *Oral Surg. Oral Med. Oral Pathol. Oral Radiol. Endodontol.* **2011**, *111*, 103–108. [CrossRef]
15. Centre for Reviews and Dissemination. *Systematic Reviews: CRD Guidance for Undertaking Reviews in Health Care*; University of York: York, UK, 2009; ISBN 978-1-900640-47-3.
16. Schwendicke, F.; Elhennawy, K.; Paris, S.; Friebertshäuser, P.; Krois, J. Deep Learning for Caries Lesion Detection in Near-Infrared Light Transillumination Images: A Pilot Study. *J. Dent.* **2019**, 103260. [CrossRef]
17. Geetha, V.; Aprameya, K.S.; Hinduja, D.M. Dental caries diagnosis in digital radiographs using back-propagation neural network. *Health Inf. Sci. Syst.* **2020**, *8*, 8–14. [CrossRef] [PubMed]
18. Casalegno, F.; Newton, T.; Daher, R.; Abdelaziz, M.; Lodi-Rizzini, A.; Schürmann, F.; Krejci, I.; Markram, H. Caries Detection with Near-Infrared Transillumination Using Deep Learning. *J. Dent. Res.* **2019**, *98*, 1227–1233. [CrossRef]
19. Moutselos, K.; Berdouses, E.; Oulis, C.; Maglogiannis, I. Recognizing Occlusal Caries in Dental Intraoral Images Using Deep Learning. In Proceedings of the 41st Annual International Conference of the IEEE Engineering in Medicine and Biology Society (EMBC), Berlin, Germany, 23–27 July 2019; pp. 1617–1620.
20. Lee, J.H.; Kim, D.H.; Jeong, S.N.; Choi, S.H. Detection and diagnosis of dental caries using a deep learning-based convolutional neural network algorithm. *J. Dent.* **2018**, *77*, 106–111. [CrossRef] [PubMed]
21. Sornam, M.; Prabhakaran, M. A new linear adaptive swarm intelligence approach using back propagation neural network for dental caries classification. In Proceedings of the IEEE International Conference on Power, Control, Signals and Instrumentation Engineering (ICPCSI), Chennai, India, 21–22 September 2017; pp. 2698–2703.

22. Singh, P.; Sehgal, P. Automated caries detection based on Radon transformation and DCT. In Proceedings of the 8th International Conference on Computing, Communication and Networking Technologies (ICCCNT), Delhi, India, 3–5 July 2017; pp. 1–6.

23. Srivastava, M.M.; Kumar, P.; Pradhan, L.; Varadarajan, S. Detection of Tooth caries in Bitewing Radiographs using Deep Learning. In Proceedings of the 31st Conference on Neural Information Processing Systems (NIPS), Long Beach, CA, USA, 4–9 December 2017; p. 4.

24. Prajapati, S.A.; Nagaraj, R.; Mitra, S. Classification of dental diseases using CNN and transfer learning. In Proceedings of the 5th International Symposium on Computational and Business Intelligence (ISCBI), Dubai, UAE, 11–14 August 2017; pp. 70–74.

25. Berdouses, E.D.; Koutsouri, G.D.; Tripoliti, E.E.; Matsopoulos, G.K.; Oulis, C.J.; Fotiadis, D.I. A computer-aided automated methodology for the detection and classification of occlusal caries from photographic color images. *Comput. Biol. Med.* **2015**, *62*, 119–135. [CrossRef]

26. Devito, K.L.; de Souza Barbosa, F.; Filho, W.N.F. An artificial multilayer perceptron neural network for diagnosis of proximal dental caries. *Oral Surg. Oral Med. Oral Pathol. Oral Radiol. Endodontol.* **2008**, *106*, 879–884. [CrossRef]

27. Kuang, W.; Ye, W. A Kernel-Modified SVM Based Computer-Aided Diagnosis System in Initial Caries. In Proceedings of the Second International Symposium on Intelligent Information Technology Application, Shanghai, China, 20–22 December 2008; pp. 207–211.

28. Bussaneli, D.G.; Boldieri, T.; Diniz, M.B.; Lima Rivera, L.M.; Santos-Pinto, L.; Cordeiro, R.D.C.L. Influence of professional experience on detection and treatment decision of occlusal caries lesions in primary teeth. *Int. J. Paediatr. Dent.* **2015**, *25*, 418–427. [CrossRef]

29. Burnham, K.P.; Anderson, D.R. *Model Selection and Multimodel Inference*; Burnham, K.P., Anderson, D.R., Eds.; Springer: New York, NY, USA, 2004; ISBN 978-0-387-95364-9.

30. Mutasa, S.; Sun, S.; Ha, R. Understanding artificial intelligence based radiology studies: What is overfitting? *Clin. Imaging* **2020**, *65*, 96–99. [CrossRef]

31. Shorten, C.; Khoshgoftaar, T.M. A survey on Image Data Augmentation for Deep Learning. *J. Big Data* **2019**, *6*, 60. [CrossRef]

32. Shokri, A.; Kasraei, S.; Lari, S.; Mahmoodzadeh, M.; Khaleghi, A.; Musavi, S.; Akheshteh, V. Efficacy of denoising and enhancement filters for detection of approximal and occlusal caries on digital intraoral radiographs. *J. Conserv. Dent.* **2018**, *21*, 162. [CrossRef]

33. Belém, M.D.F.; Ambrosano, G.M.B.; Tabchoury, C.P.M.; Ferreira-Santos, R.I.; Haiter-Neto, F. Performance of digital radiography with enhancement filters for the diagnosis of proximal caries. *Braz. Oral Res.* **2013**, *27*, 245–251. [CrossRef]

34. Kositbowornchai, S.; Basiw, M.; Promwang, Y.; Moragorn, H.; Sooksuntisakoonchai, N. Accuracy of diagnosing occlusal caries using enhanced digital images. *Dentomaxillofac. Radiol.* **2004**, *33*, 236–240. [CrossRef] [PubMed]

35. Bulman, J.S.; Osborn, J.F. Measuring diagnostic consistency. *Br. Dent. J.* **1989**, *166*, 377–381. [CrossRef]

Publisher's Note: MDPI stays neutral with regard to jurisdictional claims in published maps and institutional affiliations.

Journal of
Clinical Medicine

Review

Clinical Performance of Partial and Full-Coverage Fixed Dental Restorations Fabricated from Hybrid Polymer and Ceramic CAD/CAM Materials: A Systematic Review and Meta-Analysis

Nadin Al-Haj Husain [1,*], Mutlu Özcan [2], Pedro Molinero-Mourelle [1] and Tim Joda [3]

[1] Department of Reconstructive Dentistry and Gerodontology, School of Dental Medicine, University of Bern, 3010 Bern, Switzerland; pedro.molineromourelle@zmk.unibe.ch
[2] Division of Dental Biomaterials, Clinic for Reconstructive Dentistry, Center for Dental and Oral Medicine, University of Zurich, 8032 Zurich, Switzerland; mutlu.ozcan@zzm.uzh.ch
[3] Department of Reconstructive Dentistry, University Center for Dental Medicine Basel, University of Basel, 4058 Basel, Switzerland; tim.joda@unibas.ch
* Correspondence: nadin.al-haj-husain@zmk.unibe.ch

Received: 11 May 2020; Accepted: 1 July 2020; Published: 4 July 2020

Abstract: The aim of this systematic review and meta-analysis was to evaluate the clinical performance of tooth-borne partial and full-coverage fixed dental prosthesis fabricated using hybrid polymer and ceramic CAD/CAM materials regarding their biologic, technical and esthetical outcomes. PICOS search strategy was applied using MEDLINE and were searched for RCTs and case control studies by two reviewers using MeSH Terms. Bias risk was evaluated using the Cochrane collaboration tool and Newcastle–Ottawa assessment scale. A meta-analysis was conducted to calculate the mean long-term survival difference of both materials at two different periods ($\leq$24, $\geq$36 months(m)). Mean differences in biologic, technical and esthetical complications of partial vs. full crown reconstructions were analyzed using software package R ($p < 0.05$). 28 studies included in the systematic review and 25 studies in the meta-analysis. The overall survival rate was 99% (0.95–1.00, $\leq$24 m) and dropped to 95% (0.87–0.98, $\geq$36 m), while the overall success ratio was 88% (0.54–0.98; $\leq$24 m) vs. 77% (0.62–0.88; $\geq$36 m). No significance, neither for the follow-up time points, nor for biologic, technical and esthetical (88% vs. 77%; 90% vs. 74%; 96% vs. 95%) outcomes was overserved. A significance was found for the technical/clinical performance between full 93% (0.88–0.96) and partial 64% (0.34–0.86) crowns. The biologic success rate of partial crowns with 69% (0.42–0.87) was lower, but not significant compared to 91% (0.79–0.97) of full crowns. The esthetical success rate of partial crowns with 90% (0.65–0.98) was lower, but not significant compared to 99% (0.92–1.00) of full crowns.

Keywords: bonding; CAD/CAM; composite resin cement; dental; hybrid polymer; indirect; meta-analysis; systematic review

1. Introduction

Over the past two decades, metal-free computer-aided design/computer aided manufacturing (CAD/CAM) materials, including ceramics and composites, have been widely used in dentistry [1]. In the restorative clinical field, these materials have been gaining importance due to their biologic and esthetical properties resulting in favorable treatment outcomes in order to satisfy increased demands and expectations of patients and dentists [2,3].

The improvements in oral health during the last decades, have promoted less aggressive dental preparations changing the conventional indications and workflows of these restorations and adapting it for these metal-free materials [4,5]. The current state of the art of dental treatments accompanied by life changes in terms of time efficacy and patient care demands, have fostered the introduction of faster and cost-efficient digital clinical workflows using CAD/CAM technology facilitating high quality restorative treatments [6,7]. These workflows allow designing and manufacturing of chairside partial or full-contoured monolithic restorations, such as inlays, veneers, single crowns (SCs) or multi-spans fixed dental prostheses (FDPs), with esthetically favorable appearance, accurate marginal adaptation in a cost and time efficient production manner [3,8].

Digital technologies also enabled the development of high-performance materials like Lithium disilicate (LD), Lithium aluminosilicate ceramic reinforced with lithium disilicate glass–ceramic (LD-LAS), hybrid-polymer ceramic (HPC) and resin-matrix ceramics (RMC) including resin-based ceramics (RBC) and polymer infiltrated ceramic network (PICN) resins [9–11].

LD is one of the of the most commonly used chairside material due to its great clinical performance and high acceptance by patients, technicians and dentists. LD-LAS covers the same indication range as LD ceramics, while showing comparable flexural strength tests results, making it a high load-bearing material with excellent esthetic properties [12,13]. The group of hybrid materials (HPC, RMC, RBC and PICN) are of growing interest due their mechanical resistibility and high elasticity. These materials are based on a ceramic like hybrid ceramic also known as resin-matrix-ceramics, resin-based ceramics or nanoceramics, presenting promising results, as they follow esthetic trends combined with minimally invasive preparations in modern clinical workflows [11,14].

The gold standard in SCs and FDPs is still ceramic fused to metal. This "conventional" approach often presents esthetic shortcomings, requires a more aggressive tooth preparation and extended technical production time. Therefore, metal-free options have gradually become a favorite alternative compared to metal-ceramic restorations [15,16]. However, when using metal-free materials, clinicians should keep in mind the limited evidence that these materials present in terms of long-term performance, survival and complication rates and carefully evaluate the indication and processing technique in each unique clinical case [14].

The wide range of new hybrid polymer and ceramic CAD/CAM materials that are offered in the dental industry to manufacture tooth-borne restorations implies the need for an evidence-based study that evaluates the current clinical behavior of these materials. Therefore, the aim of this systematic review and meta-analysis was to analyze the clinical behavior of partial and full fixed restorations out of hybrid polymer and ceramic CAD/CAM materials. This present systematic review was performed in order to answer the PICO question defined as follows: In patients receiving tooth-borne partial or full crowns, are survival and clinical success rates of monolithic CAD/CAM restorations comparable to those of conventionally manufactured?

2. Experimental Section

2.1. Search Strategy

A preliminary search was conducted prior to the definition of the final PICO question, focusing on material choice (glass ceramic multiphase (e.g., Enamic); polymeric multiphase (e.g., Lava Ultimate)); Indication (tooth and implant-borne single-unit restoration and reconstruction design (crown vs. partial crown single unit).

The PICO question was then chosen as follows: *P*-population: tooth-borne partial or full crowns; I-intervention: Monolithic CAD/CAM restorations; C-control: conventionally produced/manufactured restorations (natural teeth); O-outcome: survival and clinical success (fracture, debonding, behavior); S-study designs: randomized control trials (RCT) and case–control studies.

The following MeSH terms, search terms and their combinations were used in the PubMed search: (((((((dental crowns [MeSH]) OR (dental restoration permanent [MeSH]) OR (full crown) OR (partial crown) OR (table top))))) AND (((((computer-aided design [MeSH])) OR (computer-assisted design [MeSH]) OR ((computer-aided manufacturing [MeSH])) OR (computer-assisted manufacturing

[MeSH]) OR (cerec [MeSH]) OR (CAD/CAM) OR (rapid prototyping))))) OR ((((ceramics [MeSH]) OR (dental porcelain [MeSH]) OR (polymers [MeSH]) OR (monolithic))))) AND ((((survival analysis [MeSH terms]) OR (survival rate [MeSH Terms]) OR (survival))))) OR ((((success) OR (failure) OR (dental restoration failure [MeSH terms]) OR (complications [MeSH terms]) OR (clinical behavior) OR (adverse event) OR (chipping) OR (debonding)))). The search strategy according to the focused PICOS question is presented in Table 1.

Table 1. Search strategy according to the focused question (PICO).

Focused Question (PICO)	In Patients Receiving Tooth-Borne Partial or Full Crowns, Are Monolithic CAD/CAM Restorations Comparable to Conventionally Manufactured Restorations in Terms of Survival and Clinical Success Rates?	
Search strategy	**Population**	Tooth-borne partial or full crowns. #1—((dental crowns [MeSH]) OR (dental restoration permanent [MeSH]) OR (full crown) OR (partial crown) OR (table top))
	Intervention	Monolithic CAD/CAM restorations. #2—((computer-aided design [MeSH])) OR (computer-assisted design [MeSH]) OR ((computer-aided manufacturing [MeSH])) OR (computer-assisted manufacturing [MeSH]) OR (cerec [MeSH]) OR (CAD/CAM) OR (rapid prototyping)) #3—((ceramics [MeSH]) OR (dental porcelain [MeSH]) OR (polymers [MeSH]) OR (monolithic))
	Comparison	Conventionally manufactured restorations. #4—((porcelain-fused to metal) OR (lost-wax technique)) #5—(dental alloys [MeSH])
	Outcome	Survival (rates) and/or clinical success. #6—((survival analysis [MeSH Terms]) OR (survival rate [MeSH Terms]) OR (survival)) #7—((success) OR (failure) OR (dental restoration failure [MeSH Terms]) OR (complications [MeSH Terms]) OR (clinical behavior) OR (adverse event) OR (chipping) OR (debonding))
	Search combination(s)	(#1) AND (#2 or #3) AND (#6 or #7)

The following terms were used in the EMBASE search: ('dental crowns'/exp OR 'dental restoration permanen'/exp OR 'full crown'/exp OR 'partial crown'/exp OR 'table top') AND (' computer-aided design' OR 'computer-assisted design' OR 'computer-aided manufacturing' OR ' computer-assisted manufacturing' OR 'cerec' OR 'CAD/CAM' OR 'rapid prototyping') OR ('ceramics' OR 'dental porcelain' OR 'polymers' OR 'monolithic') AND ('survival analysis' OR 'survival rate' OR 'survival') OR ('success' OR 'failure' OR 'dental restoration failure' OR 'complications' OR 'clinical behavior' OR 'adverse event' OR 'chipping' OR 'debonding') NOT [medline]/lim AND [embase]/lim.

The following terms were used in the Web of Science and IADR abstracts search: (((((((dental crowns [MeSH]) OR (dental restoration permanent [MeSH]) OR (full crown) OR (partial crown) OR (table top))))) AND ((((computer-aided design [MeSH])) OR (computer-assisted design [MeSH]) OR ((computer-aided manufacturing [MeSH])) OR (computer-assisted manufacturing [MeSH]) OR (cerec [MeSH]) OR (CAD/CAM) OR (rapid prototyping))))) OR ((((ceramics [MeSH]) OR (dental porcelain [MeSH]) OR (polymers [MeSH]) OR (monolithic))))) AND ((((survival analysis [MeSH Terms]) OR (survival rate [MeSH Terms]) OR (survival))))) OR ((((success) OR (failure) OR (dental restoration failure [MeSH Terms]) OR (complications [MeSH Terms]) OR (clinical behavior) OR (adverse event) OR (chipping) OR (debonding)))).

2.2. Information Sources

A systematic electronic literature search was conducted in PubMed MEDLINE, EMBASE and Web of Science (ISI—Web of Knowledge), including Google Scholar and IADR abstracts until 16 May 2018. The search aimed for English language clinical trials and case–control studies published in the last five years, performed on human and published in dental journals. Search syntax was categorized in a population, intervention, comparison and outcome study design; each category assembled using a combination of Medical Subject Heading [MeSH Terms].

2.3. Study Selection and Eligibility Criteria

To minimize the potential for reviewer bias, two reviewers (N.A.-H.H. and T.J.) independently conducted electronic literature searches and the study selection. Both reviewers studied the retrieved titles and abstracts and disagreements were solved by discussion. Forty-eight selected studies were then obtained in full texts, and the decision of inclusion of studies was made according to preset inclusion criteria.

The following inclusion criteria were chosen for the articles included in this systematic review: (1) RCTs and case control studies; (2) Studies with observation of a follow-up period of ≥ 1 year; (3) Studies that considered either hybrid polymers or ceramic CAD/CAM materials.

Articles meeting one or more of the following criteria were excluded: (1) In vitro or in situ studies; (2) Studies with a follow-up period less than one year; (3) Studies testing materials other than hybrid polymers or ceramic CAD/CAM materials. For quantitative analyses (meta-analysis), studies lacking a control group or standard deviation values were excluded (Figure 1).

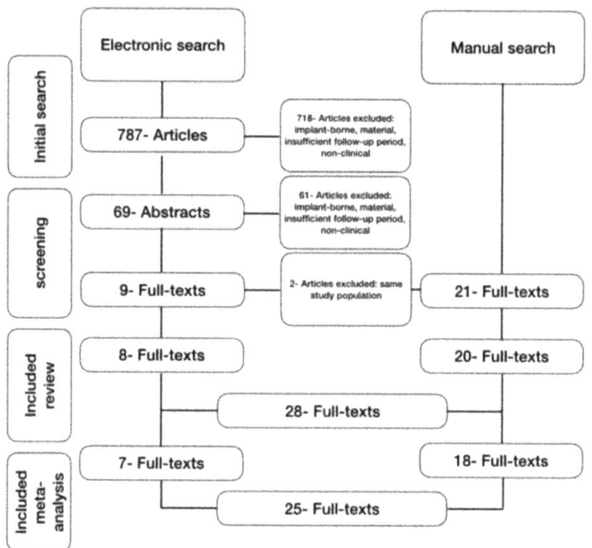

Figure 1. Flow diagram of the systematic search results.

2.4. Data Extraction and Collection

After screening the data, extracting, obtaining and screening the titles and abstracts for inclusion criteria, the selected abstracts were obtained in full texts. Titles and abstracts lacking sufficient information regarding inclusion criteria were also obtained as full texts.

Full text articles were selected in case of compliance with inclusion criteria by the two reviewers using a data extraction form. Two reviewers (N.A.-H.H. and T.J.) independently collected the following data from the included articles for further analysis: demographic information (title, authors, journal and year), study specific parameter (study type, number of treated patients, number of restorations, Ratio (restorations/patient), follow-up and drop-out), materials tested (type and commercial name, manufacturing process, luting agent, failure, survival and success rate), means and standard deviations of the clinical parameters (biologic, technical and esthetical failures).

The authors of the studies were contacted in case of unpublished data. These studies were only included if the authors provided the missing information. In order to assess the clinical performance and outcomes of the restorations, the selected studies based their evaluations on the modified United States Public Health service (USHPS) [17] criteria and the FDI World dental federation criteria [18].

For the extraction of the clinical outcomes, the relevant data of the included studies were divided into three subgroups according to their evaluated outcomes, based on the USHPS criteria and the FDI criteria: The USHPS criteria are based on an evaluation of the clinical characteristics of color, marginal adaptation, anatomic form, surface roughness, marginal staining, secondary caries and luster of restoration which is evaluated on three levels form the best to worst outcome, Alpha, Bravo and Charlie.

The FDI criteria are based on three levels that were scored into five points (Clinically very good, clinically good, clinically sufficient/satisfactory, clinically unsatisfactory, clinically poor): (A) Esthetic properties that evaluate the surface luster, the staining, color match and translucency and the esthetic anatomic form; (B) Functional properties based on the assess of fracture of material and retention, the marginal adaptation, the occlusal contour and wear, the approximal anatomic form, the radiographic examination and the patient's view; (C) Biologic properties measure the postoperative sensitivity and tooth vitality, the recurrence, the tooth integrity of caries, the periodontal response, the adjacent mucosa and the oral and general health.

2.5. Risk of Bias Assessment

The risk of bias assessment was evaluated using the Cochrane collaboration tool for randomized studies, evaluating bias risks such as sample size calculation, random sequence generation, adequate control group, materials usage following the manufacturers' instructions, tests execution by a single blinded operator, adequate statistical analysis, allocation concealment, completeness of outcome data, selective reporting and other bias. Each parameter reported by the included studies was recorded. Articles that included only one to three possible risks of bias of these items were considered at low risk for bias; four or five items, at medium risk for bias; and six to nine items, at high risk for bias.

In case of a high or unclear risk of bias the study was assigned to a judgment of risk of bias. The Newcastle–Ottawa assessment scale was applied for non-randomized studies, for the selection of the study groups, the comparability of the groups and the ascertainment of outcome or interest.

2.6. Data Analyses

The statistical analysis was performed with the software package R, Version 3.5.3 (R Core Team 2013) [19]. Both survival and success ratios were analyzed performing a meta-analysis using the logit transformation method. Results of the random effects model were reported and forest plots were drawn. Funnel plots were also produced in order to detect a possible publication bias. Overall, survival and success ratios were analyzed as well as biologic, technical and esthetical successes. The restorations instead of patients were used as the statistical unit. Studies that lacked the required information of the sample size or the follow-up time were excluded from the statistical analysis. All materials had to be pooled because of sample size considerations or missing information. The meta-analysis was done with studies reporting a follow-up time of at least 24 months.

3. Results

3.1. Study Selection

Of 795 potentially relevant studies, 48 were selected for a full-text analysis, 28 were included in the systematic review and 25 considered in the meta-analysis. Eight full text articles were selected using electronic databases and 20 further were retrieved throughout manual search. From the 25 studies included in the meta-analysis, 12 studies were randomized controlled trial, 14 prospective and 2 retrospectives (Krejci et al. 1992; Taskonak et al. 2006; Frankenberger et al. 2008; Frankenberger et al. 2009; Dukic et al. 2010; Fasbinder et al. 2010; Manhart et al. 2010; Azevedo et al. 2012; Esuivel-Opshaw et al. 2012; Murgueitio et al. 2012; Schenke et al. 2012; Taschner et al. 2012; Gehrt et al. 2013; Reich et al. 2013; Akin et al. 2014; D'all'Orologio et al. 2014; Dhima et al. 2014; Guess et al. 2014; Guess et al. 2014; Selz et al. 2014; Seydler et al. 2015; Baader et al. 2016; Botto et al. 2016; Mittal et al. 2016; Özsoy et al. 2016; Santos et al. 2016; Rauch et al. 2018) [20–46].

3.2. Study Characteristics

The characteristics of the included studies are presented in Table 2. The included articles were published between 1992 and 2018. A total of type of 28 studies including 1150 patients and 2335 reconstructions with a mean follow-up time of 4.5 years (min–max: 1–18 years) were evaluated. Materials included were composites, feldspathic ceramic, leucite reinforced glass ceramic, veneered and non-veneered lithium disilicate, veneered and monolithic zirconia and alumina. Processing techniques were stone dies incremental techniques and poured with dental stone, indirect die cast method, framework laminated with a veneering with lost-wax glaze technique, chairside and labside CAD/CAM techniques, vacuum injection mold techniques. Used luting agents were adhesive bonding systems, resin cements (Panavia, Multilink, Variolink, Tetric, Multibond) and glass ionomer luting cements (Ketac).

Table 2. Quality assessment of included studies using the Newcastle–Ottawa scale.

Study	Selection				Comparability	Outcome			Numbers of Stars (Out of 8)
	1	2	3	4	1	1	2	3	
Botto et al. 2016	–	★	–	–	★	★	★	★	5
Guess et al. 2014	★	★	★	★	★	★	★	★	8
Dhima et al. 2014	–	–	–	–	★	★	★	★	4
Dukic et al. 2010	–	–	–	–	★	★	★	★	4
Azevedo et al. 2012	★	★	★	★	★	★	★	★	8
Gehrt et al. 2013	★	★	★	★	★	★	★	★	8
Guess et al. 2014	★	★	★	★	★	★	★	★	8
Rauch et al. 2018	★	★	–	–	–	★	–	–	3
Reich et al. 2013	★	★	–	–	–	★	–	–	3
Santos et al. 2016	★	★	★	★	★	★	★	★	8
Santos et al. 2013	★	★	★	★	★	★	★	★	8
Taschner et al. 2012	★	★	★	★	★	★	★	★	8
Taskonak et al. 2006	★	★	–	★	★	★	★	–	6
Krejci et al. 1992	–	★	–	–	–	★	–	–	2

★: Each star corresponds to the subsection of quality assessment criteria.

3.3. Risks of Bias in Individual Studies

Quality and risk bias assessment of the RCTs is summarized in Figure 2 and for the case control and cohort studies reviewed in Table 1.

The Cochrane collaboration tool showed an overall low risk of bias in all the included studies. Some studies did not report enough information about the sequence generation process to allow an evaluation of either "low risk" or "high risk" (Mittal et al. 2016, Frankenberger et al. 2009). Others did not describe the allocation concealment or provide enough detail (Mittal et al. 2016, Dondi dall'Orologio et al. 2014, Ozsoy et al. 2016, Frankenberger et al. 2009). Just one study showed a high risk for the blinded outcome (Beder et al. 2016). According to the NOS scale, one study scored 2 points, two obtained 3 points, two 4 points, one 5 points, and finally seven studies obtained 8 points. These scores reflect an adequate quality of the studies included in this review.

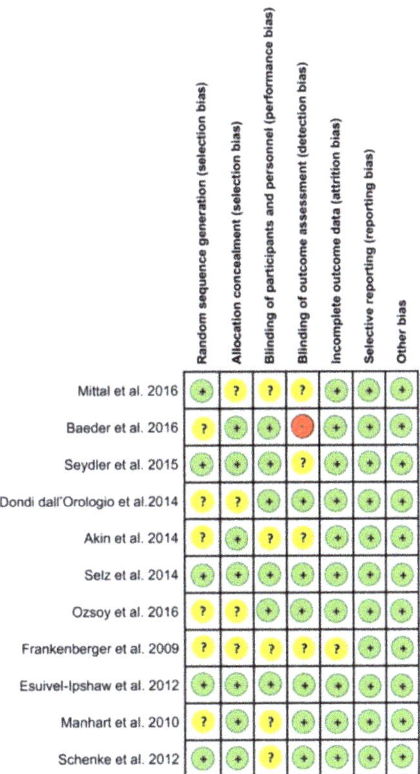

Figure 2. Summary of the Cochrane collaboration tool for assessing risk of bias for randomized controlled trials.

3.4. Meta-Analysis

Meta-analyses were performed based on 25 studies. The overall survival and success ratios of partial and full crowns were obtained using forest and funnel plots at two different time ranges: (a) ≤24 months (m); and (b) ≥36 months (m) (Table 3).

Table 3. Characteristics of included studies.

Author/ Publication Year	Journal	Study Type	Patients (N)	Restoration (n)	Ratio (n/N)	Follow-Up	Drop-Out	Material	Manufacturing Technique	Luting Agent	Failure	Survival	Success	Outcome
Mittal et al. 2016 [36]	J Clin Ped Dent	RCT	50	50	1	36 Months	0	IRC (indirect resin composite) vs. SSC (stainless steel crowns)	IRX (Composite 3-M Espe) SSC	IRC (Dual cure resin cement RelyX) SSC (luting glass ionomer cement Fuji 1)	IRC (3) SSC (2)	IRC (82.9%) SSC (90.7%)	IRC (100%) SSC (95%)	Modified FDI criteria' Dental chair side treatment time and postoperative acceptability Marginal integrity IRC < SSC Time/esthetic: IRC > SSC
Botto et al. 2016 [23]	Am J Dent	Retrospective	47	93	93/47	5–18 years		13 onlays feldspathic porcelain (Vitadur Alpha), 78 onlays, 2 inlays IPS-Empress		RelyX	6 (6.5%)	87 (93.5%)	81 (93%)	Gender, age, tooth preparation, number, type, extent, location, quality and survival of the restorations, ceramic materials, luting resin cements, parafunctional habits, secondary caries and maintenance therapy, marginal adaptation, marginal discoloration, occlusal surfaces
Baader et al. 2016 [22]	J Adhes Dent	RCT	34	68	2	6.5 years	16 patients	Vita Mark II; Cerec 3D	Indirect cast	RelyX With/without enamel etching	16:11 RXU PCCs and 5 RXU+E PCCs failed. The reasons for this were fractures of restorations (3 RXU, 4 RXU+E), debonding of PCCs with no possibility of recementation (4 RXU), one endodontic treatment followed by renewal of the restoration (1 RXU) and one renewal of the PCC due to caries at another site of the tooth, necessitating a full-crown preparation (1 RXU)	RXU of 60% and for RXU+E of 82%,	–	Modified USHPS postoperative hypersensitivity, anatomic form, marginal adaptation, marginal discoloration, surface texture and recurrent caries.

Table 3. *Cont.*

Author/ Publication Year	Journal	Study Type	Patients (N)	Restoration (n)	Ratio (n/N)	Follow-Up	Drop-Out	Material	Manufacturing Technique	Luting Agent	Failure	Survival	Success	Outcome
Seydler et al. 2015 [44]	J Prosthet Dent	RCT	60	60	1	2 years	0	veneered zirconia (VZ) group were made of zirconia frameworks veneered with CAD/CAM-produced lithium disilicate ceramic; monolithic lithium disilicate (MLD) ceramic	MLD crowns were milled (Cerec MC XL; Sirona Dental Systems) from a block (IPS e.max CAD; Ivoclar Vivadent AG) VZ crowns were milled from a zirconia blank (IPS e.max ZirCAD; Ivoclar Vivadent AG); the veneer structure was milled from an IPS e.max CAD lithium disilicate blank (both, Cerec MC XL; Sirona Dental Systems).	(Multilink; Ivoclar Vivadent AG	none	100		USHPS The quality of marginal fit, color and technical and biologic complications were recorded.
D'all' Orologio et al. 2014 [24]	Am J Dent	RCT	50	150		8 years	30 restoration, 10 patients	100 with the new restorative material, 50 with the composite as control, XP Bond ceram.x Duo Esthet.X		bonding system (XP Bond)	7% There were eight failures in the experimental group and four failures in the control group here were two key elements of failure: the presence of sclerotic dentin and the relationship between lesion and gingival margin.	93%		Retention, Sensitivity, Marginal Integrity, Caries, Contour
Akin et al. 2014 [20]	J Prosthodont	RCT	15	30	2	2 years	0	all-ceramic crowns	fabricated with CAD/CAM and heat-pressed (HP) techniques	Variolink II/Syntac; Ivoclar Vivadent	0	100		Porcelain fracture and partial debonding that exposed the tooth structure, secondary caries, extraction of abutment teeth and impaired esthetic quality or function were the main criteria for irreparable failure.

Table 3. *Cont.*

Author/ Publication Year	Journal	Study Type	Patients (N)	Restoration (n)	Ratio (n/N)	Follow-Up	Drop-Out	Material	Manufacturing Technique	Luting Agent	Failure	Survival	Success	Outcome
Guess et al. 2014 [32]	Int J Proshodont	Prospective clinical study	25	86	86/25	7 years	11 patients	all-ceramic veneers with overlap (OV) and full veneer (FV) preparation designs	Leucite-reinforced glass-ceramic veneers (IPS Empress, Ivoclar Vivadent)	(Variolink II, Ivoclar Vivadent	One OV restoration fractured (Figure 2a). cohesive ceramic fracture and crack formation within the restoration material were noted in 12 patients.	100% for FV restorations and 97.6% for OV restorations.	0.85 (CI: 0.70 to 1.00) for the FV restorations and 0.70 (CI: 0.45 to 0.95) for the OV restorations	USPHS criteria
Selz et al. 2014 [43]	Clin Oral invest	RCT	60	149	>2	5 years			In-Ceram Alumina crowns	62 Panavia, 59 Super-Bond C&B; 28 Ketac	Endodontic treatment was carried out on 7.4% of all abutment teeth and 5.4% revealed secondary caries. Unacceptable ceramic fractures were observed in 7.4%. Debonding was a rare complication (1.3%).	91.6% for Super Bond C&B-, 87.4% for Ketac Cem- and 86.3% for Panavia F-bonded	82,2 Panavia, 88.7 Super-Bond C&B; 80.1 Ketac	secondary caries, clinically unacceptable fractures, root canal treatment and debonding.
Özsoy et al. 2016 [38]	JAST	RCT	60	67	>1	2 years	2 teeth	indirect composite onlays and overlays	indirect composite (Gradia, GC, Japan)	Variolink II		100		Anatomy, marginal adaptation, marginal discoloration, color match, surface roughness, caries
Dhima et al. 2014. CAVE: Tooth & implant-borne [25]	J Prosthet Dent	Retrospective	59	226	226/59	5 years		Ceramic single crown				95%		
Dukic et al. 2010 [26]	Oper Dent	Prospective study	51	71	71/51	3 years		Ind. comp	35 Ormocer, Admira, 36 Grandio	Grandio with Voco Bifix QM	0	100	No significance Ormocer/Grandio	Modified USHPS
Azevedo et al. 2012 [21]	Braz Dent J	Prospective study	25	42	42/25	1 year	0	23 etched, non-etched, 19 etched (Filtek Supreme XT; 3M ESPE)	stone dies by the incremental technique using a LED device with power density of 1000 mW/cm^2	Etched group (ETR)—selective enamel phosphoric-acid etching + RelyX Unicem clicker; 2. Non-etched group (NER)—RelyX Unicem	0	100		More than 99% of the scores were considered clinically excellent (Alpha 1) or good (Alpha 2). Only 3 scores (0.9%) were classified as clinically sufficient (Bravo): 2 from ETR group (MS = 1, Figure 3; SE = 1) and 1 from NER group

Table 3. *Cont.*

Author/ Publication Year	Journal	Study Type	Patients (N)	Restoration (n)	Ratio (n/N)	Follow-Up	Drop-Out	Material	Manufacturing Technique	Luting Agent	Failure	Survival	Success	Outcome
Fasbinder et al. 2010 [28]	J Am Dent Assoc	Prospective study	43	62	62/43	2 years	1.6%	lithium disilicate (IPS e.max CAD, Ivoclar Vivadent, Amherst, N.Y.) all-ceramic crowns.	chairside computer-aided design/computer-aided manufacturing (CAD/CAM) system (CEREC 3, Sirona Dental Systems, Charlotte, N.C.) e.max CAD Crystall./Glaze paste (Ivoclar Vivadent) with shade tints	Multilink Automix, Ivoclar Vivadent OR: experimental self-adhesive, dual-curing cement (EC) developed by Ivoclar Vivadent.	0	100		Modified USHPS
Frankenberger et al. 2008 [29]	J Adhes Dent	Controlled clinical trial	34	96	96/34	12 years	40%	Leucite-reinforced glass ceramic IPS Empress	according to the manufacturer's instructions	4 cements: Dual Cement (*n* = 9), Variolink Low (*n* = 32), Variolink Ultra (*n* = 6) and Tetric (*n* = 49) (all Ivoclar Vivadent).	16% (15/96) without dropout	58 86%		luted with dual-cured resin composites revealed significantly fewer bulk fractures Surface roughness (loss of gloss), color match (improving with time), marginal integrity (distinct deterioration with marginal fractures in two cases with charlie scores after 12 years), tooth integrity (enamel cracks, one case rated Delta), inlay integrity (continuous deterioration over time, predominantly chipping of the ceramic, two charlie and two delta scores) and hypersensitivity

Table 3. *Cont.*

Author/ Publication Year	Journal	Study Type	Patients (N)	Restoration (n)	Ratio (n/N)	Follow-Up	Drop-Out	Material	Manufacturing Technique	Luting Agent	Failure	Survival	Success	Outcome
Frankenberger et al. 2009 [30]	Dent Mater	RCT	39	98	98/39	4 years	3%	Cergogold glass ceramic inlays	One dental ceramist produced all inlays according to the manufacturer's instructions and recommendations within 2 weeks after impression taking.	Multibond and Definite Ormocer resin composite Definite Multibond/Definite ($n = 45$) Syntac/Variolink Ultra ($n = 53$)	21 restorations had to be replaced due to inlay fracture ($n = 11$), tooth fracture ($n = 4$), hypersensitivities ($n = 3$) or marginal gap formation ($n = 3$).	77 survival rate 89.9%,	significantly changed over time: color match, marginal integrity, tooth integrity, inlay integrity, sensitivity, hypersensitivity and X-ray control Color match was inferior for Variolink, but only at the 2-year recall (Mann–Whitney U-test, $p < 0.05$), marginal integrity was inferior for Variolink, but only at the 0.5 and 1-year recall (Mann–Whitney U-test, $p < 0.05$) and proximal contacts were inferior in the definite group, but only at baseline	criteria marginal integrity, tooth integrity and inlay integrity
Gehrt et al. 2013 [31]	Clin Oral invest	prospective study	41	104	104/41	9 years	4 patients, 10 crowns	lithium-disilicate crowns	frameworks were laminated by a prototype of a veneering material combined with an experimental glaze. lost-wax technique	adhesively luted (69.2%) or inserted with glass–ionomer cement (30.8%). adhesively luted (IPS Ceramic etchant/Monobond S/dual-cured Variolink II, Ivoclar Vivadent) and 32 (30.8%) crowns were inserted with glass–ionomer cement (Vivaglass, Ivoclar Vivadent)	4 (4.3%)	97.4% after 5 years and 94.8% after 8 years	There were five rated technical complications (5.3%). Three crowns (3.3%) suffered from minor chipping of the veneering material. Major chippings did not occur. There were four biologic complications (4.3%). Two anterior crowns (2.1%) had to be treated endodontically 94.7 months after insertion.	Biologic complications such as loss of vitality joined by declined endodontic condition, endodontic disease and occurrence of caries & Technical complications such as loss of retention, minor chipping

Table 3. *Cont.*

Author/ Publication Year	Journal	Study Type	Patients (N)	Restoration (n)	Ratio (n/N)	Follow-Up	Drop-Out	Material	Manufacturing Technique	Luting Agent	Failure	Survival	Success	Outcome
Guess et al. 2014 [32]	Int J Proshtodont	Prospective Study	25	80	80/25	7 years	42 restorations	40 lithium disilicate pressed PCRs (IPS e.max-Press, Ivoclar Vivadent) and 40 leucite-reinforced glass–ceramic CAD/CAM PCRs (ProCAD, Ivoclar Vivadent).	computer-aided design/computer-assisted manufacture (CAD/CAM) ProCAD, Ivoclar Vivadent; Cerec 3 InLab, Sirona	hybrid composite resin material (Tetric/Syntac Classic, Ivoclar Vivadent)	1 restoration	100% for pressed PCRs and 97% for CAD/CAM PCR	No secondary caries, endodontic complications or postoperative complaints were ob- served. Minimal cohesive ceramic fractures (Figure 2a,b) were noted in 5 patients, but all affected restorations remained in situ 0.84 (CI: 0.70–0.98) for the pressed PCRs and 0.58 for the CAD/CAM PCRs (CI: 0.38–0.78).	modified United States Public Health Service (USPHS)
Murgueitio et al. 2012 [37]	J Prosthodont	Prospective study	99	210	210/99	3 years	?	leucite-reinforced IPS Empress Onlays and Partial Veneer Crowns	the manufacturer's instructions using the vacuum injection mold technique for leucite-reinforced ceramic material (IPS Empress).	Variolink II, Ivoclar Vivadent	The mode of failure was classified and evaluated as (1) adhesive, (2) cohesive, (3) combined failure, (4) decementation, (5) tooth sensitivity and (6) pulpal necrosis 33%	96.66%	Increased material thickness produced less probability of failures. Vital teeth were less likely to fail than nonvital teeth. Second molars were five times more susceptible to failure than first molars. Tooth sensitivity postcementation and the type of opposing dentition were not statistically significant in this study.	USPHS

Table 3. *Cont.*

Author/ Publication Year	Journal	Study Type	Patients (*N*)	Restoration (*n*)	Ratio (*n*/N)	Follow- Up	Drop- Out	Material	Manufacturing Technique	Luting Agent	Failure	Survival	Success	Outcome
Esuivel-Ipshaw et al. 2012 [27]	J Prosthodont	RCT	32	37	37/32	3 years	1 restoration	(1) metal-ceramic crown (MC) made from a Pd–Au–Ag–Sn–In alloy (Argedent 62) and a glass- ceramic veneer (IPS d.SIGN veneer); (2) non-veneered (glazed) lithium disilicate glass–ceramic crown (LDC) (IPS e.max Press core and e.max Ceram Glaze); and (3) veneered lithia disilicate glass–ceramic crown (LDC/V) with glass–ceramic veneer (IPS Empress 2 core and IPS Eris).		Variolink II, Ivoclar Vivadent	0?	100?	between years 2 and 3, gradual roughening of the occlusal surface occurred in some of the ceramic-ceramic crowns, possibly caused by dissolution and wear of the glaze. Statistically **significant** differences in surface texture ($p = 0.0013$) and crown wear ($p = 0.0078$) were found at year 3 between the metal-ceramic crowns and the lithium-disilicate-based crowns.	tissue health, marginal integrity, secondary caries, proximal contact, anatomic contour, occlusion, surface texture, cracks/chips (fractures), color match, tooth sensitivity and wear (of crowns and opposing enamel). Numeric rankings ranged from 1 to 4, with 4 being excellent and 1 indicating a need for immediate replacement.
Manhart et al. 2010 [35]	Quintessence Int	RCT	89	155	155/89	3 years	Artglass inlays (35%) and Charisma inlays (21%)	Resin composite	The inlays were postcured in a light oven (Uni-XS, Heraeus Kulzer)	adhesive system Solid Bond (Heraeus Kulzer)	five Artglass and 10 Charisma inlays failed mainly because of postoperative symptoms, bulk fracture and loss of marginal integrity	5 Artglass and ten Charisma inlays had to be (3 years)	Small Charisma inlays exhibited a statistically **significant better** performance for the "integrity of the restoration" parameter ($p = 0.022$).	Modified USPHS
Rauch et al. 2018 [39]	Clin Oral invest	Prospective	34	41	41/34	10 years	15 restorations	monolithic lithium disilicate crowns	chairside CAD/CAM technique.	Multilink Sprint, Ivoclar Vivadent	5 five failures occurred due to one crown fracture, an abutment fracture, one endodontic problem, a root fracture and a replacement of one crown caused by a carious	24/29	Due to the small amount of technical complications and failures, the clinical performance of monolithic lithium disilicate crowns was completely satisfying.	Modified USHPS

Table 3. *Cont.*

Author/ Publication Year	Journal	Study Type	Patients (*N*)	Restoration (*n*)	Ratio (*n/N*)	Follow- Up	Drop- Out	Material	Manufacturing Technique	Luting Agent	Failure	Survival	Success	Outcome
Reich et al. 2013 [40]	Clin Oral invest	Prospective clinical trial	34	41	41/34	4 years	12 restoration	lithium disilicate crowns	chairside CAD/CAM technique (Cerec)	Multilink Sprint (Ivoclar-Vivadent)	1 failure 96.3% after 4 years according to Kaplan–Meier	28	The complication-free rate comprising all events after 4 years was 83%, whereas the rate dropped down to 71% after 4.3 years	Modified USHPS
Santos et al. 2016 [41]	Clin Oral invest	Prospective clinical trial	35	86	86/35	5 year	17.91% restoration	sintered Duceram (Dentsply Degussa) and pressable IPS Empress (Ivoclar Vivadent).	poured with dental stone type IV (Durone, Dentsply).	Variolink II, Ivoclar Vivadent	8 failures Four IPS restorations were fractured, two restorations presented secondary caries (one from IPS and one from Duceram) and two restorations showed unacceptable defects at the restoration margin and needed replacement (one restoration from each ceramic system).	56	87% significant differences in relation to marginal discoloration, marginal integrity and surface texture between the baseline and five-year recall for both systems	Modified USHPS
Schenke et al. 2012 [42]	Clin Oral invest	RCT	29	58	58/29	2 years	0	ceramic blocks (Vita 3D Master CEREC Mark II, CAD/CAM designed and machined with the CEREC III system (Sirona CEREC III Software Version 3.0 (600/800), Sirona, Bensheim, Germany)	an indirect method on a die cast	RelyX Unicem with/without enamel etching	4 failures	54	Statistically significant changes were observed for marginal adaptation (MA) and marginal discoloration (MD) between BL and 2 years, but not between the two groups (RXU, RXU+E). Percentage of alfa values at BL for MA (RXU, 97% and RXU+E, 100%) and for MD (RXU, 97% and RXU+E, 97%) decreased to RXU, 14% and RXU+E, 28% for MA and to RXU, 50% and RXU+E, 59% for MD after 24 months.	Modified USHPS

Table 3. *Cont.*

Author/ Publication Year	Journal	Study Type	Patients (N)	Restoration (n)	Ratio (n/N)	Follow- Up	Drop- Out	Material	Manufacturing Technique	Luting Agent	Failure	Survival	Success	Outcome
Taschner et al. 2012 [45]	Dent Mater	Prospective controlled clinical study	30	83	83/30	2 years	0	IPS-Empress	at a commercial dental laboratory according to manufacturer's instructions	Group 1: 43 inlays/onlays were luted with RX; group 2: 40 inlays/onlays were luted with Syntac/Variolink II low viscosity (SV, Ivoclar Vivadent).	1	82/83 restorations	Indirect restorations luted with RX showed lower tooth and marginal integrity compared to the multistep approach.	Surface roughness, Color match, Anatomic form, Marginal integrity, Integrity tooth, Integrity inlay, Proximal contact, Changes in sensitivity, Radiographic check, Subjective satisfaction
Taskonak et al. 2006 [46]	Dent Mater	Prospective clinical trial	15	40	40/15	2 years		lithia-disilicate-based all-ceramic (Empress II) FDP/Crowns (20 FDPs/ 20 crowns)			10 (50%) catastrophic failures of FPDs occurred			marginal adaptation, color match, secondary caries and visible fractures in the restorations
Krejci et al. 1992 [34]	Quintessence Int	Prospective clinical trial	10	10	1	1.5 years	0	IPS/Empress Inlays	According to manufacturer's instruction	Dual curing composite, Dual cement, Vivadent, Inc.	0	100	1 hypersensitivity, Discoloration at the marginal	Modified USHPS
Azevdo et al. 2012 [21]	Braz Dent J	Prospective clinical trial	25	42	42/25	1 year	0	Indirect resin composite	The composite resin restorations were built over plaster casts using the incremental technique with a LED device for light-curing the increments	1. Etched group (ETR)—selective enamel phosphoric-acid etching + RelyX Unicem clicker; 2. Non-etched group (NER)–RelyX Unicem RelyX	0	100	More than 99% of the scores were considered clinically excellent (Alpha 1) or good (Alpha 2) (Figure 2). Only 3 scores (0.9%) were classified as clinically sufficient (Bravo): 2 from ETR group (MS = 1, Figure 3; SE = 1) and 1 from NER group (SE).	Modified USHPS

3.5. Survival Ratios

As for the survival ratios it could be observed that at the time frame up to 24 m the estimated survival is 99%, while after at least 36 m it dropped to 95%. Forest and funnel plots ≤24 m revealed homogeneous results (heterogeneity $I^2 = 47\%$, $p = 1.00$) and low suspicion for a publication bias, while forest and funnel plots ≥36 m demonstrated heterogeneous results (heterogeneity $I^2 = 93\%$, $p < 0.01$) and a slight suspicion of a publication bias (Figures 3–7).

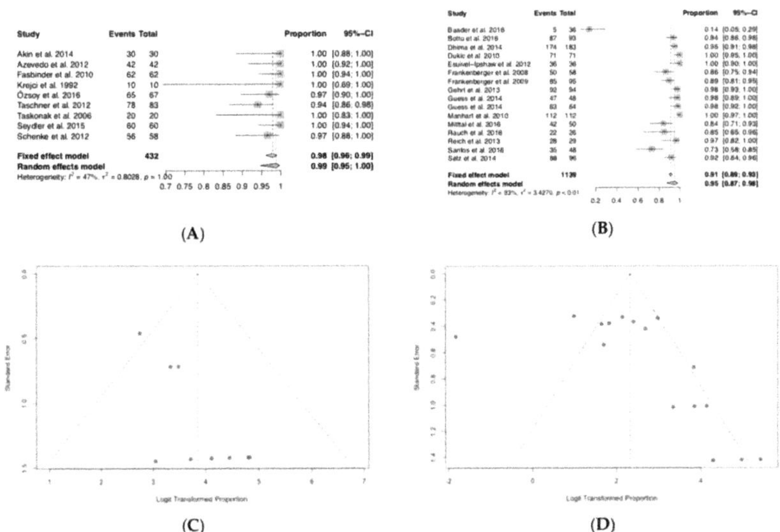

Figure 3. Survival ratios of all included specimens. (**A**) Forest plot ≤24 months; (**B**) forest plot ≥36 months; (**C**) funnel plot ≤24 months; (**D**) funnel plot ≥36 months.

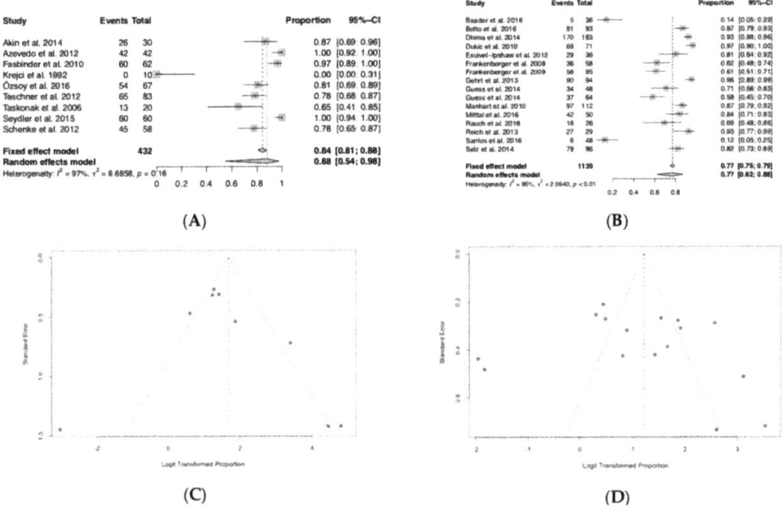

Figure 4. Success ratios of all biologic, technical and esthetical aspects. (**A**) Forest plot ≤24 months; (**B**) forest plot ≥36 months; (**C**) funnel plot ≤24 months; (**D**) funnel plot ≥36 months.

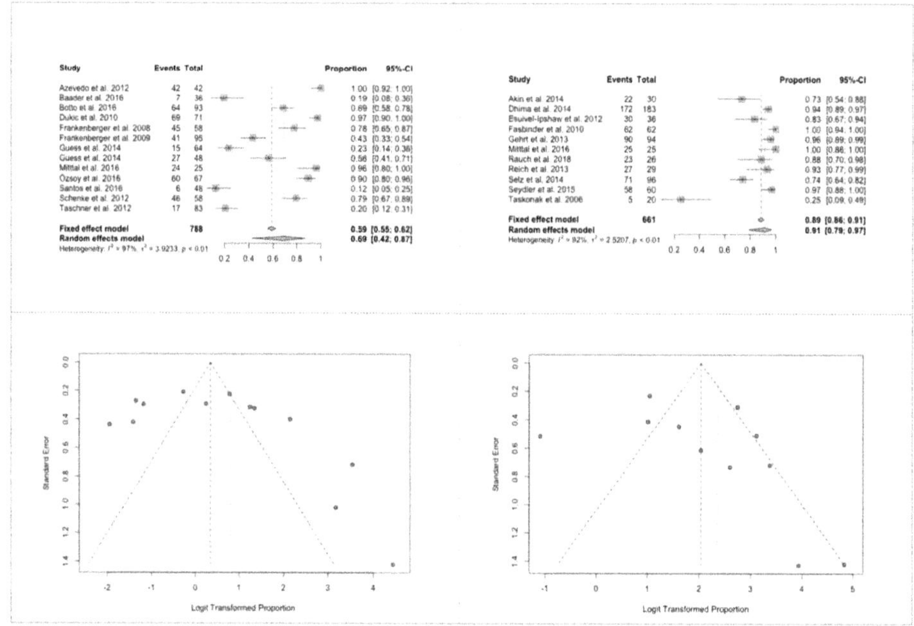

Figure 5. Success ratios of all biologic aspects. (**A**) Forest plot for partial and (**B**) full crowns; (**C**) funnel plot for partial and (**D**) full crowns.

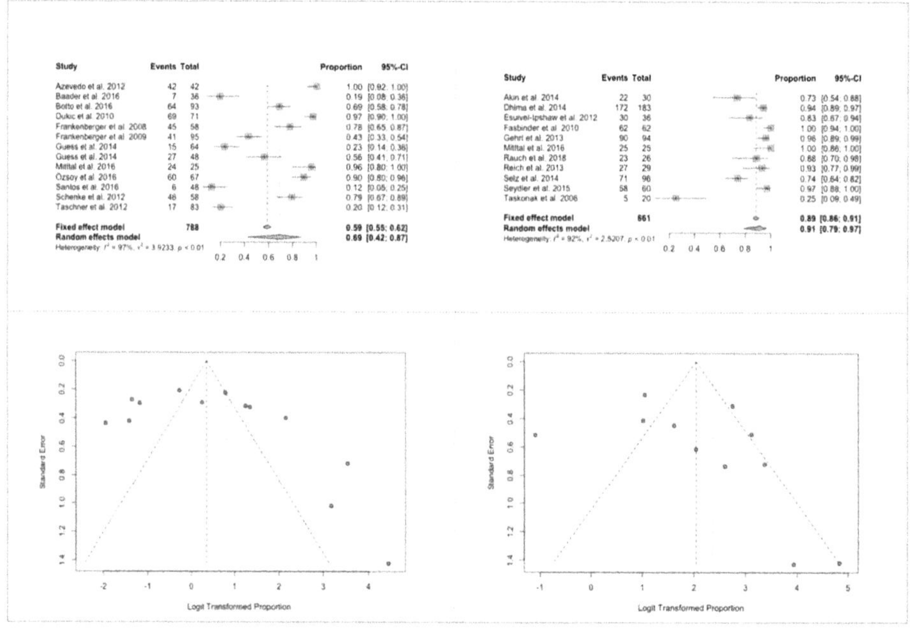

Figure 6. Success ratios of all technical aspects. (**A**) Forest plot for partial and (**B**) full crowns; (**C**) funnel plot for partial and (**D**) full crowns.

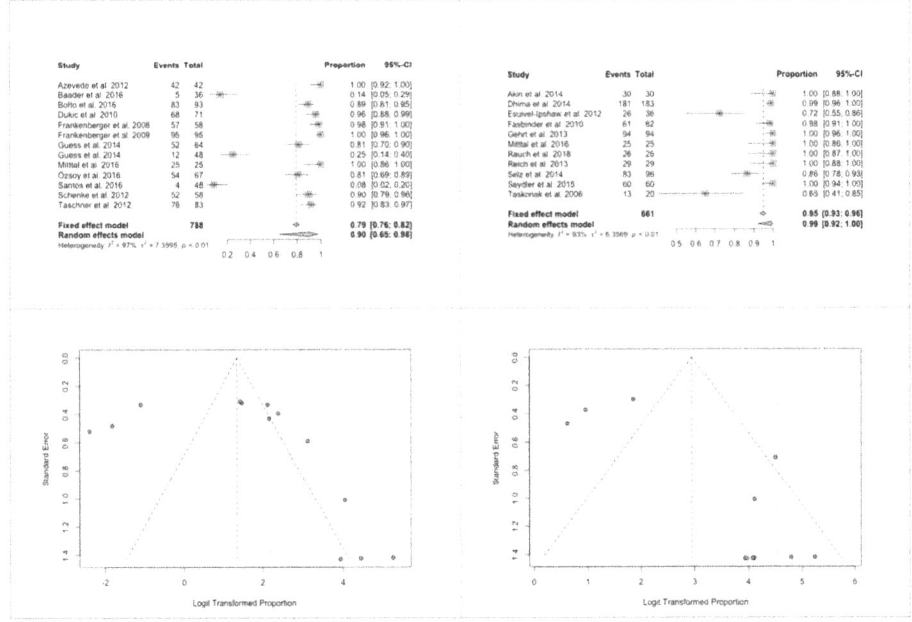

Figure 7. Success ratios of all esthetical aspects. (**A**) Forest plot for partial and (**B**) full crowns; (**C**) funnel plot for partial and (**D**) full crowns.

3.6. Success Ratios of All Biologic, Technical and Esthetical Aspects

The estimated success ratio at ≤24 m was 88% (95% COI: 0.54–0.98), while after at least 36 m it dropped to 77% (95% COI: 0.62–0.88). Forest plot ≤24 m revealed not strongly homogeneous results (heterogeneity I^2 = 97%, p = 0.16). However, heterogeneity is not statistically significant. Funnel plot ≤24 m showed very small and extremely large values. Forest plot ≥36 m demonstrated highly heterogeneous results (I^2 = 95%, $p < 0.01$). The plot illustrates the studies with the remarkably noticeable results. The wide range and heterogeneity of included material types (composites, feldspathic ceramic, leucite reinforced glass ceramic, veneered and non-veneered lithium disilicate, veneered and monolithic zirconia and alumina), processing techniques and luting agents did not allow any further statistical analysis as regards to an analysis for the material type only.

3.7. Success Ratios of All Biologic Criteria

The estimated success ratio at ≤24 m was 88% (95% COI: 0.58–0.97), while after at least 36 m it dropped to 75% (95% COI: 0.56–0.88). Results of the forest Plot <24 m presented very heterogeneous results (I^2 = 96%, $p < 0.01$). The funnel Plot <24 m showed, apart from the before mentioned two studies the distribution of published results, a slight skew in favor of high success rates, indicating a possible publication bias.

For forest plot >36 m (I^2 of 97%, $p < 0.01$) these study results were also very heterogeneous, and a large dispersion could be observed. In general, the results of the funnel Plot >36 m presented great variability among the published studies.

3.8. Success Ratios of All Technical Criteria

After 2 years the estimated success ratio was 90% (95% COI: 0.74–0.97), while after 3 years it dropped to 74% (95% COI: 0.50–0.89). Forest plot <24 m presented (I^2 of 93%, $p < 0.01$) heterogenous results and after 3 years (I^2 of 97%, $p < 0.01$). The funnel plot after 2 years showed a tendency towards overproportioned high success rates studies.

3.9. Success Ratios of All Esthetical Criteria

The success ratios are very high at 24 m 96% (95% COI: 0.87–0.99) and dropped very slightly after 36 m 95% (95% COI: 0.78–0.99). Forest plot <24 m presented (I^2 of 86%, $p = 0.08$) statistically insignificant heterogenous results and after 3 years (I^2 of 97%, $p < 0.01$) heterogenous results, because of 3 studies showing only 8%–25% success rates, while all other included studies presented ≥72%. Funnel plot did not show any bias during the first 2 years, while the 3 mentioned studies presented very low success rates, many others shower too high success rates. The overall results did not show any bias.

The biologic success rates of full crowns were much higher than those of partial crowns. Forest plot of partial (I^2 of 97%, $p < 0.01$) and full (I^2 of 92%, $p < 0.01$) crowns showed very heterogeneous studies, while funnel plots exhibited a possibility of publication bias for partial and low possibility of bias for full crowns, even though there was a slight hint of too high success rates.

The technical success rates of full crowns were much higher and significantly different ($p < 0.05$) compared to partial crowns. Forest plot showed heterogeneous results for partial crowns (I^2 of 98%, $p < 0.01$) and homogeneous results for full crowns (I^2 of 66%, $p = 0.63$). Funnel plot for partial crowns showed a rather unlikely publication bias, the variation is very high, for full crowns the results were all in the expected range, with an asymmetric distribution. Higher success rates were often demonstrated as statistically expected. A publication bias seems to be possible.

The esthetical success of partial crowns was also higher compared to full crowns, but not as high as it was for biologic and technical success rates. Forest plot of partial crowns (I^2 of 97%, $p < 0.01$) revealed heterogeneous results with three studies showing low success rates, the funnel plot exhibited at both sides a high prevalence of studies in the upper and lower end of the graph with more studies presenting high results. The forest plot of full crowns (I^2 of 93%, $p < 0.01$) showed also heterogeneous results, because of the two studies Esquivel-Ipshaw et al. and Taskonak et al. reporting low results. The funnel plot showed many results with high success rates and three with low results. Because of the sample size it was not possible to conclude if a bias was possible or not.

4. Discussion

This systematic review including meta-analysis was conducted to evaluate the clinical short- and long-term survival rates and biologic, technical and esthetical success ratios of partial and full crowns using hybrid polymer and ceramic CAD/CAM materials.

Some data were reported on CAD/CAM processing methods regarding survival and clinical survival rates. However, to best of author's knowledge, no similar systematic review based on hybrid polymer and ceramic materials on survival and complications rates has been published yet. Since these materials have been developed recently, their indications and clinical applicability are still being studied. In the present review, the existence of a great variety and heterogeneity of hybrid polymer and ceramic materials and their indications has been observed.

The meta-analysis of this study was performed for mean long-term survival rates and for biologic, technical and esthetic complication ratios for partial vs. full crown reconstructions at two different follow-up periods. Due to the variety of the CAD/CAM materials, their differing compositions and the lack of homogeneity, the variable "material" could not be included in the meta-analysis. This finding was also observed in the systematic review by Alves de Carvalho et al. [47]. investigating clinical survival rates in single restorations using CAD/CAM technologies with a minimum follow-up of three years, describing a great variety of studies analyzing different materials. Their results are in agreement with the present systematic review related to the heterogeneity caused by the variety of the materials assessed [47]. The review of Rodrigues et al. included studies on CAD/CAM materials for single crown, multiple- unit or partial ceramic crown with a 24 to 84-month follow-up based on the longevity and failures rates, suggesting that the longevity of CAD/CAM restorations is lower compared to the conventionally fabricated restorations [48], as they presented a 1.84 higher failure rate during a follow-up period of 24 to 84 months. However, the results of the present systematic review showed that when partial and full crown reconstructions made of hybrid polymer and ceramic CAD/CAM materials were analyzed, the overall survival rate was 99% (0.95–1.00) up to 24 months and dropped to 95% (0.87–0.98) at ≥36 months.

These results were assessed based on the restoration type, given higher success rates for the overall clinical performance in full crown reconstructions compared to partial crowns. Similar data were found for survival rates of full crowns, estimated 5-year survival rate for leucite or lithium-disilicate reinforced glass ceramic (96.6%) and sintered alumina and zirconia (96%) were similar [16]. For partial restorations, our results are also in agreement with the literature, Sampaio FBWR et al. found estimated survival rates for CAD/CAM of 97% after five years [49].

Current trends for material selection in tooth-supported single restorations showed that, both clinicians and patients are favoring esthetic and nonmetallic restorations. However, for full crowns, literature is still supporting the porcelain-fused-to-metal crowns as the gold standard, with results of 5-year survival rates exceeding 95% [16,50]. Furthermore, in terms of longevity, the literature showed that full and partial CAD/CAM ceramic crowns have lower long-term survival compared to the ones produced through conventional techniques [48]. Analyzing the results of other studies of full ceramic crowns, the literature provided data on leucite or disilicate reinforced ceramics survival rates of 96.6% and 95%, respectively [16], these results are comparable to those found in this review.

The other large CAD/CAM processed material group was zirconia, showing a 5-year survival of 91.2% (82.8–95.6%) [16]. Digital developments, new materials and advanced processing techniques enabled the minimal invasive approach in dentistry throughout partial restorations. Partial crowns have been widely used for years, as composite resins were a less predictable treatment option for direct restorations. Among other factors, the longevity of partial restorations depended on the restorative material, the patient and the experience of the clinician. Previous reviews show survival rates of 92% and 95% at five years and 91% at 10 years, (Morimoto et al.) or in a more recent study the survival rate data for inlays was 90.89% and 93.50% in a follow-up period of one to five years [51].

Gold alloys have served as gold standard for partial crowns for years [52]. However, the increasing price of gold and the high esthetic demands of patients have caused advancement of materials such as hybrid polymer and ceramic CAD/CAM materials. The current evidence of gold restorations is limited, suggesting a survival rate of 95.4% observed in a retrospective, clinical study studying 1314 gold restorations; whereas inlays had a failure rate of 4.7% after more than 20 years [53]. Another study evaluated 391 posterior gold inlays during a mean follow-up period of 11.6 years and observed 82.9% of success rate and a 6.4% failure rate [52].

The development, evolution and improvement of composite resins, high strength ceramics and adhesive techniques have allowed the development of hybrid materials to compensate the deficiencies and limitations of gold alloys. In this regard, a systematic review evaluating 5811 restorations showed a survival rate of feldspathic porcelain and glass–ceramics for five-year follow-up of 95% and at the 10-year follow-up of 2154 restorations, a survival rate of was 91% [54].

In addition to ceramics and gold alloys composite resin materials have been increasingly used due to improvements in the composition and thereby related mechanical properties. Previous reviews on resins were inconclusive whether longevity and survival rates of resins are higher compared to ceramics [55]. However, a recent review on CAD/CAM materials for full and partial crowns that included resin-matrix ceramic showed an estimated survival rate after five years of 82.5% [47,49].

Survival rates are a reliable indicator to assess clinical performance. However, after placement and during exposure to the oral cavity restorations can present complications compromising their longevity, survival and clinical success. The clinical performance based on the overall success ratio of biologic, technical and esthetical aspects was 88% (0.54–0.98; ≤24 m) vs. 77% (0.62–0.88; ≥36 m) for the different follow-up periods. The meta-analysis could not find any significance regarding both follow-up time (≤24 m or ≥36 m) and their biologic, technical and esthetical (88% vs. 77%; 90% vs. 74%; 96% vs. 95%) outcome. However, it presented a significant difference in the technical clinical performance between full 93% (0.88–0.96) and partial 64% (0.34–0.86) crowns, in favor of full crown reconstructions ($p < 0.05$). Biologic and esthetical success rates of full crowns (91% (0.79–0.97) vs. 99% (0.92–1.00)) were comparable to those of partial crowns (69% (0.42–0.87) vs. 90% (0.65–0.98)). This meta-analysis suggests that in case of possible technical failure a full crown reconstruction should be preferred compared to a partial crown.

Restoration failures are considered as such when they need repair or replacement, the general assessment of these failures can also be considered in terms of success rates. The success rates, assessed by biologic, technical and esthetical aspects showed a decrease in success from 24 to 36 months. Compared to previous reviews the present data were higher compared to ceramic, zirconia and CAD/CAM single crown reconstructions reported in previous studies [16,48,56].

This study assessed the failures as either biologic, technical and esthetic complications, although during the analysis of the included studies, the lack of homogeneity of the results did not allow for its specific analysis resulting in an overall complications analysis. Considering tooth-supported restorations complications, the success ratio of biologic complications decreased in case of caries occurrence, loss of pulp vitality, endodontic treatment, tooth fracture and hypersensitivity. The present study showed a biologic success rate of 88% at the follow-up period ≤24 m and 75% at ≥36 m. The most frequent biologic complication reported in the literature was caries and loss of pulp vitality. Comparing full and partial restorations higher biologic complications rates (21% more) were observed in partial reconstructions. Considering the characteristics of partial restorations, in terms of indications and dental preparation, full crowns could hide biologic complications. Therefore, caries can be diagnosed more easily in partial crowns compared to full crowns and could explain the results obtained in this study. The biologic complications for full crowns were lower in metal-ceramic restorations than in full ceramic reconstructions [16,57].

Technical complications include ceramic fracture, cracks, core failure, chipping, problems with microleakage and the loss of retention. Ceramic chipping has been described as the most common technical complication, finding similar ranges for metal ceramics and fully ceramic crowns with no statistic differences between materials. However, the overall technical complication rates in the present study were higher compared to conventional and other CAD/CAM materials [16,57].

Missing clinical workflows and lacking experience with these newly developed materials could have an influence in the complications derived from bonding techniques and microleakage, factors such as polymerization of resin cement, degradation of adhesive, enzymatic degradation of bonding of these materials composition could explain the higher failure rates compared to conventional groups or metal-ceramic restorations regarding biologic and technical complication rates [51].

The technical complications in partial restorations are increasing during the follow-up assessment and between groups showing less complications for full coverage restorations. Considering the design and the manufacturing process, the complications could have been due to defects of the thickness and the roughness of the final preparations milled by CAD/CAM chairside units. Some partial crowns are designed and milled using chairside devices, lacking a verification of material thickness throughout the technician. Technical complications may also result in esthetical problems, such as discoloration or wear of glace. The results of the review for esthetic were higher at 36 months and however lower compared to the other studies. Considering the posterior localization of the restorations, it is possible that the results are due to the fact that materials are biomimetic, and patients do notice esthetical failures less than in the anterior sites.

Given these data, the results for the CAD/CAM crowns of hybrid polymer and ceramics are comparable regarding the 5-year success rates performance with other materials.

A tendency for lower failure rate for glass-matrix ceramics and polycrystalline ceramics compared to leucite and feldspathic ceramic could be observed. The high survival rate of glass-matrix ceramics—followed by resin-matrix ceramics and polycrystalline ceramics—should, however, be considered with caution due to shorter follow-up periods of the latter materials.

Dual curing agents are preferred for ceramic and resin-matrix ceramic inlays in order to compensate for the light transmission throughout the restoration and to allow complete polymerization even at the bottom of the cavity, where the access of LED curing light is limited [58]. Despite the wide diversity of included materials, most studies used chemically polymerized or LED polymerized dual curing agents. In studies where chemical and dual curing cements were compared, the dual curing systems achieved better results and presented lower failure rates compared to only chemical luting agents.

According to the findings of this systematic review, a great heterogeneity of the methodological data between studies with lack of properly comparisons (control and study groups), no homogeneous restoration material type groups and a short follow-up examination was observed. More homogeneous studies with the more comparable materials, manufacturing techniques and CAD/CAM software system with a control groups in a split-mouth randomized controlled study design should be conducted.

The density of published high survival rates is statistically slightly conspicuously high. In the lower section, there is the study by Baader et al. 2016, which stands out regarding the low survival ratios. However, further small studies, which published a low outcome are lacking.

5. Conclusions

Summary for success rates and different follow-up times including all biologic, technical and esthetical parameters could be listed as follows:

- All success rates decreased after 36 or more months compared to 24 months;
- The esthetic success rates were greatest, followed by the almost identical rate of technical and biologic success rates;
- There were no significant differences at the 95% level between the two follow-up times nor between the biologic, technical and esthetic aspects;
- Both the biologic, technical and esthetic success rates were higher for full crowns than for partial crowns;
- The technical success rate of full crowns was statistically significantly higher than that of partial crowns;
- The esthetic success rates are greater than the biologic or technical ones, but neither for the full crowns nor for the partial crowns these comparisons were of significance.

Author Contributions: Conceptualization, N.A.-H.H., M.Ö. and T.J.; methodology, N.A.-H.H., M.Ö. and T.J.; software, N.A.-H.H., M.Ö., P.M.-M. and T.J.; validation, N.A.-H.H., M.Ö., P.M.-M. and T.J.; formal analysis, N.A.-H.H., M.Ö., P.M.-M. and T.J.; investigation, N.A.-H.H., M.Ö. and T.J.; resources, N.A.-H.H., M.Ö. and T.J.; data curation, N.A.-H.H., M.Ö. and T.J.; writing—original draft preparation, N.A.-H.H., M.Ö., P.M.-M. and T.J.; writing—review and editing, N.A.-H.H., M.Ö., P.M.-M. and T.J.; visualization, N.A.-H.H., M.Ö., P.M.-M. and T.J.; supervision, N.A.-H.H., M.Ö., P.M.-M. and T.J.; project administration, N.A.-H.H., M.Ö., P.M.-M. and T.J.; funding acquisition, none. All authors have read and agreed to the published version of the manuscript.

Funding: This research received no external funding.

Conflicts of Interest: The authors declare no conflicts of interest.

References

1. Ruse, N.D.; Sadoun, M.J. Resin-composite blocks for dental CAD/CAM applications. *J. Dent. Res.* **2014**, *93*, 1232–1234. [CrossRef] [PubMed]
2. Mainjot, A.K.; Dupont, N.M.; Oudkerk, J.C.; Dewael, T.Y.; Sadoun, M.J. From artisanal to CAD-CAM blocks: State of the art of indirect composites. *J. Dent. Res.* **2016**, *95*, 487–495. [CrossRef] [PubMed]
3. Spitznagel, F.A.; Boldt, J.; Gierthmuehlen, P.C. CAD/CAM Ceramic Restorative Materials for Natural Teeth. *J. Dent. Res.* **2018**, *97*, 1082–1091. [CrossRef] [PubMed]

4.   Abt, E.; Carr, A.B.; Worthington, H.V. Interventions for replacing missing teeth: Partially absent dentition. *Cochrane Database Syst. Rev.* **2012**, *15*, CD003814. [CrossRef]

5.   Schneider, C.; Zemp, E.; Zitzmann, N.U. Oral health improvements in Switzerland over 20 years. *Eur. J. Oral Sci.* **2017**, *125*, 55–62. [CrossRef]

6.   Joda, T.; Brägger, U. Digital vs. conventional implant prosthetic work-flows: A cost/time analysis. *Clin. Oral Implants Res.* **2015**, *26*, 1430–1435. [CrossRef]

7.   Berrendero, S.; Salido, M.P.; Ferreiroa, A.; Valverde, A.; Pradíes, G. Comparative study of all-ceramic crowns obtained from conventional and digital impressions: Clinical findings. *Clin. Oral Investig.* **2019**, *23*, 1745–1751. [CrossRef]

8.   Tsirogiannis, P.; Reissmann, D.R.; Heydecke, G. Evaluation of the marginal fit of single-unit, complete-coverage ceramic restorations fabricated after digital and conventional impressions: A systematic review and meta-analysis. *J. Prosthet. Dent.* **2016**, *116*, 328–335. [CrossRef]

9.   Coldea, A.; Swain, M.V.; Thiel, N. Mechanical properties of polymer- infiltrated ceramic-network materials. *Dent. Mater.* **2013**, *29*, 419–426. [CrossRef]

10.   Mörmann, W.H.; Stawarczyk, B.; Ender, A.; Sener, B.; Attin, T.; Mehl, A. Wear characteristics of current aesthetic dental restorative CAD/CAM materials: Two-body wear, gloss retention, roughness and martens hardness. *J. Mech. Behav. Biomed. Mater.* **2013**, *20*, 113–125. [CrossRef]

11.   Gracis, S.; Thompson, V.P.; Ferencz, J.L.; Silva, N.R.; Bonfante, E.A. A new classification system for all-ceramic and ceramic-like restorative materials. *Int. J. Prosthodont.* **2015**, *28*, 227–235. [CrossRef] [PubMed]

12.   Belli, R.; Petschelt, A.; Hofner, B.; Hajtó, J.; Scherrer, S.S.; Lohbauer, U. Fracture rates and lifetime Estimations of CAD/CAM All-ceramic Restorations. *J. Dent. Res.* **2016**, *95*, 67–73. [CrossRef] [PubMed]

13.   Homsy, F.R.; Özcan, M.; Khoury, M.; Majzoub, Z.A.K. Comparison of fit accuracy of pressed lithium disilicate inlays fabricated from wax or resin patterns with conventional and CAD-CAM technologies. *J. Prosthet. Dent.* **2018**, *120*, 530–536. [CrossRef] [PubMed]

14.   Aslan, Y.U.; Coskun, E.; Ozkan, Y.; Dard, M. Clinical evaluation of three types of CAD/CAM inlay/onlay materials after 1-Year clinical follow up. *Eur. J. Prosthodont. Restor. Dent.* **2019**, *27*, 131–140.

15.   Zarone, F.; Russo, S.; Sorrentino, R. From porcelain-fused-to-metal to zirconia: Clinical and experimental considerations. *Dent. Mater.* **2011**, *27*, 83–96. [CrossRef]

16.   Sailer, I.; Makarov, N.A.; Thoma, D.S.; Zwahlen, M.; Pjetursson, B.E. All-ceramic or metal-ceramic tooth-supported fixed dental prostheses (FDPs)? A systematic review of the survival and complication rates. Part I: Single crowns (SCs). *Dent. Mater.* **2015**, *31*, 603–623. [CrossRef]

17.   Ryge, G.; Snyder, M. Evaluating the clinical quality of restorations. *J. Am. Dent. Assoc.* **1973**, *87*, 369–377. [CrossRef]

18.   Hickel, R.; Peschke, A.; Tyas, M.; Mjör, I.; Bayne, S.; Peters, M.; Hiller, K.A.; Randall, R.; Vanherle, G.; Heintze, S.D. FDI World Dental Federation—Clinical criteria for the evaluation of direct and indirect restorations. Update and clinical examples. *J. Adhes. Dent.* **2010**, *12*, 259–272. [CrossRef]

19.   R Foundation for Statistical Computing, Vienna, Austria. Available online: http://www.R-project.org (accessed on 27 June 2020).

20.   Akin, A.; Toksavul, S.; Toman, M. Clinical Marginal and Internal Adaptation of Maxillary Anterior Single Ceramic Crowns and 2-year Randomized Controlled Clinical Trial. *J. Prosthodont.* **2015**, *24*, 345–350. [CrossRef]

21.   Azevedo, C.G.S.; De Goes, M.F.; Ambrosano, G.M.B.; Chan, D.C.N. 1-year Clinical Study of Indirect Resin Composite REstorations Luted with a Self-Adhesive Resin Cement: Effect of Enamel Etching. *Braz. Dent. J.* **2012**, *23*, 97–103. [CrossRef]

22.   Baader, K.; Hiller, K.A.; Buchalla, W.; Schmalz, G.; Federlin, M. Self-adhesive Luting of Partial Ceramic Crowns: Selective Enamel Etching Leads to Higher Survival after 6.5 Years in Vivo. *J. Adhes. Dent.* **2016**, *18*, 69–79. [PubMed]

23.   Botto, E.B.; Baró, R.; Borgia Botto, J.L. Clinical performance of bonded ceramic inlays/onlays: A 5- to 18-year retrospective longitudinal study. *Am. J. Dent.* **2016**, *29*, 187–192.

24.   D'all'Orologio, G.D.; Lorenzi, R. Restorations in abrasion/erosion cervical lesions: 8-year results of a triple blind randomized controlled trial. *Am. J. Dent.* **2014**, *27*, 245–250.

25.   Dhima, M.; Paulusova, V.; Carr, A.B.; Rieck, K.L.; Lihse, C.; Salinas, T.J. Practice-based clinical evaluation of ceramic single crowns after at least five years. *J. Prosthet. Dent.* **2014**, *3*, 124–130. [CrossRef]

26. Dukic, W.; Dukic, O.L.; Milardovic, S.; Delija, B. Clinical evaluation of indirect composite restorations at baseline and months after placement. *Oper. Dent.* **2010**, *35*, 156–164. [CrossRef]

27. Esuivel-Opshaw, J.; Rose, W.; Oliveira, E.; Yang, M.; Clark, A.E.; Anusavice, K. Randomized, controlled clinical trial of bilayer ceramic and metal-ceramic crown performance. *J. Prosthodont.* **2012**, *22*, 166–173. [CrossRef]

28. Fasbinder, D.J.; Dennison, J.B.; Heys, D.; Neiva, G. A Clinical Evaluation of Chairside Lithium Disilicate CAD/CAM Crowns: A two-year report. *J. Am. Dent. Assoc.* **2010**, *141*, 10S–14S. [CrossRef]

29. Frankenberger, R.; Taschner, M.; Garcia-Godoy, F.; Petschelt, A.; Krämer, N. Leucite-reinforced glass ceramic inlays and onlays after 12 years. *J. Adhes. Dent.* **2008**, *10*, 393–398.

30. Frankenberger, R.; Reinelt, C.; Petschelt, A.; Krämer, N. Operator vs. material influence on clinical outcome of bonded ceramic inlays. *Dent. Mater.* **2009**, *25*, 960–968. [CrossRef]

31. Gehrt, M.; Wolfart, S.; Rafai, N.; Reich, S.; Edelhoff, D. Clinical results of lithium-disilicate crowns after up to 9 years of service. *Clin. Oral Investig.* **2013**, *17*, 275–284. [CrossRef]

32. Guess, P.C.; Selz, C.F.; Voulgarakis, A.; Stampf, S.; Stappert, C.F. Prospective clinical study of press-ceramic overlap and full veneer restorations: 7-year result. *Int. J. Prosthodont.* **2014**, *27*, 355–358. [CrossRef] [PubMed]

33. Guess, P.C.; Selz, C.F.; Steinhart, Y.N.; Stampf, S.; Strub, J.R. Prospective clinical split-mouth study of pressed and CAD/CAM all-ceramic partial-coverage restorations: 7-year results. *Int. J. Prosthodont.* **2013**, *26*, 21–26. [CrossRef] [PubMed]

34. Krejci, I.; Krejci, D.; Lutz, F. Clinical evaluation of a new pressed glass ceramic inlay material over 1.5 years. *Oper. Dent.* **1992**, *23*, 181–186.

35. Manhart, J.; Chen, H.Y.; Mehl, A.; Hickel, R. Clinical study of indirect composite resin inlays in posterior stress-bearing preparations placed by dental students: Results after 6 months and 1, 2, and 3 years. *Quintessence Int.* **2010**, *41*, 399–410.

36. Mittal, H.C.; Goyal, A.; Gauba, K.; Kapur, A. Clinical Performance of Indirect Composite Onlays as Esthetic Alternative to Stainless Steel Crowns for Rehabilitation of a Large Crarious Primary Molar. *J. Clin. Pediatr. Dent.* **2016**, *40*, 345–352. [CrossRef]

37. Murgueitio, R.; Bernal, G. Three-Year Clinical Follow-Up of Posterior Teeth Restored with Leucite-Reinforced IPS Empress ONlays and Partial Veneer Crowns. *J. Proshtodont.* **2012**, *21*, 340–345. [CrossRef]

38. Özsoy, A.; Kuşdemir, M.; Öztürk-Bozkurt, F.; Toz Akalın, T.; Özcan, M. Clinical performance of indirect composite onlays and overlays: 2-year follow up. *J. Adhes. Sci. Technol.* **2016**, *30*, 1808–1818.

39. Rauch, A.; Reich, S.; Dalchau, L.; Schierz, O. Clinical survival of chair-side generated monolithic lithium disilicate crowns: 10-year results. *Clin. Oral Investig.* **2018**, *22*, 1763–1769. [CrossRef]

40. Reich, S.; Schierz, O. Chair-side generated posterior lithium disilicate crowns after 4 years. *Clin. Oral Investig.* **2013**, *17*, 1765–1772. [CrossRef]

41. Santos, M.J.; Freitas, M.C.; Azeedo, L.M.; Santos, G.C., Jr.; Navarro, M.F.; Francisschone, C.E.; Mondelli, R.F. Clinical evaluation of ceramic inlays and onlays fabricated with two systems: 12-year follow-up. *Clin. Oral Investig.* **2016**, *20*, 1683–1690. [CrossRef]

42. Schenke, F.; Federlin, M.; Hiller, K.A.; Moder, D.; Schmalz, G. Controlled, prospective, randomized, clinical evaluation of partial ceramic crowns inserted with RelyX Unicem with or without selective enamel etching. Results after 2 years. *Clin. Oral Investig.* **2012**, *16*, 451–461. [CrossRef] [PubMed]

43. Selz, C.F.; Strub, J.R.; Cach, K.; Guess, P.C. Long-term performance of posterior InCeram Alumina crowns cemented with different luting agents: A prospective, randomized clinical split-mouth study over 5 years. *Clin. Oral Investig.* **2014**, *18*, 1695–1703. [CrossRef] [PubMed]

44. Seydler, B.; Schmitter, M. Clinical Performance of two different CAD/CAM-fabricated ceramic crowns: 2-Year results. *J. Prosthet. Dent.* **2015**, *114*, 212–216. [CrossRef] [PubMed]

45. Taschner, M.; Krämer, N.; Lohbauer, U.; Pelka, M.; Breschi, L.; Petschelt, A.; Frankenberger, R. Leucite-reinforced glass ceramic inlays luted with self-adhesive resin cement: A 2-year in vivo study. *Dent. Mater.* **2012**, *28*, 535–540. [CrossRef] [PubMed]

46. Taskonak, B.; Sertgöz, A. Two-year clinical evaluation of lithia-discilicate-based all-ceramic crowns and fixed partial dentures. *Dent. Mater.* **2006**, *22*, 1008–1113. [CrossRef]

47. Alves de Carvalho, I.F.; Santos Marques, T.M.; Araujo, F.M.; Azevedo, L.F.; Donato, H.; Correia, A. Clinical Performance of CAD/CAM Tooth-Supported Ceramic Restorations: A systematic Review. *Int. J. Periodontics. Restorative. Dent.* **2018**, *38*, e68–e78. [CrossRef]

48. Rodrigues, S.B.; Franken, P.; Celeste, R.K.; Leitune, V.C.B.; Collares, F.M. CAD/CAM or conventional ceramic materials restorations longevity: A systematic review and meta-analysis. *J. Prosthodont. Res.* **2019**, *63*, 389–395. [CrossRef]

49. Sampaio, F.B.W.R.; Özcan, M.; Gimenez, T.C.; Moreira, M.S.N.A.; Tedesco, T.K.; Morimoto, S. Effects of manufacturing methods on the survival rate of ceramic and indirect composite resotorations: A systematic review and meta-analysis. *J. Esthet. Restor. Dent.* **2019**, *31*, 561–571. [CrossRef]

50. Rekow, E.D.; Silva, N.R.; Coelho, P.G.; Zhang, Y.; Guess, P.; Thompson, V.P. Performance of dental ceramics: Challenges for improvement. *J. Dent. Res.* **2011**, *90*, 937–952. [CrossRef]

51. Vagropoulou, G.I.; Klifopoulou, G.L.; Vlahou, S.G.; Hirayama, H.; Michalakis, K. Complications and survival rates of inlays and onlays vs complete coverage restorations: A systematic review and analysis of studies. *J. Oral Rehabil.* **2018**, *5*, 903–920. [CrossRef]

52. Mulic, A.; Svendsen, G.; Kopperud, S.E. A retrospective clinical study on the longevity of posterior class II cast gold inlays/onlays. *J. Dent.* **2018**, *70*, 46–50. [CrossRef] [PubMed]

53. Donovan, T.; Simonsen, R.J.; Guertin, G.; Tucker, R.V. Retrospective clinical evaluation of 1314 cast gold restorations in service from 1 to 52 years. *J. Esthet. Restor. Dent.* **2014**, *16*, 194–204. [CrossRef] [PubMed]

54. Morimoto, S.; Rebello de Sampaio, F.B.; Braga, M.M.; Sesma, N.; Özcan, M. Survival Rate of Resin and Ceramic Inlays, Onlays and Overlays: A systematic Review and Meta-analysis. *J. Dent. Res.* **2016**, *95*, 985–994. [CrossRef] [PubMed]

55. Grivas, E.; Roudsari, R.V.; Satterthwaite, J.D. Composite inlays: A systematic review. *Eur. J. Prosthodont. Restor. Dent.* **2014**, *22*, 117–124.

56. Angeletaki, F.; Gkogkos, A.; Papazoglou, E.; Kloukos, D. Direct versus indirect inlay/onlay composite restorations in posterior teeth. A systematic review and meta-analysis. *J. Dent.* **2016**, *53*, 12–21. [CrossRef]

57. Poggio, C.E.; Ercoli, C.; Rispoli, L.; Maiorana, C.; Esposito, M. Metal-free materials for fixed prosthodontic restorations. *Cochrane Database Syst. Rev.* **2017**, *20*, 12CD009606. [CrossRef]

58. Hofmann, N.; Papsthart, G.; Hugo, B.; Klaiber, B. Comparison of photo-activation versus chemical or dual-curing of resin-based luting cements regarding flexural strength, modulus and surface hardness. *J. Oral Rehabil.* **2001**, *28*, 1022–1028. [CrossRef]

Journal of
Clinical Medicine

Article

Influence of Preparation Design, Marginal Gingiva Location, and Tooth Morphology on the Accuracy of Digital Impressions for Full-Crown Restorations: An In Vitro Investigation

Selina A. Bernauer [1], Johannes Müller [2], Nicola U. Zitzmann [1] and Tim Joda [1,*]

1   Department of Reconstructive Dentistry, UZB University Center for Dental Medicine Basel, University of
    Basel, 4058 Basel, Switzerland; selina.bernauer@unibas.ch (S.A.B.); n.zitzmann@unibas.ch (N.U.Z.)
2   Private Practice, 80634 Munich, Germany; dr.johannes.a.mueller@gmail.com
*   Correspondence: tim.joda@unibas.ch

Received: 18 November 2020; Accepted: 7 December 2020; Published: 9 December 2020

Abstract: (1) Background: Intraoral optical scanning (IOS) has gained increased importance in prosthodontics. The aim of this in vitro study was to analyze the IOS accuracy for treatment with full crowns, considering possible influencing factors. (2) Methods: Two tooth morphologies, each with four different finish-line designs for tooth preparation and epi- or supragingival locations, were digitally designed, 3D-printed, and post-processed for 16 sample abutment teeth. Specimens were digitized using a laboratory scanner to generate reference STLs (Standard Tessellation Language), and were secondary-scanned with two IOS systems five times each in a complete-arch model scenario (Trios 3 Pod, Primescan AC). For accuracy, a best-fit algorithm (Final Surface) was used to analyze deviations of the abutment teeth based on 160 IOS-STLs compared to the reference STLs (16 preparations × 2 IOS-systems × 5 scans per tooth). (3) Results: Analysis revealed homogenous findings with high accuracy for intra- and inter-group comparisons for both IOS systems, with mean values of 80% quantiles from 20 ± 2 μm to 50 ± 5 μm. Supragingival finishing lines demonstrated significantly higher accuracy than epigingival margins when comparing each preparation ($p < 0.05$), whereas tangential preparations exhibited similar results independent of the gingival location. Morphology of anterior versus posterior teeth showed slightly better results in favor of molars in combination with shoulder preparations only. (4) Conclusion: The clinical challenge for the treatment with full crowns following digital impressions is the location of the prospective restoration margin related to the distance to the gingiva. However, the overall accuracy for all abutment teeth was very high; thus, the factors tested are unlikely to have a strong clinical impact.

Keywords: fixed prosthodontics; full crown; tooth preparation; intraoral optical scanning (IOS); digital dentistry

1. Introduction

Continuous technical development has expanded opportunities in reconstructive dentistry and prosthodontics [1]. In particular, intraoral optical scanning (IOS), computer-aided design, and computer-aided manufacturing (CAD/CAM) have fostered complete digital workflows for the treatment of fixed dental prostheses (FDPs) [2]. IOS has become indispensable in everyday dental practice, in university education [3,4], and in dental laboratories [5–7].

Digital impressions have been proven to be more time-efficient compared to conventional impressions, and a majority of patients have preferred the digital impression technique rather than the conventional approach with plastic materials [8–10]. At the same time, IOS has simplified the process chain between dentist and dental technician. Complete digital workflows have rendered various work steps superfluous such as tray preparation, disinfecting, shipping of the conventional impression, and further preparations for the fabrication of gypsum dental casts [11]. IOS technology offers new possibilities for clinical routine in selected indications, especially in the field of fixed prosthodontics. By taking a digital impression, the intraoral situation is visually recorded with neither the mucosa nor the teeth needing to be physically touched. This prevents possible gingival displacement or tooth movement from the application of conventional elastomeric impression material [12–14]. IOS is also advantageous for the treatment of periodontally compromised dentitions with recessions, enlarged interdental spaces, and dental undercuts (such as pontics or cantilevers), which make an accurate impression difficult [15]. Additionally, IOS opens the door to chairside CAD/CAM systems that could offer treatment protocols with single-unit restorations in one clinical session [16]. In contrast to conventional impressions, technical factors must be taken into account when using a digital approach. IOS requires a direct line of sight on the object in order to create 3D surface files, which are known as standard tessellation language (STL) [17]. Additional studies have shown that a supragingival preparation margin in the impression is more accurate [17,18].

Besides the technical development related to digital impressions, the finish-line design for tooth preparation has remained a crucial aspect for the abutment tooth [19,20]. Are the same finish-line designs applicable for full-crown restorations, or are adjustments required to facilitate the application of IOS? The challenge is now to analyze the existing parameters of tooth preparation in order to identify the best design for an accurate IOS and further STL processing for clinically acceptable restorations, while considering minimal invasiveness combined with modern materials and adhesive luting technology [21].

The aim of this in vitro study was to analyze the influence of different finish lines for complete crown preparations, their locations related to the gingival margin, and tooth morphology on the accuracy of digital impressions. The null hypotheses tested were that the IOS accuracy does not depend on the finish-line design (tangential, narrow chamfer, wide chamfer, and shoulder), the gingival positioning of the finishing line (epi- and supragingival), or on tooth morphology (incisor and molar); secondly, there is no difference in performance between the IOS systems used (Trios 3 Pod, 3Shape, Copenhagen, Denmark and Cerec Primescan AC, Dentsply Sirona, Bensheim, Germany).

2. Materials and Methods

A maxillary dental training model was used as reference (Dental Model AG-3, Frasaco, Tettnang, Germany). A maxillary central incisor (FDI 11) was selected to represent the anterior tooth morphology, while a first maxillary molar (FDI 16) was chosen to represent posterior sites. Based on a standardized complete crown preparation, two typodonts were manually prepared with a supragingival finishing line, 0.4 mm chamfer, and a 4–6° convergence angle. Substance removal was incisal 2.0 mm, palatal 1.0 mm, and labial 1.0–1.5 mm for tooth 11, and occlusal 1.5 mm and labial 1.0 mm, palatal 1.0 mm, and interdental 1.0 mm for the molar. All practical work steps were performed by the same operator (S.B.), a postgraduate prosthodontic resident, and each step was supervised by a senior clinician and board-certified prosthodontist (J.M.).

Both prepared typodonts were digitized with a laboratory desktop scanner (Series 7, Institute Straumann AG, Basel, Switzerland), and these served as the basis for the digital designs of the virtual modifications to create the test specimens, involving four different finish-line designs for both morphologies. These designs were digitally computed with the software Geomatic Design X (3D Systems, Rock Hill, SC, USA) and saved as STL files. The following finish-line designs were applied: tangential, narrow chamfer (0.4 mm), wide chamfer (0.8 mm), and shoulder (0.8 mm). Each design was applied in an epigingival or a 1.0 mm supragingival position, resulting in a total of eight tooth preparations for the anterior and another eight for the posterior region. Figure 1 displays the study setup with 16 different specimens (Figure 1).

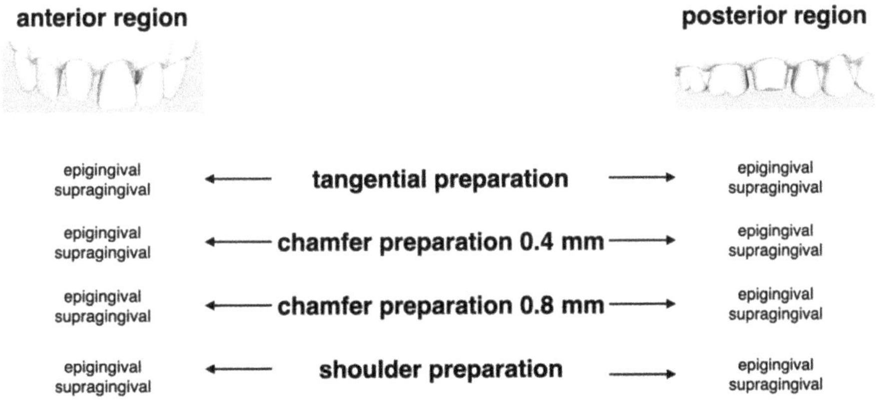

anterior region

posterior region

epigingival supragingival	⟵ **tangential preparation** ⟶	epigingival supragingival
epigingival supragingival	⟵ **chamfer preparation 0.4 mm** ⟶	epigingival supragingival
epigingival supragingival	⟵ **chamfer preparation 0.8 mm** ⟶	epigingival supragingival
epigingival supragingival	⟵ **shoulder preparation** ⟶	epigingival supragingival

Figure 1. Trial setting: Four different preparation designs in anterior and posterior regions separated for epi- and supragingival finishing lines.

Finally, the 16 virtual tooth preparations were 3D-printed for the production of standardized replicas (3D-Printer Objet260 Connex2, Stratasys, Eden Prairie, MN, USA). The color of the rubber-like material used was a mixture of Vero White Plus RGD 835 and Tango Black Plus FLX 980. This mixture resulted in the color DM8515 (Stratasys, Eden Prairie, MN, USA). All 3D-printed teeth were mounted in the reference model and manually finalized with diamond burs (Intensive SA, Montagnola, Switzerland) and Sof-Lex discs (3M ESPE AG, Saint Paul, MN, USA) to achieve an exact finishing line and a smooth surface. In order to avoid potential deviations due to printing errors and to visualize the manual corrections, all teeth were removed from the reference model and digitized with the same laboratory desktop scanner that was used for the initial digitalization.

Successively, all 16 3D-printed teeth were remounted in the reference model and scanned by one experienced operator (S.B.) with two IOS systems (Trios 3 Pod and Cerec Primescan AC). Each preparation, including adjacent teeth, was captured five times in order to minimize potential scanning errors. The scans were carried out according to the manufacturer's recommendations.

For accuracy of analysis, a total of 160 IOS-STLs (16 specimens × 2 IOS-systems × 5 scans = 160 STLs) were then superimposed to the corresponding original reference STLs with the software Final Surface (GFaI e.V., Berlin, Germany). A best-fit algorithm was applied for deviation analysis in order to minimize the distances between the two surfaces being compared. Here, the distance to the surface of the IOS-STL to be examined was considered for all surfaces of the matching reference STL. Scanning data beyond 2 mm from the finishing lines were digitally cut to guarantee an accurate fine registration. Trimmed scan data obtained from five scans by each IOS were paired, and these pairs were inspected (STL-1 vs. STL-2, STL-1 vs. STL-3, STL-1 vs. STL-3, etc.). Deviations between polygons formed by the point cloud constituting the two superimposed scans were calculated, and the distance data of all superimposed pairs were summarized [22].

Numerical variables of interest were descriptively analyzed with sample means for 80% quantiles including standard deviation. Since the IOS-STLs under investigation appeared with a plus or minus of data points compared to the reference STLs, the use of 80% quantiles ensured error minimization regarding "too small" and "too large" areas for deviation analysis. Statistics were carried out using R 4.0.3 (The R Project for Statistical Computing, Vienna, Austria), and a significance level was set at 0.05.

3. Results

Trios 3 Pod and Primescan AC could successfully capture all selected preparations as tangential, narrow and wide chamfer, and shoulder, respectively (Figure 2). Intra- and intergroup analyses comparing both IOS systems revealed homogenous results with high accuracy representing mean values of 80% quantiles ranging from 20 ± 2 μm to 50 ± 5 μm throughout all tested abutment teeth (Tables 1 and 2). Supragingival finishing lines demonstrated significantly higher accuracy than epigingival margins when comparing preparation designs against each other ($p < 0.05$), whereas tangential preparations exhibited similar results independent of the gingival location of the finishing line. Morphology of anterior versus posterior teeth showed slightly better results in favor of molars in combination with shoulder preparations only.

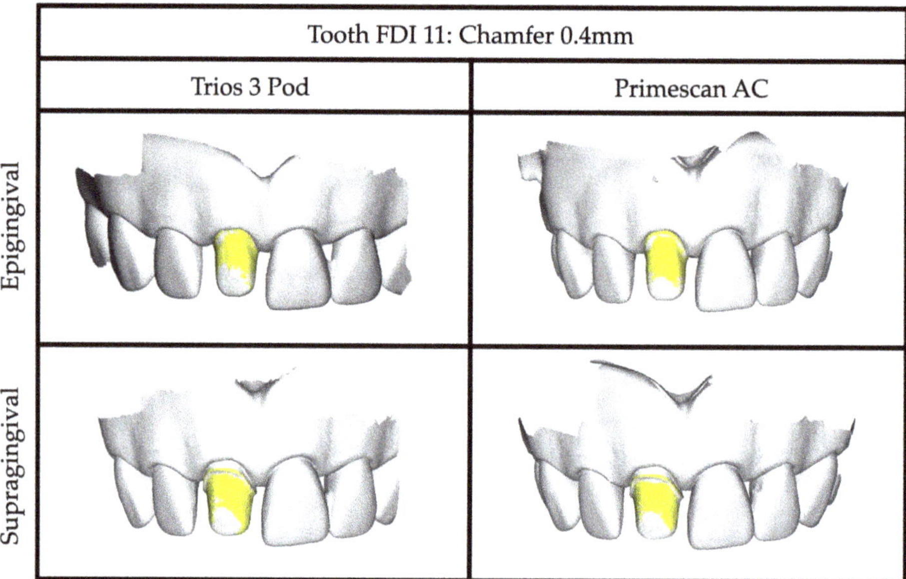

Figure 2. 3D color mapping depicting sample abutment teeth in position FDI 11 with epi-/supragingival 0.4 mm chamfer captured with Trios 3/Primescan AC and superimposition to the corresponding references (Final Surface, GFaI e.V., Berlin, Germany).

Table 1. Anterior tooth morphology: Deviation (in μm) of IOS-STLs compared to the reference STLs summarizing mean values of 80% quantiles, including standard deviations (SD) of the different preparation designs separated for epi- and supragingival finishing lines ([a–f] $p < 0.05$).

		Trios 3 Pod	Primescan AC
Epigingival	Tangential	34 ± 6	35 ± 5
	Chamfer 0.4 mm	[a] 38 ± 4	[b] 40 ± 6
	Chamfer 0.8 mm	[c] 42 ± 5	[d] 45 ± 6
	Shoulder	[e] 48 ± 5	[f] 50 ± 5
Supragingival	Tangential	30 ± 1	31 ± 2
	Chamfer 0.4 mm	[a] 28 ± 3	[b] 26 ± 2
	Chamfer 0.8 mm	[b] 29 ± 3	[d] 30 ± 3
	Shoulder	[e] 40 ± 6	[f] 39 ± 5
([a] $p = 0.0036$, [b] $p = 0.001$, [c] $p = 0.0013$, [d] $p = 0.0008$, [e] $p = 0.0008$, [f] $p = 0.0025$)			

Table 2. Posterior tooth morphology: Deviation (in µm) of IOS-STLs compared to the reference STLs summarizing mean values of 80% quantiles, including standard deviations (SD) of the different preparation designs separated for epi- and supragingival finishing lines ([a–f] $p < 0.05$).

		Trios 3 Pod	Primescan AC
Epigingival	Tangential	30 ± 4	31 ± 4
	Chamfer 0.4 mm	[a] 40 ± 6	[b] 39 ± 4
	Chamfer 0.8 mm	[c] 39 ± 4	[d] 41 ± 5
	Shoulder	[e] 34 ± 4	[f] 36 ± 5
Supragingival	Tangential	29 ± 3	30 ± 3
	Chamfer 0.4 mm	[a] 28 ± 3	[b] 32 ± 3
	Chamfer 0.8 mm	[c] 27 ± 2	[d] 27 ± 1
	Shoulder	[e] 21 ± 2	[f] 20 ± 2
([a] $p = 0.0018$, [b] $p = 0.0128$, [c] $p = 0.0018$, [d] $p = 0.001$, [e] $p = 0.0013$, [f] $p = 0.0006$)			

4. Discussion

The aim of this in vitro study was to analyze IOS accuracy for complete crown restorations, considering maxillary incisor and molar tooth morphologies with four different finish-line designs (tangential, narrow chamfer, wide chamfer, and shoulder) in epi- and supragingival margin positions. The results demonstrated that all specimens were successfully digitized with high accuracy independently of the IOS device used. However, the supragingival finishing lines were captured significantly better than the epigingivally located margins. Therefore, the hypothesis that IOS accuracy does not depend on any of the factors listed above was partially rejected.

IOS technology offers new possibilities for clinical routine in selected indications, especially in the field of fixed prosthodontics with all the advantages mentioned above. In the present study, the position of the finishing line with respect to the gingiva showed differences between epi- and supragingival margins. IOS recorded the supragingival preparations more precisely. Two further investigations have also demonstrated higher reproducibility for supragingival finishing lines [17,18]. Divergent literature states that the supra- and epigingival margins can be scanned without significant differences. In that mentioned in vitro study, supragingival finishing lines were made visible by gingival retraction [23]. The significant difference between the epi- and supragingival margins could be attributed to the absence of gingival retraction. Sufficient soft-tissue management is a crucial success factor and, therefore, should be ensured in clinical routine.

Based on the results of this in vitro investigation, it can be recommended that, to ensure a higher predictability of digital impression-taking in clinical routine, the finishing line must be clearly visible, with healthy gingiva surrounded a full 360°. Therefore, complete-crown finishing lines should be prepared supragingivally whenever possible using IOS [19,24], including proper soft-tissue management during impression taking, which remains a crucial success factor for any kind of impression technique [25] until future technology can provide novel IOS possibilities for scanning through tissue and liquids.

Today, IOS requires a direct line of sight on the object being scanned, and a minimal distance of 0.5 mm between adjacent teeth seems to be the critical threshold for the optical resolving power [17]. Otherwise, the IOS software takes over to calculate the preparation margins virtually, instead of capturing the intraoral situation with optical precision. Not all surfaces of a tooth seem to be recorded with the same accuracy; for example, distal and lingual surfaces have shown the lowest accuracy [26,27]. Finally, the complexity of the geometry to be scanned has an impact on the accuracy as well. Supragingival complete crown preparations have demonstrated significantly better results than intracoronal inlay preparations using different IOS systems [28]. For more complex preparations, e.g., for post copings or adhesive attachments, capturing with IOS is currently not feasible.

However, what are the limitations of digital impressions for treatment with complete crown restorations? Do any influencing factors affect the successful use of IOS in clinical routine?

Based on the results of the present trial, conventional crown preparation designs can be applied with a digital capturing by IOS while considering minimal invasiveness. It is possible to focus on the desired requirements for single-unit restorations in everyday clinical practice. The anatomical position and morphology of the area to be restored must be analyzed first. Moreover, the selected material has to be considered when selecting the preparation characteristics. Basically, the following parameters have been summarized for metal-based complete crowns: (i) convergence angle between two opposing prepared axial surfaces in the range of 10° to 22°; (ii) retentive vertical surfaces of at least 3 mm and a height-to-diameter ratio of at least 0.4 to provide adequate resistance form; (iii) teeth should be reduced uniformly to facilitate esthetic dental work, as well as anatomically to keep the teeth's characteristic geometric shape and to avoid pulp trauma [19].

The translation from in vitro to in vivo always involves difficulties. The presented trial setting reflects ideal and constant conditions. The clinical real-world scenario has to tackle multi-factor challenges such as irregular tooth preparations in terms of design and distance to the gingival margin, different dental surfaces, perfused soft tissue, saliva and sulcus fluids, limited access in the oral cavity, and patient movement. It was also not possible to work with gingival retraction in this in vitro setting. This study was carried out in a stable single-jaw setting using a typodont model with ideally prepared artificial abutment teeth. The absence of saliva, tongue, mouth opening, and individual patient anatomy simplified IOS scanability [29]. The impact of mouth opening, in particular, needs be further investigated for mandibular impressions in vivo. During IOS, patients must maintain an extensive mouth opening for a longer time compared to the conventional approach. This could lead to slight deformations of the mandible [30,31]. With the conventional method, the mouth must only initially be opened wide; however, during the setting time of the material, the patient can almost rest in a relaxing position. In vivo, this could lead to deviations in scanning accuracy between anterior and posterior areas, which could not be detected in this in vitro setting. Additionally, only two IOS devices were used, which reduces the power of generalization, and the operators could not be blinded for the intervention and the type of scanner used. For further studies, it would be useful to include a greater variety of IOS scanners and also to perform subgingival preparations in vivo, where appropriate soft-tissue management could be applied.

5. Conclusions

Within the limitations of this study, the following can be concluded:

(1)   the overall accuracy for all abutment teeth was very high, without significant differences in the performance of 3Shape Trios 3 Pod versus Cerec Primescan AC;
(2)   the supragingival finishing lines were captured significantly better than the epigingivally located margins using IOS. If the clinical situation allows, a supragingival margin should be chosen accordingly;
(3)   the tooth morphology seems to be a negligible factor for IOS accuracy in terms of single-unit complete crown restorations.

Author Contributions: Conceptualization, S.A.B., J.M. and T.J.; methodology, S.A.B., J.M. and T.J.; software, S.A.B. and T.J.; validation, S.A.B. and T.J.; investigation, S.A.B. and J.M.; resources, J.M. and T.J.; data curation, S.A.B. and T.J.; writing—original draft preparation, S.A.B.; writing—review and editing, S.A.B., N.U.Z. and T.J.; visualization, S.A.B. and T.J.; supervision, J.M. and T.J. All authors have read and agreed to the published version of the manuscript.

Funding: This research received no external funding.

Acknowledgments: The authors thank Institute Straumann AG, Basel, Switzerland, for their support of the study by donating the prepared teeth. Thanks also go to Dentsply Sirona for making the Cerec Primescan AC scanner available.

Conflicts of Interest: The authors declare no conflict of interest.

References

1. Joda, T.; Ferrari, M.; Gallucci, G.O.; Wittneben, J.; Brägger, U. Digital technology in fixed implant prosthodontics. *Periodontolgy 2000* **2016**, *73*, 178–192. [CrossRef] [PubMed]
2. Joda, T.; Zarone, F.; Ferrari, M. The complete digital workflow in fixed prosthodontics: A systematic review. *BMC Oral Health* **2017**, *17*, 1–9. [CrossRef] [PubMed]
3. Zitzmann, N.U.; Matthisson, L.; Ohla, H.; Joda, T. Digital Undergraduate Education in Dentistry: A Systematic Review. *Int. J. Environ. Res. Public Health* **2020**, *17*, 3269. [CrossRef] [PubMed]
4. Zitzmann, N.U.; Kovaltschuk, I.; Lenherr, P.; Dedem, P.; Joda, T. Dental Students' Perceptions of Digital and Conventional Impression Techniques: A Randomized Controlled Trial. *J. Dent. Educ.* **2017**, *81*, 1227–1232. [CrossRef]
5. Joda, T.; Gallucci, G.O. The virtual patient in dental medicine. *Clin. Oral Implant. Res.* **2014**, *26*, 725–726. [CrossRef]
6. Schoenbaum, T.R. Dentistry in the digital age: An update. *Dent. Today* **2012**, *31*, 12–13.
7. Eaton, K.A.; Reynolds, P.A.; Grayden, S.K.; Wilson, N.H.F. A vision of dental education in the third millennium. *Br. Dent. J.* **2008**, *205*, 261–271. [CrossRef]
8. Schepke, U.; Meijer, H.J.A.; Kerdijk, W.; Cune, M.S. Digital versus analog complete-arch impressions for single-unit premolar implant crowns: Operating time and patient preference. *J. Prosthet. Dent.* **2015**, *114*, 403–406. [CrossRef]
9. Joda, T.; Lenherr, P.; Dedem, P.; Kovaltschuk, I.; Bragger, U.; Zitzmann, N.U. Time efficiency, difficulty, and operator's preference comparing digital and conventional implant impressions: A randomized controlled trial. *Clin. Oral Implant. Res.* **2017**, *28*, 1318–1323. [CrossRef]
10. Yuzbasioglu, E.; Kurt, H.; Turunc, R.; Bilir, H. Comparison of digital and conventional impression techniques: Evaluation of patients' perception, treatment comfort, effectiveness and clinical outcomes. *BMC Oral Health* **2014**, *14*, 10. [CrossRef]
11. Christensen, G.J. Impressions are changing: Deciding on conventional, digital or digital plus in-office milling. *J. Am. Dent. Assoc. (1939)* **2009**, *140*, 1301–1304. [CrossRef] [PubMed]
12. Gintaute, A.; Straface, A.; Zitzmann, N.U.; Joda, T. Die Modellgussprothese 2.0: Digital von A bis Z? *Swiss. Dent. J.* **2020**, *130*, 229–235. [PubMed]
13. Masri, R.; Driscoll, C.F.; Burkhardt, J.; von Fraunhofer, A.; Romberg, E. Pressure generated on a simulated oral analog by impression materials in custom trays of different designs. *J. Prosthodont.* **2002**, *11*, 155–160. [CrossRef] [PubMed]
14. Al-Ahmad, A.; Masri, R.; Driscoll, C.F.; Von Fraunhofer, J.; Romberg, E. Pressure Generated on a Simulated Mandibular Oral Analog by Impression Materials in Custom Trays of Different Design. *J. Prosthodont.* **2006**, *15*, 95–101. [CrossRef]
15. Schlenz, M.A.; Schubert, V.; Schmidt, A.; Wöstmann, B.; Ruf, S.; Klaus, K. Digital versus Conventional Impression Taking Focusing on Interdental Areas: A Clinical Trial. *Int. J. Environ. Res. Public Health* **2020**, *17*, 4725. [CrossRef]
16. Baroudi, K.; Ibraheem, S.N. Assessment of Chair-side Computer-Aided Design and Computer-Aided Manufacturing Restorations: A Review of the Literature. *J. Int. Oral Health* **2015**, *7*, 96–104.
17. Ferrari, M.; Keeling, A.; Mandelli, F.; Giudice, G.L.; Garcia-Godoy, F.; Joda, T. The ability of marginal detection using different intraoral scanning systems: A pilot randomized controlled trial. *Am. J. Dent.* **2018**, *31*, 272–276.
18. Keeling, A.; Wu, J.; Ferrari, M. Confounding factors affecting the marginal quality of an intra-oral scan. *J. Dent.* **2017**, *59*, 33–40. [CrossRef]
19. Goodacre, C.J. Designing tooth preparations for optimal success. *Dent. Clin. N. Am.* **2004**, *48*, 359–385. [CrossRef]
20. Ahmed, W.M.; Shariati, B.; Gazzaz, A.Z.; Sayed, M.E.; Carvalho, R.M. Fit of tooth-supported zirconia single crowns—A systematic review of the literature. *Clin. Exp. Dent. Res.* **2020**. [CrossRef]
21. Balevi, B. Limited evidence on the best position for prosthetic margins. *Evidence-Based Dent.* **2013**, *14*, 103–104. [CrossRef] [PubMed]

22. Mühlemann, S.; Greter, E.A.; Park, J.M.; Hämmerle, C.H.F.; Thoma, D.S. Precision of digital implant models compared to conventional implant models for posterior single implant crowns: A within-subject comparison. *Clin. Oral Implant. Res.* **2018**, *29*, 931–936. [CrossRef] [PubMed]

23. Koulivand, S.; Ghodsi, S.; Siadat, H.; Alikhasi, M. A clinical comparison of digital and conventional impression techniques regarding finish line locations and impression time. *J. Esthet. Restor. Dent.* **2020**, *32*, 236–243. [CrossRef] [PubMed]

24. Valderhaugw, J.; Birkeland, J. Periodontal conditions in patients 5 years following insertion of fixed prostheses: Pocket depth and loss of attachment. *J. Oral Rehabil.* **1976**, *3*, 237–243. [CrossRef]

25. Shetty, K. Gingival tissue management: A necessity or a liability. *Triv. Dent. J.* **2011**, *2*, 112–119.

26. Roperto, R.; Oliveira, M.; Porto, T.; Ferreira, L.; Melo, L.; Akkus, A. Can Tooth Preparation Design Affect the Fit of CAD/CAM Restorations? *Compend. Contin. Educ. Dent. (Jamesburg, NJ: 1995)* **2017**, *38*, e13–e17.

27. Chiu, A.; Chen, Y.W.; Hayashi, J.; Sadr, A. Accuracy of CAD/CAM Digital Impressions with Different Intraoral Scanner Parameters. *Sensors* **2020**, *20*, 1157. [CrossRef]

28. Ashraf, Y.; Sabet, A.; Hamdy, A.; Ebeid, K. Influence of Preparation Type and Tooth Geometry on the Accuracy of Different Intraoral Scanners. *J. Prosthodont.* **2020**. [CrossRef]

29. Fluegge, T.V.; Schlager, S.; Nelson, K.; Nahles, S.; Metzger, M.C. Precision of intraoral digital dental impressions with iTero and extraoral digitization with the iTero and a model scanner. *Am. J. Orthod. Dentofac. Orthop.* **2013**, *144*, 471–478. [CrossRef]

30. Chen, D.C.; Lai, Y.L.; Chi, L.Y.; Lee, S.Y. Contributing factors of mandibular deformation during mouth opening. *J. Dent.* **2000**, *28*, 583–588. [CrossRef]

31. Law, C.; Bennani, V.; Lyons, K.; Swain, M.V. Mandibular Flexure and Its Significance on Implant Fixed Prostheses: A Review. *J. Prosthodont.* **2012**, *21*, 219–224. [CrossRef] [PubMed]

Publisher's Note: MDPI stays neutral with regard to jurisdictional claims in published maps and institutional affiliations.

Journal of
Clinical Medicine

Article

Analysis of The Reproducibility of Subgingival Vertical Margins Using Intraoral Optical Scanning (IOS): A Randomized Controlled Pilot Trial

Edoardo Ferrari Cagidiaco [1,2], **Fernando Zarone** [3], **Nicola Discepoli** [4], **Tim Joda** [5] and **Marco Ferrari** [6,*]

1   Department of Periodontics, Complutense University, 28001 Madrid, Spain;
    edoardo.ferrari.cagidiaco@gmail.com
2   Department of Prosthodontic and Dental Materials, University of Siena, 53100 Siena, Italy
3   Department of Neurosciences, Reproductive and Odontostomatological Sciences,
    University of Naples Federico II, 80131 Napoli, Italy; fernandozarone@mac.com
4   Department of Periodontics, University of Siena, 53100 Siena, Italy; ndiscepoli@me.com
5   Department of Reconstructive Dentistry, University Center for Dental Medicine Basel,
    CH-4058 Basel, Switzerland; tim.joda@unibas.ch
6   Department of Prosthodontics and Dental Materials, University of Siena, 53100 Siena, Italy
*   Correspondence: ferrarm@gmail.com; Tel.: +39-340-3277096

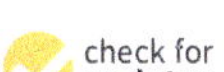

check for
updates

Citation: Ferrari Cagidiaco, E.;
Zarone, F.; Discepoli, N.; Joda, T.;
Ferrari, M. Analysis of The
Reproducibility of Subgingival
Vertical Margins Using Intraoral
Optical Scanning (IOS): A
Randomized Controlled Pilot Trial. *J.
Clin. Med.* **2021**, *10*, 941. https://
doi.org/10.3390/jcm10050941

Academic Editor: Luca Testarelli

Received: 4 February 2021
Accepted: 22 February 2021
Published: 1 March 2021

Publisher's Note: MDPI stays neutral
with regard to jurisdictional claims in
published maps and institutional affiliations.

Abstract: Background: The aim of this randomized controlled trial was to evaluate the capability of an IOS (Intra Oral Scanner) device, used in standardized conditions, to detect margins of abutments prepared with knife-edge finishing line located at three different levels in relation to the gingival sulcus. Methods: sixty abutment teeth for treatment with full crowns were selected and randomly divided in three groups accordingly to the depth of the finishing line: Group A: supragingival margin; Group B: 0.5–1.0 mm into the sulcus; Group C: 1.5–2.0 mm into the sulcus. Temporary crowns were placed for two weeks and then digital impressions (Aadva IOS 100, GC, Japan) were made of each abutment. As controls, analog impressions were taken, poured, and scanned using a laboratory scanner (Aadva lab scanner, GC, Japan). Two standard tessellation language (STL) files were generated for each abutment, subsequently processed, and superimposed by Exocad software (Exocad GmbH, Darmstadt, Germany), applying the "best-fit" algorithm in order to align the scan of the conventional with the digital impressions. The distances between each preparation margin and the adjacent gingival tissue were measured. Four measures were taken, two interproximally and buccally, for a total of six measures of each abutment considering three modes of impressions. The data were statistically evaluated using two-way analysis of variance (ANOVA) for each site and the Bonferroni test. Results: there was no difference between the two kinds of impression in Group A in both sites, in Group B a difference of 0.483 mm and 0.682 mm at interproximal and buccal sites, respectively, and in Group C 0.750 mm and 0.964 mm at interproximal and buccal sites, respectively. The analysis performed on a site level (mesial/distal/vestibular) for the depth of both vertical preparations revealed significant differences ($p < 0.0001$). After a post hoc analysis (Bonferroni), vestibular sites of the shallow vertical preparations resulted in significantly lower values compared to the other sites prepared deeply. Conclusions: the results showed that the location of the margin is an important factor in making a precise and complete impression when IOS (Intra Oral Scanner) is used. Moreover, deep preparation into the sulcus is not recommended for IOS (Intra Oral Scanner) impressions.

Keywords: knife-edge preparation; IOS; superimposition; digital impression; subgingival margins

1. Introduction

Key factors for long-term clinical success in fixed prosthodontics are respect of function, biocompatibility, marginal and internal fit, fracture resistance, and appealing esthetics.

In particular, a marginal gap, at the level of the restorative finish line, has a highly detrimental effect on the quality of the restoration, inducing micro-leakage, cement dissolution by oral fluids, and biofilm accumulation, with consequences such as caries or endodontic and periodontal problems [1–4]. Up to now, the precision of marginal fit has been reported up to 200 microns and beyond [5–8], although a precise, scientifically validated evaluation of the maximum acceptable marginal gap has never been provided; the threshold of 120 microns, defined by McLean, has been considered as a reference in dental literature since 1971 [9]. It is generally accepted that all subsequent clinical and laboratory work steps influence the overall success of a fixed restoration, from tooth preparation to cementation [10]. Here, the final impression is one of the most important steps to achieving the final marginal adaptation of the restoration, independent on the material and technique selected. In conventional impression procedures, the final result is strongly affected by dimensional distortions of impression materials and gypsum [11,12], to the extent that half of misfits have been considered to be ascribed to the impression procedure and to the production of the gypsum cast, the other half being mainly related to the production techniques of the prosthesis [13,14]. The introduction of the digital impression by using intraoral scanning (IOS) has changed the restorative scenario in prosthodontics by the acquisition of anatomic information without the use of physical impression materials, transforming shapes into digital files [15–18].

One of the most critical steps during impression taking, both conventional and digital, is detecting the finish line, in particular for subgingival tooth preparation. In this context, adequate soft tissue management without inflammation is mandatory for a successful impression, supported by gingival displacement to expose the finish. In the conventional impression procedure, this is usually obtained using gingival retraction cords or materials which temporarily modify the marginal soft tissue, with the purpose of detecting the necessary sub-gingival anatomic information and of widening the gingival sulcus without tearing the subtle light material margin, due to its low consistency [19]. Following the digital impression technique, it is not different to the conventional approach. In both cases, the detection of the finish line relies on a clean, healthy gingival sulcus, proper soft tissue displacement, and clear visibility of the prepared tooth anatomy.

The aim of this randomized controlled clinical trial was to test the capability of an IOS device (Aadva IOS 100, GC, Japan) used in standardized conditions, to detect margins of abutments prepared with knife-edge finishing line located at three different levels in relation to the gingival sulcus.

The null hypothesis was that there was no difference in the capability of the IOS based on the vertical position of the prepared finish line.

2. Experimental Section

In this study, 60 patients (28 female and 32 male) with a mean age of 45 ($\pm$20.5) years (range 18–69) in need of a tooth-borne single crown in posterior sites were recruited. The present prospective clinical trial was approved by the Ethical Committee of the University of Siena (n.18895). For each included individual, a signed written consent was obtained after clear information about the study. Guidelines of the CONSORT statement were followed.

Inclusion Criteria: age $\geq$ 18 years; single full crown in posterior sites (maxilla or mandible); periodontally healthy or successfully treated; general good health.

Exclusion criteria: presence of any active infection; severe periodontal inflammation; presence of chronic systemic disease; smoking more than 15 cigarettes per day; bruxism habits.

2.1. Randomization/Allocation Concealment/Masking of Examiners

Included patients were recruited between May and November of 2018 in the Department of Fixed Prosthodontics at the University of Siena and randomly divided into three groups of twenty each (3 × *n* = 20) according to the location depth of the finishing line made on the prepared abutments in relation to the sulcus:

Group A: supragingival margin.
Group B: margin 0.5–1.0 mm into the sulcus.
Group C: margin 1.5–2.00 mm into the sulcus.

Treatment assignment was noted in a detailed registration and treatment assignment form. Allocation concealment was performed by opaque, sealed, and sequentially numbered envelopes. The statistician generated the allocation sequence by means of a computer-generated random list and instructed a different subject to assign a sealed envelope containing the type of IOS. The opaque envelope was opened before IOS selection and communicated to the operator (EFC—Edoardo Ferrari Cagidiaco). Blinding of the examiner was maintained throughout all experimental procedures (Figure 1).

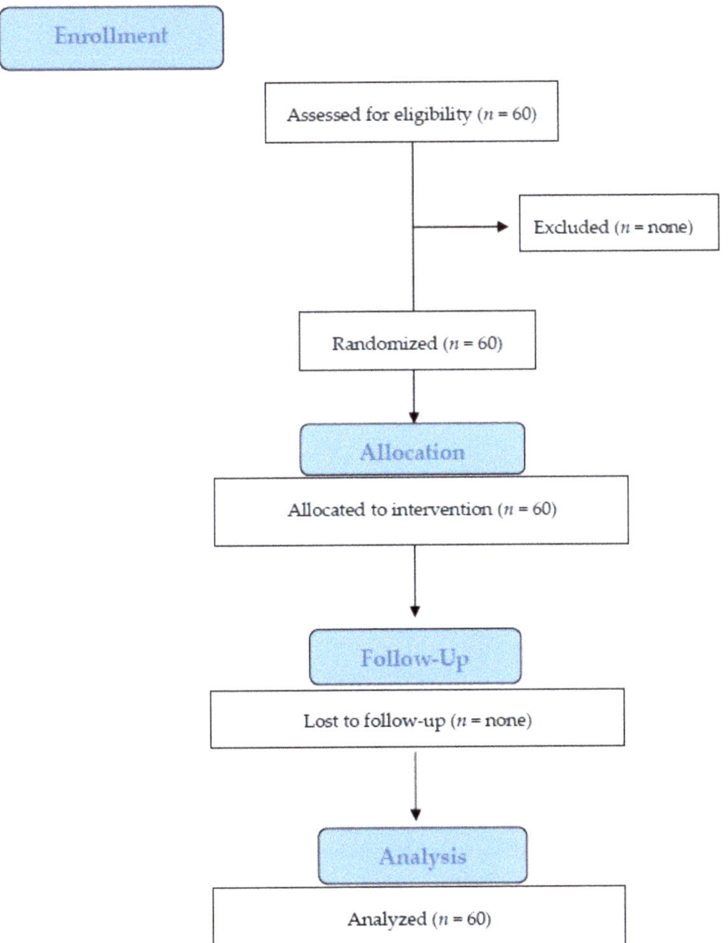

Figure 1. Flow diagram.

2.2. Clinical Setting

Abutment tooth preparations of Group A were performed following the generally accepted recommendations for CAD/CAM (Computer-Aided Design/Computer-Aided Manufacturing)-restorations with supragingivally located margins in order to remain visible [20]. In Group B, the margins were placed 0.5–1.0 mm into the sulcus and in Group C, the margins were placed around 1.5–2.0 mm in depth. Clinical pictures were taken of each quadrant and the corresponding preparations (Figure 2).

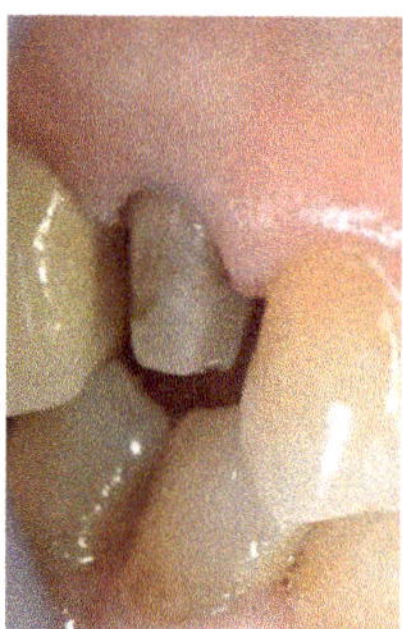

Figure 2. The abutment after preparation.

All abutments received a temporary crown for 2 weeks [21,22] and then the final IOS impressions were made. The impression site was prepared according to the double retraction cord technique: the first, thinner cord (Ultrapack #00; Ultradent, South Jordan, UT, USA) was gently placed into the gingival sulcus, followed by the insertion of a second, wider-diameter cord (Ultrapack #1; Ultradent, South Jordan, UT, USA) at a more coronal level, visible around the preparation margins. IOS was initially performed according to the manufacturer's guidelines (Aadva IOS 100, GC Co., Tokyo, Japan): firstly, the upper arch was scanned, followed by the lower arch, and then the bite registration was performed. A total of twenty scans of each group (A, B, and C) were collected and saved in the standard tessellation language (STL) format (Figure 3a).

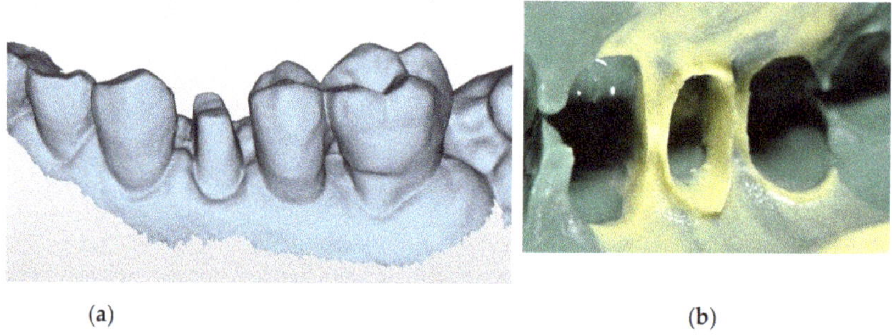

(a) (b)

Figure 3. Digital impression (**a**)Analogic impression. The deep preparation is evident (**b**).

Any scanning shot considered incorrect or showing evident defects was discarded.

As the control, a conventional impression was made using polyvinyl siloxane (Ex'lance, GC) (Figure 3b).

The viscoelastic properties of the material facilitate the detection of the area below the gingival margins. Impressions were cleansed, disinfected, poured in Type IV Dental Die Stone (FujiRock, GC, Tokyo, Japan), and finally scanned by a laboratory scanner (Aadva lab scanner, GC, Tokyo, Japan), generating STL files of the control protocol.

2.3. Software Measurements

Each STL file generated by both the IOS and the lab scanner was processed by the same dental master technician, using the Exocad software (Exocad GmbH, Darmstadt,

Germany), applying the "best-fit" algorithm in order to align the scan of the conventional with the digital impression (Figure 4a).

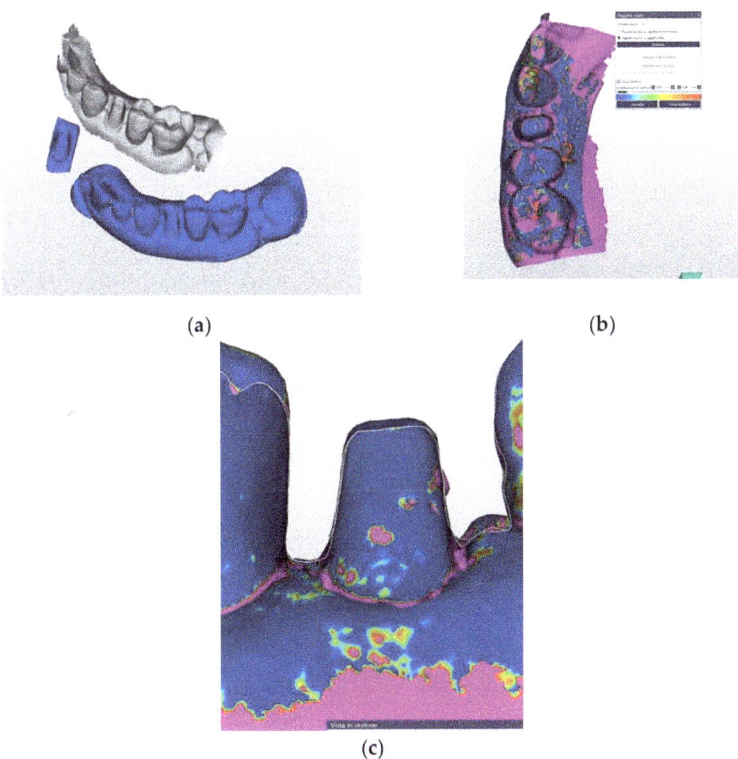

(a)

(b)

(c)

Figure 4. The two digital casts before being superimposed (**a**). The two digital casts after being superimposed (**b**). The abutment after being sovraimposed (**c**).

The superimposition of the STL files allowed measurement of the distance between each preparation margin and the adjacent gingival tissue, after making a section of each abutment in either the mesial-distal or buccal-lingual direction (Figure 4b,c).

The straight distance between the most coronal part of the gingival margin and the apical finish line of the preparation were used as distances to be recorded, and both vertical distances (made by conventional and digital impressions) were measured and recorded. The most coronal part of the gingival tissue was always the same, and the most apical part into the sulcus varied accordingly for each impression. Four measures were taken, two interproximally (mesial and distal) and buccally (buccal), for a total of six measures of each abutment considering three modes of impressions.

2.4. Statistical Analysis

All the data were collected and processed statistically. Descriptive statistics (means, standard deviations, 95% confidence intervals) were performed on the studied parameters using Stata 15-IC (IBM, NY, USA). The Wilcoxon rank sum test was used to analyze the media each measure.

Two-way analysis of variance (ANOVA) for each site and the Bonferroni test were conducted to assess the overall statistical significance of the differences among the groups ($p > 0.05$).

3. Results

Table 1 shows the results for the mean distance of the prepared root that cannot be detected with the digital impression compared to the conventional one.

Table 1. Statistical results for the mean distance of the prepared root that cannot be detected with the digital impression compared to the conventional one.

$n = 20$	Juxtagengival Margins Group A	Subgengival Margins (within 1.5 mm) Group B	Deepest Margins (1.5–2.0 mm) Group C
Interproximal margins	0	0.682	0.964
Buccal margins	0	0.483	0.750

There was no difference between the two kinds of impression in Group A in both sites, in Group B a difference of 0.483 mm and 0.682 mm at the interproximal and buccal site respectively, and in Group C 0.750 mm and 0.964 mm at the interproximal and buccal site respectively (Figure 5a,b and Figure 6a,b).

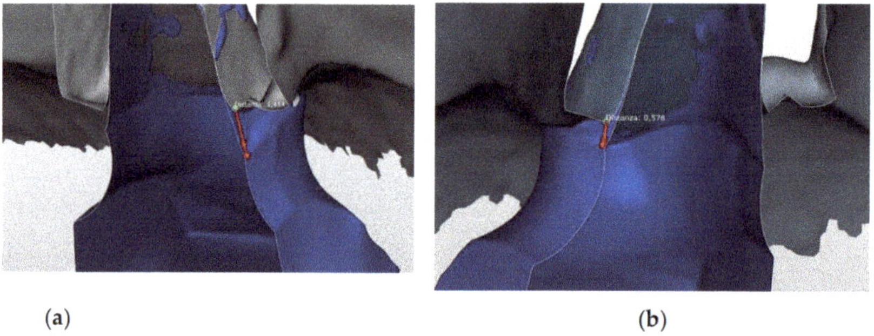

(a)   (b)

Figure 5. Differences in preparation reading of the two impressions in the mesial and distal areas (**a**,**b**).

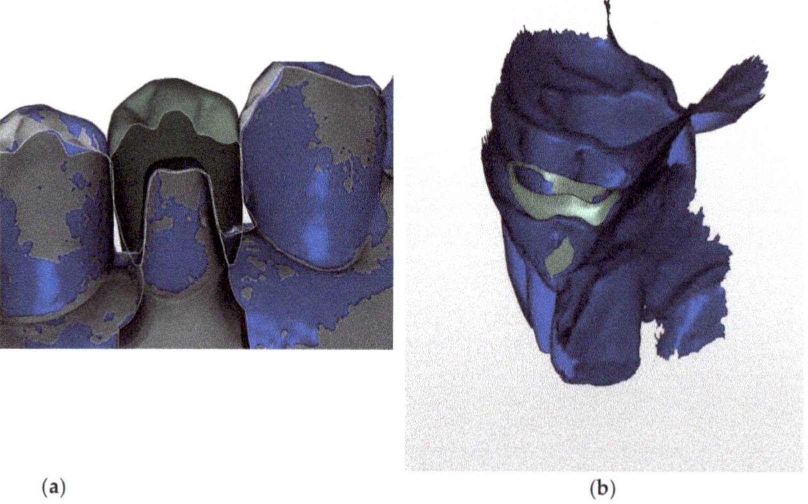

(a)   (b)

Figure 6. After waxing up the two crowns it is evident the difference in depth (**a**,**b**).

The difference between the depth of the sulci, analyzed according to the two vertical preparations (Group B /<1 mm vs. Group C /1.5–2.0 mm), was statistically significant, with a difference of 0.28 mm (SE—Standard Error: 0.5; IC—Interval of Confidence: 95% −0.4–0.2) ($p < 0.00$).

The analysis performed on a site level (mesial/distal/vestibular) on the depth of both vertical preparations revealed significant differences (F = 12.15; $p < 0.0001$) (Tables 2 and 3). After a post hoc analysis (Bonferroni) the vestibular site of the Group B vertical preparation was always statistically inferior to the other sites prepared deeply (Group C) (Tables 4 and 5).

The number of intraoral scans rejected from the study due to evident errors was 2 for Group A, 3 for Group B and 4 for Group C, respectively; and, essentially, were the first scanning shots made by the operator. However, 20 scanning shots for each group were finally performed and evaluated.

Table 2. The analysis performed on a site level (mesial/distal/vestibular) on the depth of both vertical preparations.

Site	Mean	Std. Dev.	Freq.
Bmesial	0.66	0.27	20
Bdistal	0.73	0.28	20
Bbuccal	0.48	0.12	20
Cmesial	1.01	0.30	20
Cdistal	0.92	0.27	20
Cbuccal	0.78	0.15	20
Total	0.76	0.29	120

One-way measure site, bonferroni tabulate: B (Group B) and C (Group C). Bmesial: Mesial site group B, Bdistal: Distal site group B; Bbuccal: Buccal site group B; Cmesial: Mesial site group C; Cdistal: dDistal site group C; Cbuccal: Buccal site group C; Std.Dev: Standard Deviation; Freq.:Frequency.

Table 3. Analysis of variance.

Source	SS	df	MS	F	Prob > F
Between groups	3.55666457	5	0.711332913	12.15	0.0000
Within groups	6.6717138	114	0.058523805		
Total	10.2283784	119	0.058523805		

Bartlett's test for equal variances: $chi^2(5)$ = 21.8654 Prob > chi^2 = 0.001.

Table 4. A post hoc analysis (Bonferroni).

Row Mean−| Column Mean	Bmesial	Bdistal	Bbuccal	Cmesial	Cdistal
Bdistal	0.07				
	1.000				
Bbuccal	−0.18	−0.24			
	0.357	0.027			
Cmesial	0.36	0.29	0.53		
	0.000	0.004	0.000		
Cdistal	0.26	0.19	0.44	−0.10	
	0.014	0.211	0.000	1.000	
Cbuccal	0.12	0.05	0.30	−0.23	−0.14
	1.000	1.000	0.002	0.045	1.000

Measure, by depth. Two-sample *t* test with equal variances.

Table 5. Statistical data about differences between the two types of impressions.

Group	Obs	Mean	Std. Err.	Std. Dev	[95% Confidence.	Interval]
Group B	60	0.62295	0.0325056	0.2517873	0.5579064	0.6879936
Group C	60	0.9044833	0.0340743	0.263938	0.8363009	0.9726658
Combined	120	0.7637167	0.0267633	0.293177	0.7107227	0.8167106
Difference		−0.2815333	0.0470921		−0.3747886	−0.1882781

diff = mean(Group B)−mean(Group C); t = −5.9784. Ho: diff = 0 degrees of freedom = 118 Ha: diff < 0 Ha: diff = 0 Ha: diff > 0. Pr(T < t) = 0.0000 Pr(|T| > |t|) = 0.0000 Pr(T > t) = 1.0000; Std.Err—standard error; Std.Dev—standard deviation.

4. Discussion

The restorative finishing line of full crowns can be designed according to various geometries, mainly horizontally or vertically oriented, and as shoulder, chamfer, and knife edge preparations, with mixed typologies based on the angulation of the marginal zone. When a partial crown is prepared for an esthetic restoration a horizontal margin is usually prepared, such as a shoulder design, with a sharp external angle. The presence of this sharp angle facilitates the check of the distance between the finish line and adjacent tooth, as well as the distance between the finish line and the soft tissues. However, the preparation of an abutment for a digital impression must consider limitations due to the digital impression device [23].

Based on the results of this clinical trial, the null hypothesis, that there was no difference in the capability of the IOS independent of the vertical position of the prepared finish line, was rejected ($p < 0.005$). It was pointed out that the deeper into the sulcus the position of the margin is, more of the part of the prepared root will be lost during the digital impression.

Several clinical parameters were kept under control to ensure uniformity in order to reduce the risk of bias in this RCT. All the soft tissues around preparation margins were in similarly healthy condition; the operator was a long-time experienced user of IOS and each patient received detailed instructions before performing the digital impression.

The accuracy of digital impression systems has been extensively studied in recent years [20,23]. However, the wide majority of studies were performed in vitro and designed to detect differences among different scanners [23].

The problem is that the in vitro laboratory conditions often differ from real, daily clinical situations [24]. The clinical use of IOS can be heavily complicated by factors such as: humidity of the oral environment, saliva flow, soft tissue presence and health condition, possible movements of the patient, scanning procedure and technique, limited access of the scanning probe to posterior teeth (for instance, hampered by lips and cheeks), and the varying translucency of enamel and dentine [25]. However, the results of this study showed that when all the aforementioned factors were controlled as fully as possible during impression taking, the depth of the finishing line inside the sulcus can negatively influence the final quality and accuracy of the digital impression.

A possible explanation for this finding is related to the discrete nature of intraoral scans. Unlike conventional impressions, which record a continuous surface, digital scans sample the surface at discrete intervals. A continuous surface is then generated in the software by 'joining the dots' according to the "stitching" algorithm. If the sample density of information is too low relative to the topology of the region (e.g., in a small patch of the impression near the gingival crevice and containing an angular crown margin too), the generated 3D surface will not replicate the true anatomy.

The results of this study clearly pointed out limitations in taking a predictable digital impression when a margin placed 1.5–2 mm into the sulcus was used and showed the need for a coronally positioned finishing line in order to catch the margins.

It was stated that low quality of impressions and insufficient preparations were the greatest obstacles for the production of high-end dental restorations [26]. In this context, IOS seems to be a logical step to prevent many possible errors.

However, it must be considered that performing a preparation is a common procedure in general dental practice, as a necessary prerequisite for the fabrication of fixed prosthetic restoration, and influences overall success substantially. During preparation, biological and technical necessities often oppose each other and therefore sometimes make it a difficult procedure for the dentist. Additionally, in daily practice the cervical margin is often located equigingivally and/or subgingivally and the positioning of the margin can be a serious obstacle to taking a perfect digital impression [27].

When the finishing line is located in the sulcus and the IOS is used, a certain amount of prepared root can't be captured [28]. The prepared root which is not captured in the digital impression and that remains uncovered by the margin of the crown will be covered by a long epithelium attachment the same type of periodontal attachment formed after scaling and root planning [29].

The skill of the operator and the role of temporary crowns may help to address margins positioned more in depth into the sulcus.

However, few scientific data are available regarding the capability of IOS to catch margins located deeply into the sulcus. Consequently, the results of this randomized clinical trial strongly suggest the use of IOS in combination with supragingival preparations only.

It has to be emphasized that only one IOS device has been evaluated in this study; therefore, these results cannot be directly translated to other trials using different IOS devices. Similar clinical studies with a wider number of IOS are desirable.

5. Conclusions

Based on the results of this clinical study, the following conclusions can be drawn:

1. The deeper the position of the finishing line into the sulcus, it is more difficult to capture the margin using IOS.

2. Digital impression is not recommended when crowns' margins are positioned deep (1.5–2 mm) into the sulcus.

Author Contributions: Conceptualization, E.F.C. and M.F.; methodology, E.F.C., M.F., and F.Z.; software, N.D.; validation, N.D. and T.J.; formal analysis, N.D.; investigation, E.F.C. and M.F.; resources, E.F.C.; writing—original draft preparation, E.F.C. and F.Z.; writing—review and editing, M.F. and T.J.; supervision, M.F.; funding acquisition, M.F. All authors have read and agreed to the published version of the manuscript. Authorship was limited to those who have contributed substantially to the work reported.

Funding: This research received no external funding.

Institutional Review Board Statement: The study was conducted according to the guidelines of the Declaration of Helsinki, and approved by the Institutional Review Board (or Ethics Committee) of NAME OF INSTITUTE (protocol n.18895, code PSO001, Siena 16.01.2020).

Informed Consent Statement: Written informed consent has been obtained from the patients to publish this paper.

Data Availability Statement: The data presented in this study are available on request from the corresponding author. The data are not publicly available due to privacy of patients.

Conflicts of Interest: The authors declare no conflict of interest.

References

1. Gardner, F.M. Margins of complete crowns—Literature review. *J. Prosthet. Dent.* **1982**, *48*, 396–400. [CrossRef]
2. Anusavice, K.J.; Carroll, J.E. Effect of incompatibility stress on the fit of metal-ceramic crowns. *J. Dent. Res.* **1987**, *66*, 1341–1345. [CrossRef] [PubMed]
3. Yeo, I.S.; Yang, J.H.; Lee, J.B. In vitro marginal fit of three all-ceramic crown systems. *J. Prosthet. Dent.* **2003**, *90*, 459–464. [CrossRef]
4. Larson, T.D. The clinical significance of marginal fit. *Northwest Dent.* **2012**, *91*, 22–29.

5. Contrepois, M.; Soenen, A.; Bartala, M.; Laviole, O. Marginal adaptation of ceramic crowns: A systematic review. *J. Prosthet. Dent.* **2013**, *110*, 447–454. [CrossRef] [PubMed]
6. Boitelle, P.; Mawussi, B.; Tapie, L.; Fromentin, O. A systematic review of CAD/CAM fit restoration evaluations. *J. Oral. Rehabil.* **2014**, *41*, 853–874. [CrossRef]
7. Yildirim, G.; Uzun, I.H.; Keles, A. Evaluation of marginal and internal adaptation of hybrid and nanoceramic systems with microcomputed tomography: An in vitro study. *J. Prosthet. Dent.* **2017**, *118*, 200–207. [CrossRef]
8. Papadiochou, S.; Pissiotis, A.L. Marginal adaptation and CAD-CAM technology: A systematic review of restorative material and fabrication techniques. *J. Prosthet. Dent.* **2018**, *119*, 545–551. [CrossRef] [PubMed]
9. Mclean, J.W. The estimation of cement film thickness by an in vivo technique. *Br. Dent. J.* **1971**, *131*, 107–111. [CrossRef]
10. Martignoni, M.; Schonenberger, A. *Precision Fixed Prosthodontics: Clinical and Laboratory Aspects*; Quintessence Publishing: Berlin, Germany, 1990.
11. Johnson, G.H.; Craig, R.G. Accuracy of four types of rubber impression materials compared with time of pour and a repeat pour of models. *J. Prosthet. Dent.* **1985**, *53*, 484–490. [CrossRef]
12. Millstein, P.L. Determining the accuracy of gypsum casts made from type IV dental stone. *J. Oral Rehab.* **1992**, *19*, 239–243. [CrossRef]
13. Schneider, A.; Kurtzman, G.M.; Silverstein, L.H. Improving implant framework passive fit and accuracy through the use of verification stents and casts. *J. Dent. Technol.* **2001**, *18*, 23–25.
14. Heckmann, S.M.; Karl, M.; Wichmann, M.G.; Winter, W.; Graef, F.; Taylor, T.D. Cement fixation and screw retention: Parameters of passive fit: An in vitro study of three-unit implant-supported fixed partial dentures. *Clin. Oral Impl. Res.* **2004**, *15*, 466–473. [CrossRef] [PubMed]
15. Kihara, H.; Hatakeyama, W.; Komine, F.; Takafuji, K.; Takahashi, T.; Yokota, J.; Oriso, K.; Kondo, H. Accuracy and practicality of intraoral scanner in dentistry: A literature review. *J. Prosthodont. Res.* **2020**, *64*, 109–113. [CrossRef] [PubMed]
16. Aswani, K.; Wankhade, S.; Khalikar, A.; Deogade, S. Accuracy of an intraoral digital impression: A review. *J. Indian Prosthodont. Soc.* **2020**, *20*, 27–37. [CrossRef]
17. Zarone, F.; Ruggiero, G.; Ferrari, M.; Mangano, F.; Joda, T.; Sorrentino, R. Accuracy of a chairside intraoral scanner compared with a laboratory scanner for the completely edentulous maxilla: An in vitro 3-dimensional comparative analysis. *J. Prosthet. Dent.* **2020**, *124*, 767. [CrossRef]
18. Mangano, F.; Lerner, H.; Margiani, B.; Solop, P.; Latuta, N.; Admakin, O. Congruence between Meshes and Library Files of Implant Scanbodies: An In Vitro Study Comparing Five Intraoral Scanners. *J. Clin. Med.* **2020**, *9*, 2174. [CrossRef] [PubMed]
19. Safari, S.; Vossoghi Sheshkalani, M.; Hoseini Ghavam, F.; Hamedi, M. Gingival Retraction Methods for Fabrication of Fixed Partial Denture: Literature Review. *J. Dent. Biomater.* **2016**, *3*, 205–213.
20. Goodacre, C.J.; Campagni, W.V.; Aquilino, S.A. Tooth preparations for complete crowns: An art form based on scientific principles. *J. Prosthet. Dent.* **2001**, *85*, 363–376. [CrossRef]
21. Di Fiore, A.; Vigolo, P.; Monaco, C.; Graiff, L.; Ferrari, M.; Stellini, E. Digital impression of teeth prepared with a subgingival vertical finish line: A new clinical approach to manage the interim crown. *J. Osseointegr.* **2019**, *11*, 544–547.
22. Schmitz, J.; Valenti, M. Interim restoration technique for gingival displacement with a feather edge preparation design and digital scan. *J. Prosthet. Dent.* **2019**, *123*. [CrossRef] [PubMed]
23. Guth, J.F.; Keul, C.; Stimmelmayr, M.; Beuer, F.; Edelhoff, D. Accuracy of digital models obtained by direct and indirect data capturing. *Clin. Oral Invest.* **2013**, *17*, 1201–1208. [CrossRef] [PubMed]
24. Ferrari, M.; Keeling, A.; Mandelli, F.; Lo Giudice, G.; Garcia-Godoy, F.; Joda, T. The ability of marginal detection using different intraoral scanning systems: A pilot randomized controlled trial. *Am. J. Dent.* **2018**, *31*, 272–276. [PubMed]
25. Abduo, J. Accuracy of Intraoral Scanners: A Systematic Review of Influencing Factors. *Eur. J. Prosth. Rest. Dent.* **2018**, *26*, 101–121.
26. Cho, S.H.; Schaefer, O.; Thompson, G.A.; Guentsch, A. Comparison of accuracy and reproducibility of casts made by digital and conventional methods. *J. Prosthet. Dent.* **2015**, *113*, 310–315. [CrossRef]
27. Bader, J.D.; Rozier, R.G.; McFall, W.T., Jr.; Ramsey, D.L. Effect of crown margins on periodontal conditions in regularly attending patients. *J. Prosthet. Dent.* **1991**, *65*, 75–79. [CrossRef]
28. Mandelli, F.; Ferrini, F.; Gastaldi, G.; Gherlone, E.; Ferrari, M. Improvement of a Digital Impression with Conventional Materials: Overcoming Intraoral Scanner Limitations. *Int. J. Prosthodont.* **2017**, *30*, 373–376. [CrossRef]
29. Caton, J.G.; Zander, H.A. The attachment between tooth and gingival tissues after periodic root planning and soft tissue curettage. *J. Periodontology* **1979**, *9*, 462–466. [CrossRef]

Journal of
Clinical Medicine

Article

Marginal and Internal Fit of Ceramic Restorations Fabricated Using Digital Scanning and Conventional Impressions: A Clinical Study

Jeong-Hyeon Lee [1,2,†], Keunbada Son [2,3,†] and Kyu-Bok Lee [1,2,*]

1   Department of Prosthodontics, School of Dentistry, Kyungpook National University,
    2177 Dalgubeol-daero, Jung-gu, Daegu 41940, Korea; prossn@naver.com
2   Advanced Dental Device Development Institute (A3DI), Kyungpook National University,
    2177 Dalgubeol-daero, Jung-gu, Daegu 41940, Korea; sonkeunbada@gmail.com
3   Department of Dental Science, Graduate School, Kyungpook National University,
    2177 Dalgubeol-daero, Jung-gu, Daegu 41940, Korea
*   Correspondence: kblee@knu.ac.kr
†   These authors contributed equally to this work (co-first author).

Received: 26 November 2020; Accepted: 11 December 2020; Published: 14 December 2020

Abstract: This clinical study was designed with the aim of fabricating four ceramic crowns using the conventional method and digital methods with three different intraoral scanners and evaluate the marginal and internal fit as well as clinician satisfaction. We enrolled 20 subjects who required ceramic crowns in the upper or lower molar or the premolar. Impressions were obtained using digital scans, with conventional impressions (polyvinyl siloxane and desktop scanner) and three different intraoral scanners (EZIS PO, i500, and CS3600). Four lithium disilicate glass-ceramic crowns were fabricated for each patient. In the oral cavity, the proximal and occlusal adjustments were performed, and the marginal fit and internal fit were evaluated using the silicone replica technique. The clinician satisfaction score of the four crowns was evaluated as per the evaluations of the proximal and occlusal contacts made during the adjustment process and the marginal and internal fit. For statistical analysis, the differences among the groups were analyzed with one-way analysis of variance and Tukey HSD test as a post-test; Pearson correlation analysis was used for analyzing the correlations ($\alpha = 0.05$). There was a significant difference in the marginal and internal fit of the ceramic crowns fabricated using three intraoral scanner types and one desktop scanner type ($p < 0.001$); there was a significant difference in the clinician satisfaction scores ($p = 0.04$). The clinician satisfaction score and marginal fit were significantly correlated (absolute marginal discrepancy and marginal gap) ($p < 0.05$). An impression technique should be considered for fabricating a ceramic crown with excellent goodness-of-fit. Further, higher clinician satisfaction could be obtained by reproducing the excellent goodness-of-fit using the intraoral scanning method as compared to the conventional method.

Keywords: marginal and internal fit; intraoral scanner; conventional method; ceramic crown; digital workflow

---

1. Introduction

In the processes of fabricating and restoring prostheses, it is crucial to take impressions accurately [1–3]. The use of conventional impression material for taking impressions causes the patient discomfort, such as gagging; further, there may be various problems, such as the possibility of deformation of the impression material and contamination by the saliva and blood in the oral cavity [3,4]. In contrast, the use of an intraoral canner for taking a digital impression is a method that obtains impressions via direct scanning [5,6]. Moreover, it is possible to correct it and check the bite, looking at the three-dimensional (3D) virtual cast displayed in real time on the monitor [7]. In the conventional method that uses impression material, dental stone, investing material, and alloy, etc., there are differences in the expansion and contraction rate of each material; therefore, the goodness-of-fit of the prostheses may differ, depending on the proficiency of the dental technician [8]. However, prosthesis fabrication using an intraoral scanner offers the advantage in that the work process can be standardized [9]. Furthermore, digital impression taking is unlikely to cause the deformation of the impression material, and the additional scan and work are easy [9]. In addition, compared to conventional impression taking, this method involves a lower cost and less time; thus, this method tends to be used increasingly [10].

An intraoral scanner is an essential tool in chairside computer-aided design and computer-aided manufacturing (CAD/CAM) system because it allows the acquisition of a virtual cast directly from the patient's oral cavity without the need for any additional work process [11–15]. In the dental clinic, where quick fabrication is necessary, lithium disilicate ceramic material is preferred because it takes less milling time and less crystallization time [16]. The lithium disilicate ceramic crown is superior to a zirconia crown in terms of better aesthetics, faster fabrication, and easier post-processing process [17,18].

The accuracy of the impression obtained from the patient's oral cavity is important for the fabrication of a well-fitting prosthesis [13–16]. The deformation of the impression may lead to poor marginal and internal fit of the prostheses along with inaccurate working cast fabrication [13,14]. This may cause issues, such as plaque deposits in the oral cavity, secondary caries, cement dissolution, and periodontal disease [13]. Furthermore, the poor internal fit may cause loss of the retention force of the prostheses and lower fracture resistance [14,16]. In general, it is judged that the clinically allowable marginal fit of the fixed prostheses is 100–120 μm [17–20]. Previous studies fabricated zirconia coping, using an intraoral scanner, and reported about 100 μm as the marginal fit [21,22]. Many earlier trials have shown various results of the marginal fit [23–26]; however, few studies have studied the marginal and internal fit of the crowns fabricated using various intraoral scanners.

Several studies have compared the conventional method to the digital method [27–31]. However, the results were different and inconsistent [29–31]. Some studies have demonstrated superior accuracy of the conventional method [23–25], and some have shown better results using the digital method [27,28]. However, most previous studies were in vitro trials that performed extraoral evaluations, and various conditions that might be reflected in the oral cavity (obstacles, such as saliva, limited scan space, tongue, and cheek) have not been reflected [23–25]. Thus, in vivo studies conducted directly in the oral cavity with the conventional method of impression taking and intraoral scanning are continuously needed. In addition, most previous studies have compared the marginal and internal fit of the crowns fabricated using the conventional and digital method. There is insufficient research on the comparison of the ceramic crowns fabricated with different intraoral scanners [23–25].

In the present clinical study, we fabricated four ceramic crowns for each patient using the conventional method and the digital method using three different intraoral scanners and compare the marginal and internal fit of the ceramic crowns with respect to clinician satisfaction. The first null hypothesis is that there would be no differences in the marginal and internal fit of the prostheses fabricated using the conventional and digital methods. The second null hypothesis is that the marginal and internal fit of the ceramic crowns and clinician satisfaction were not correlated.

2. Materials and Methods

The present clinical test was conducted after obtaining approval from the IRB of the Kyungpook National University Dental Hospital (Approval Number: KNUDH-2019-02-02-02). The present clinical study was conducted from April 2019 to April 2020. Twenty subjects (10 women and 10 men) who required a ceramic crown on the upper or lower molar or premolar were enrolled. Of the participants, those with poor oral hygiene or more than one crown; those with parafunctional activities, such as bruxism, clenching, and grinding; those with acute or chronic temporomandibular joint dysfunction sensation or mental abnormalities; and those with serious medical conditions; as well as those who were a pregnant or breastfeeding were excluded. Based on the pilot experiment, training was provided for all the processes involved in the fabrication of ceramic crowns; using power software (G*Power version 3.1.9.2; Heinrich-Heine-Universität Düsseldorf, Düsseldorf, Germany), 20 participants were selected for the analyses (actual power = 96.1%; power = 95%; α = 0.05).

All the subjects were trained for intraoral scanning, crown design using CAD software, and all digital workflows, including the CAM process in advance. All the intraoral processes were prepared by one skilled dentist. The dentist was blinded to the information about the type of crown. Abutment teeth were ground as per the standard crown treatment guidelines [32]. All the abutment teeth were ground, using diamond rotary cutting instruments (852.FG.010; Jota AG, Rüthi, SG, Switzerland) to have 0.5-mm supragingival finish line, 2-mm occlusal reduction, and 6-degree convergence angle; the line range was rounded.

Impressions were obtained, using a conventional impression (desktop scanner) and three intraoral scanner types for each patient (Figure 1). All the scanners used in the present clinical study were calibrated immediately before performing scanning, and scanning was done under uniform conditions of ambient light and surface condition of the dried tooth by a single skilled operator. The same skilled dentist (J.-H.L.) conducted all the clinical tests, and four ceramic crowns were fabricated for each patient.

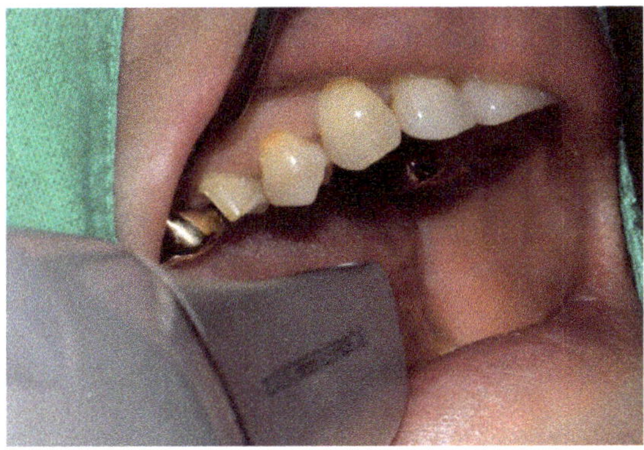

Figure 1. Intraoral scanning.

Group 1: Conventional impression (polyvinyl siloxane (PVS) (Aquasil Ultra; Dentsply Sirona, Bensheim, Germany) (PVS group)
Group 2: Intraoral scanner (EZIS PO; DDS, Seoul, South Korea) (EZIS PO group)
Group 3: Intraoral scanner (i500; MEDIT, Seoul, South Korea) (i500 group)
Group 4: Intraoral scanner (CS3600; Carestream Dental, Atlanta USA) (CS3600 group)

For the PVS Group, an impression was taken, using PVS impression and a double-arch tray (Dual Arch Impression Tray; 3M, MN, USA) in the oral cavity. For the material of the obtained impression, a working cast was fabricated, using Type IV dental stone (FUGIROCK; GC, Leuven, Belgium). The fabricated working cast was scanned with a desktop scanner (E1; 3Shape, Copenhagen, Denmark) and converted to an STL file. All the processes of the working cast fabrication, desktop scanning, and ceramic crown fabrication were performed by a skilled dental technician. For Intraoral Scan Groups, three intraoral scanner types were used, including EZIS PO, i500, and CS3600. All the digital scans were performed as per the manufacturers' instructions. All the intraoral scanning processes were performed by a skilled dentist (J.-H.L.).

In the scan file obtained from each group, the cement space was set at 80 μm in the dental CAD software (EZIS VR; DDS, Seoul, Korea), and the crown was designed for the anatomical shape (Figure 2). After the design preparation was complete, crowns were fabricated, using four-axis milling equipment (EZIS HM; DDS, Seoul, Korea). For the crown material, lithium disilicate glass-ceramic block (IPS e.max CAD; Ivoclar Vivadent AG, Schaan, Liechtenstein) was used. The milled ceramic crown was cleaned using the method recommended by the manufacturer, and it was crystallized and finished. Four ceramic crowns were fabricated for each patient, and 80 ceramic crowns were fabricated for 20 patients.

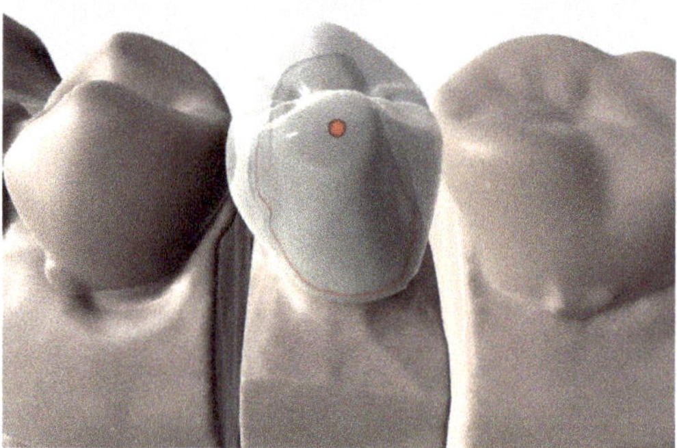

Figure 2. Computer-aided design of a crown.

Four crowns were tried for each patient intraorally and adjusted to enable optimum proximal and occlusal contacts. The marginal and internal fit for the adjustment and cementation in the oral cavity was subjected to clinical evaluation. The marginal fit was checked for appropriateness by probing with a dental explorer (5 XTS™ EXPLORER; Hu-Friedy, Chicago, IL, USA), and the internal fit was checked with silicone paste (Fit Checker; GC, Tokyo, Japan). After trying that in by putting the silicone paste, it was hardened under the patient's bite force (Figure 3), and the transparent region of the silicon was marked on the intaglio surface of the crown, using graphite and adjusted with a diamond rotary cutting instrument. Finally, the occlusal contacts of the ceramic crown were adjusted. Using articulating paper (AccuFilm II; Parkell, Inc., Farmingdale, NY, USA), the regions of earlier contacts or interference were checked during the centric occlusion and eccentric occlusion and carefully removed, using diamond rotary cutting instruments. Using shim stock foil, proximal, and occlusal contacts were checked, and the final grinding was performed as per the recommendation of the manufacturer of the lithium disilicate glass-ceramic.

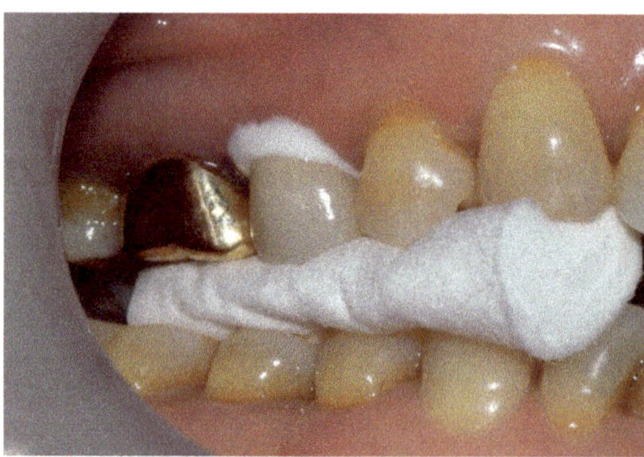

Figure 3. Taking silicone film for checking the fit.

For each patient, the marginal and internal fit of four ceramic crowns was evaluated, using the silicone replica technique. After all the adjustments were made, silicone indicator paste (Fit Checker; GC, Tokyo, Japan) was injected into the intaglio surface of each crown, and the position of the crown was maintained at the patient's bite force till the end of the silicon polymerization process. After the hardening of the silicone indicator paste, the ceramic crown was removed, and a light-body PVS impression was (Aquasil Ultra; Dentsply Sirona, Bensheim, Germany) was injected in the intaglio surface of the crown and hardened for five minutes to support the thin silicon layer. The silicon in which the space between the ceramic crown and the abutment tooth was duplicated was cut from the crown in the medial-mesiodistal and buccolingual directions, and the gap was evaluated at 60× magnification with an industrial video microscope system (IMS 1080P; SOMETECH, Seoul, Korea). With respect to the position of measurement, marginal fit (absolute marginal discrepancy and marginal gap) and internal fit (chamfer, axial, angle, and occlusal gap) were measured (Figure 4). In the internal fit, the chamfer gap was evaluated at the central point of the chamfer region, and the angle gap was assessed at the central point of the angle region. The axial gap was evaluated at the central point between the chamfer gap and the angle gap, and the occlusal gap was measured at the center of the occlusal region and the central point of the axial gap.

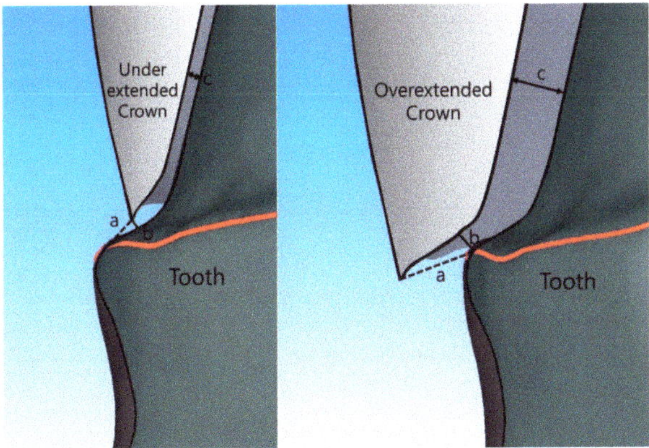

Figure 4. Schematic showing measurement positions for marginal and internal fit. a, Absolute marginal discrepancy. b, Marginal gap. c, Internal gap.

One prosthetic dentist evaluated the quality of the four crowns for each patient based on the clinician satisfaction score. The order of the four crowns was set based on the evaluations of the proximal and occlusal contacts performed in the adjustment process and the marginal and internal fit. The crown showing the best quality was assigned four points and that with the lowest quality was scored one point. The best ceramic crown was chosen and cemented as per the standard prosthetic protocol.

All the data were analyzed using SPSS statistical software (IBM, Armonk, NY, USA). First, the normality of the data was investigated using Shapiro–Wilk test. The data were normal, and the equality of dispersion was evaluated using the Levene test. As per the result, the differences among the groups were analyzed using One-way ANOVA and Tukey HSD test as a post-test ($\alpha = 0.05$).

Moreover, in order to analyze the correlations between the marginal and internal fit and the clinician satisfaction score, Pearson correlation analysis was used. The correlations were divided as per the size of the Pearson correlation coefficient (PCC), as reported previously [33]. The results of the correlation analysis among the variables were explained through the following criteria by the previous studies [14,16,33]: perfect (PCC = +1 or −1), strong (PCC = +0.7–+0.9 or −0.7––0.9), moderate (PCC = +0.4–+0.6 or −0.4––0.6), and weak (PCC = +0.1–+0.3 or −0.1––0.3) ($\alpha = 0.05$).

3. Results

There were significant differences in the marginal and internal fit of the ceramic crowns fabricated using three intraoral scanner types and one desktop scanner type ($p < 0.001$; Figure 5; Table 1). There was no significant difference in the marginal gap as per the three intraoral scanner types ($p > 0.05$; Figure 5; Table 1). There was a higher value for the gap of the marginal fit (absolute marginal discrepancy and marginal gap) in the desktop scanner as compared to that in the three intraoral scanner types ($p < 0.001$; Figure 5; Table 1). There was a significant difference in the internal fit based on the three intraoral scanner types and the desktop scanner ($p < 0.001$); however, no special tendency was observed among the groups (Figure 5; Table 1).

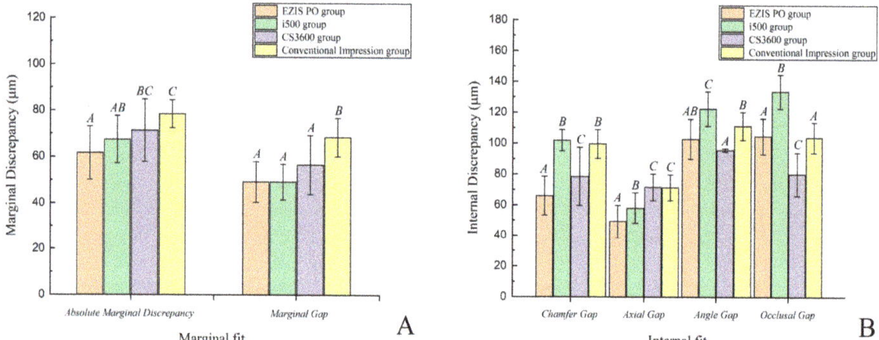

Figure 5. Comparison of the marginal and internal fit of ceramic restorations. (**A**) Marginal fit. (**B**) Internal fit. Same uppercase letters (A, B, C) are not significantly different ($p > 0.05$).

Table 1. Comparison of the discrepancy (μm) of ceramic crowns fabricated with the intraoral scanners and conventional impression technique.

Measurement Position	Discrepancy (Mean ± SD)				F	*p*
	Intraoral Scanner Group			Conventional Impression Technique		
	EZIS PO	i500	CS3600			
Absolute Marginal Discrepancy	61.6 ± 11.5 [A]	67.4 ± 10.2 [A,B]	71.3 ± 13.5 [B,C]	78.4 ± 6 [C]	8.758	<0.001 *
Marginal Gap	49.1 ± 8.8 [A]	49.1 ± 7.7 [A]	56.5 ± 12.7 [A]	68.4 ± 8.3 [B]	17.771	<0.001 *
Chamfer Gap	65.9 ± 12.7 [A]	101.9 ± 6.9 [B]	78.5 ± 18.8 [C]	99.5 ± 9.3 [B]	36.483	<0.001 *
Axial Gap	49.2 ± 10.6 [A]	58 ± 10 [B]	71.6 ± 8.6 [C]	71.3 ± 8.3 [C]	26.547	<0.001 *
Angle Gap	102.8 ± 12.9 [A,B]	122.4 ± 11.2 [C]	95.7 ± 12.8 [A]	111.2 ± 9 [B]	19.505	<0.001 *
Occlusal Gap	104.5 ± 11.6 [A]	133.5 ± 11 [B]	80 ± 13.9 [C]	103.7 ± 9.9 [A]	69.547	<0.001 *

* $p < 0.05$; significance was determined using one-way ANOVA. Different letters (A, B, C) indicate that the difference between the groups was significant, as determined using Tukey's HSD post hoc test ($p < 0.05$).

There were significant differences in the clinician satisfaction score of the ceramic crowns fabricated with three intraoral scanner types and one desktop scanner type ($p = 0.04$; Figure 6; Table 2). The value of the clinician satisfaction score was lower in the desktop scanner as compared to that in the three intraoral scanner types ($p < 0.001$; Figure 6; Table 2), while there were no significant differences as per the three intraoral scanner types ($p > 0.05$; Figure 6; Table 2).

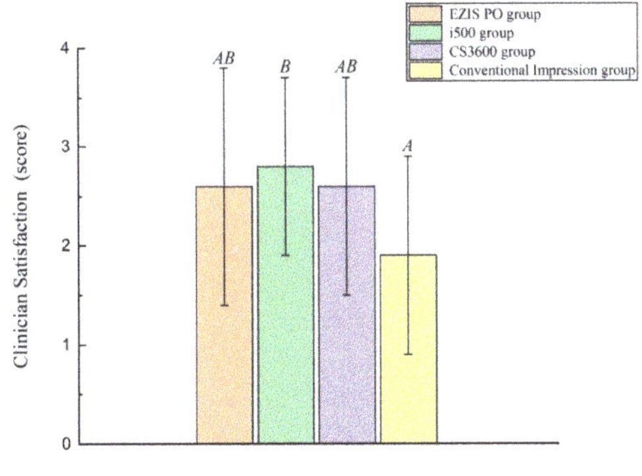

Figure 6. Comparison of clinician satisfaction score of ceramic restorations. Same uppercase letters denote that the difference was not significant ($p > 0.05$).

Table 2. Comparison of the marginal and internal fit of ceramic crowns fabricated with the intraoral scanners and conventional impression technique.

	Intraoral Scanner Group			Conventional Impression Technique	F	*p*
	EZIS PO	**i500**	**CS3600**			
Clinician satisfaction (Mean ± SD, Score)	2.6 ± 1.2 A,B	2.8 ± 0.9 B	2.6 ± 1.1 A,B	1.9 ± 1 A	2.909	0.04 *

* $p < 0.05$; significance was determined using one-way ANOVA. Different letters (A, B, C) indicate that the difference between the groups was significant, as determined by Tukey's HSD post hoc test ($p < 0.05$).

Clinician satisfaction score and marginal fit (absolute marginal discrepancy and marginal gap) had a significant correlation ($p < 0.05$; Table 3). The clinician satisfaction score and absolute marginal discrepancy showed a weak negative correlation ($p = 0.015$; PCC = −0.271; Table 3); the clinician satisfaction score and marginal gap showed a general negative correlation ($p < 0.001$; PCC = −0.403; Table 3).

Table 3. Correlation coefficient between the clinical satisfaction score and the marginal and internal fit.

	Absolute Marginal Discrepancy		Marginal Gap		Chamfer Gap	
	p	PCC	*p*	PCC	*p*	PCC
Clinician Satisfaction	0.015	−0.271	<0.001	−0.403	0.207	-
	Axial Gap		Angle Gap		Occlusal Gap	
	p	PCC	*p*	PCC	*p*	PCC
	0.166	-	0.526	-	0.457	-

4. Discussion

The present clinical study aimed to fabricate ceramic crowns using the conventional method and the digital method with three different intraoral scanners and compare the marginal and internal fit of the ceramic crowns with clinician satisfaction. Thus, the first null hypothesis stated that there would be no differences between the marginal and internal fit of the ceramic crowns fabricated using the four methods and clinician satisfaction; however, this hypothesis was false for all the ceramic crowns ($p < 0.001$). The second null hypothesis stated that there would be no correlation between the marginal and internal fit of the ceramic crowns and clinician satisfaction; this hypothesis was partially dismissed only with respect to the correlation between the marginal fit and clinician satisfaction ($p < 0.001$).

In Chairside CAD/CAM workflow, the use of an intraoral scanner is essential [11,12]. However, previous studies did not investigate the impact of the type of intraoral scanner in the clinical environment on the marginal and internal fit of the ceramic crown. The present results suggest that conventional impression and intraoral scanner type may affect the marginal and internal fit of the ceramic prostheses and the clinician's satisfaction score; thus, clinicians should consider the method for acquiring a virtual cast for the fabrication of an excellent ceramic crown.

Most previous studies report the goodness-of-fit of the prostheses based on the type of the restoring material [13,14]. Moreover, these studies mostly evaluated if the marginal fit could be applied in the clinical setting [15,16]. In many previous studies, the clinically allowable range of the marginal fit is assumed to be a value from 100 μm to 120 μm [17–19], and a range of 50–100 μm is recommended for the internal fit [20–22]. In the present trial, all the ceramic crowns that were fabricated using the conventional method and the digital method with three different intraoral scanners were in the clinically allowable range of marginal fit. However, the internal fit (angle and occlusal gap) had a value exceeding the gap of 100 μm except for CS3600 Group.

Many previous studies have compared the marginal fit and internal fit of the prostheses that was fabricated as per various dental CAD/CAM workflows. Ortorp A. et al. reported poorer marginal fit in the digital workflow (222.5 ± 124.6 μm) as compared to that in the conventional workflow (118 ± 49.7 μm) [23]. Varol S. et al. reported poorer marginal fit in the digital workflow (86.17 ± 27.61 μm) as compared to that in the conventional workflow (77.26 ± 29.23 μm) [24]. In a similar manner, Bayramoglu E. et al. reported poorer marginal fit in the digital workflow (120.4 ± 54.5 μm) as compared to that in the conventional workflow (75.4 ± 16.6 μm) [25]. However, Massignan Berejuk H. et al. reported poorer marginal fit in the conventional workflow (11.56 ± 8.74 μm) as compared to that in the digital workflow (1.85 ± 1.50 μm) [26]. Previous studies have shown different results. Several earlier researches have reported superior marginal fit of prostheses fabricated using the conventional method in a working cast with a physical impression material than that with an intraoral scanner [23–25]. However, recently, the use of chairside CAD/CAM workflow has increased, and many intraoral scanners have recently been developed [27,28]. Thus, most recent studies have reported better marginal fit of prostheses fabricated using an intraoral scanner than that of those fabricated using the conventional method [27,28]. In keeping with these results, in the present study, the method for fabrication with three kinds of intraoral scanner showed better marginal and internal fit of the prostheses than the conventional method. As per a systematic review of the multi-unit fixed dental prosthesis fabricated using the digital workflow, Russo LL et al. reported that studies of a single crown fabricated with the digital workflow are generally conducted; however, few studies of a multi-unit fixed dental prosthesis have been performed, and it is important to perform additional studies to confirm the clinical reliability of the findings [29]. Thus, in addition to the evaluation of the single ceramic crown conducted in the present study, it is necessary to perform a study on the multi-unit fixed dental prosthesis.

In the present clinical study, we found a significant correlation between the clinician satisfaction score and the marginal fit (absolute marginal discrepancy and marginal gap) ($p < 0.05$). However, there was no correlation between the clinician satisfaction score and the internal fit ($p > 0.05$) because the marginal fit was recognized as the most important factor in the process wherein clinicians check the prostheses in the oral cavity. Many previous studies have reported that marginal fit is an important element that influences the prognosis of the fixed dental prosthesis [13,20–26]. Based on the present results, clinicians can use the intraoral scanning method rather than the conventional method for the fabrication of ceramic crowns with excellent goodness-of-fit and realize high clinician satisfaction by reproducing the excellent goodness-of-fit obtained using the intraoral scanning method.

Previous studies have shown a difference in the scanning accuracy based on the intraoral scanner used [11,12]. In the present study, there were significant differences in the marginal fit and internal fit of the ceramic crowns that were fabricated, based on the three intraoral scanner types ($p < 0.001$); however, all values were within the clinically allowable range (within 120 μm), and there were no big differences. Moreover, there were no significant differences in the clinicians' satisfaction among the ceramic crowns fabricated with the three intraoral scanner types ($p > 0.05$). It is judged that there was no impact on the clinician satisfaction because all the ceramic crowns fabricated with the three intraoral scanner types were in the clinically allowable range (within 120 µm). Further, it is necessary to conduct additional studies to evaluate the three intraoral scanner types used in the present study and examine the impact of the scanning accuracy on the marginal fit and the internal fit.

The present clinical study has certain limitations. It is necessary to conduct additional studies on various intraoral scanners other than those used in the present study. Moreover, it is crucial to perform additional studies using materials other than the lithium disilicate glass-ceramic material that was used in the present clinical study, such as zirconia. We believe that it is important to conduct a study of the multi-unit fixed dental prosthesis. Finally, additional studies on the prognosis should be conducted as a continuation of the present clinical study.

5. Conclusions

Based on the findings of this clinical study, the following conclusions were drawn:

1. There was an impact on the marginal and internal fit of the ceramic crowns based on the type of intraoral scanner that was used; however, there was no difference in the clinicians' satisfaction with the prostheses.
2. The ceramic crowns fabricated using an intraoral scanner showed superior marginal fit and internal fit as well as higher clinician satisfaction than those fabricated using the conventional method with PVS impression.
3. The excellent marginal fit of the fabricated ceramic crowns can achieve high clinician satisfaction.
4. Thus, clinicians should consider the use of the impression method for fabricating a ceramic crown with excellent goodness-of-fit and can realize high clinician satisfaction by reproducing excellent goodness-of-fit using the intraoral scanning method rather than the conventional method.

Author Contributions: Conceptualization, J.-H.L. and K.S.; funding acquisition, K.-B.L.; methodology, J.-H.L. and K.S.; validation, J.-H.L. and K.S.; formal analysis, J.-H.L.; investigation, J.-H.L.; data curation, J.-H.L.; software, J.-H.L. and K.S.; writing—original draft, J.-H.L. and K.S.; visualization, K.S.; supervision, K.-B.L.; project administration, K.-B.L. All authors have read and agreed to the published version of the manuscript.

Funding: This research was supported by the Ministry of Trade, Industry & Energy (MOTIE, Korea) under the Industrial Technology Innovation Program (No. 10062635).

Acknowledgments: The authors thank the researchers from the Advanced Dental Device Development Institute, Kyungpook National University, for their time and contributions to the study.

Conflicts of Interest: The authors declare no conflict of interest. The funders had no role in the design of the study; in the collection, analyses, or interpretation of the data; in the writing of the manuscript; or in the decision to publish the results.

References

1. Falahchai, M.; Hemmati, Y.B.; Asli, H.N.; Emadi, I. Marginal gap of monolithic zirconia endocrowns fabricated by using digital scanning and conventional impressions. *J. Prosthet. Dent.* **2020**. [CrossRef] [PubMed]
2. Porrelli, D.; Berton, F.; Piloni, A.C.; Kobau, I.; Stacchi, C.; Di Lenarda, R.; Rizzo, R. Evaluating the stability of extended-pour alginate impression materials by using an optical scanning and digital method. *J. Prosthet. Dent.* **2020**. [CrossRef] [PubMed]
3. Sahin, V.; Jodati, H.; Evis, Z. Effect of storage time on mechanical properties of extended-pour irreversible hydrocolloid impression materials. *J. Prosthet. Dent.* **2020**, *124*, 69–74. [CrossRef] [PubMed]
4. Bohner, L.O.L.; Canto, G.D.L.; Marció, B.S.; Laganá, D.C.; Sesma, N.; Neto, P.T. Computer-aided analysis of digital dental impressions obtained from intraoral and extraoral scanners. *J. Prosthet. Dent.* **2017**, *118*, 617–623. [CrossRef]
5. Michelinakis, G.; Apostolakis, D.; Tsagarakis, A.; Kourakis, G.; Pavlakis, E. A comparison of accuracy of 3 intraoral scanners: A single-blinded in vitro study. *J. Prosthet. Dent.* **2019**. [CrossRef]

6. Park, J.M.; Kim, R.J.Y.; Lee, K.W. Comparative reproducibility analysis of 6 intraoral scanners used on complex intracoronal preparations. *J. Prosthet. Dent.* **2020**, *123*, 113–120. [CrossRef]

7. Braian, M.; Wennerberg, A. Trueness and precision of 5 intraoral scanners for scanning edentulous and dentate complete-arch mandibular casts: A comparative in vitro study. *J. Prosthet. Dent.* **2019**, *122*, 129–136. [CrossRef]

8. Stimmelmayr, M.; Groesser, J.; Beuer, F.; Erdelt, K.; Krennmair, G.; Sachs, C.; Güth, J.F. Accuracy and mechanical performance of passivated and conventional fabricated 3-unit fixed dental prosthesis on multi-unit abutments. *J. Prosthodont. Res.* **2017**, *61*, 403–411. [CrossRef]

9. Hayama, H.; Fueki, K.; Wadachi, J.; Wakabayashi, N. Trueness and precision of digital impressions obtained using an intraoral scanner with different head size in the partially edentulous mandible. *J. Prosthodont. Res.* **2018**, *62*, 347–352. [CrossRef]

10. Kihara, H.; Hatakeyama, W.; Komine, F.; Takafuji, K.; Takahashi, T.; Yokota, J.; Kondo, H. Accuracy and practicality of intraoral scanner in dentistry: A literature review. *J. Prosthodont. Res.* **2020**, *64*, 109–113. [CrossRef]

11. Park, G.H.; Son, K.; Lee, K.B. Feasibility of using an intraoral scanner for a complete-arch digital scan. *J. Prosthet. Dent.* **2019**, *121*, 803–810. [CrossRef] [PubMed]

12. Son, K.; Lee, K.B. Effect of Tooth Types on the Accuracy of Dental 3D Scanners: An In Vitro Study. *Materials* **2020**, *13*, 1744. [CrossRef] [PubMed]

13. Son, K.; Lee, S.; Kang, S.H.; Park, J.; Lee, K.B.; Jeon, M.; Yun, B.J. A comparison study of marginal and internal fit assessment methods for fixed dental prostheses. *J. Clin. Med.* **2019**, *8*, 785. [CrossRef] [PubMed]

14. Jang, D.; Son, K.; Lee, K.B. A Comparative study of the fitness and trueness of a three-unit fixed dental prosthesis fabricated using two digital workflows. *Appl. Sci.* **2019**, *9*, 2778. [CrossRef]

15. Kang, B.H.; Son, K.; Lee, K.B. Accuracy of five intraoral scanners and two laboratory scanners for a complete arch: A comparative in vitro study. *Appl. Sci.* **2020**, *10*, 74. [CrossRef]

16. Lee, K.; Son, K.; Lee, K.B. Effects of Trueness and Surface Microhardness on the Fitness of Ceramic Crowns. *Appl. Sci.* **2020**, *10*, 1858. [CrossRef]

17. Alajaji, N.K.; Bardwell, D.; Finkelman, M.; Ali, A. Micro-CT Evaluation of Ceramic Inlays: Comparison of the Marginal and Internal Fit of Five and Three-Axis CAM Systems with a Heat Press Technique. *J. Esthet. Restor. Dent.* **2017**, *29*, 49–58. [CrossRef]

18. Roperto, R.; Assaf, H.; Soares-Porto, T.; Lang, L.; Teich, S. Are different generations of CAD/CAM milling machines capable to fabricate restorations with similar quality? *J. Clin. Exp. Dent.* **2016**, *8*, e423.

19. Sachs, C.; Groesser, J.; Stadelmann, M.; Schweiger, J.; Erdelt, K.; Beuer, F. Full-arch prostheses from translucent zirconia: Accuracy of fit. *Dent. Mater.* **2014**, *30*, 817–823. [CrossRef]

20. Colpani, J.T.; Borba, M.; Della Bona, Á. Evaluation of marginal and internal fit of ceramic crown copings. *Dent. Mater.* **2013**, *29*, 174–180. [CrossRef]

21. Mously, H.A.; Finkelman, M.; Zandparsa, R.; Hirayama, H. Marginal and internal adaptation of ceramic crown restorations fabricated with CAD/CAM technology and the heat-press technique. *J. Prosthet. Dent.* **2014**, *112*, 249–256. [CrossRef] [PubMed]

22. Lins, L.; Bemfica, V.; Queiroz, C.; Canabarro, A. In vitro evaluation of the internal and marginal misfit of CAD/CAM zirconia copings. *J. Prosthet. Dent.* **2015**, *113*, 205–211. [CrossRef] [PubMed]

23. Örtorp, A.; Jönsson, D.; Mouhsen, A.; von Steyern, P.V. The fit of cobalt–chromium three-unit fixed dental prostheses fabricated with four different techniques: A comparative in vitro study. *Dent. Mater.* **2011**, *27*, 356–363. [CrossRef] [PubMed]

24. Varol, S.; Kulak-Özkan, Y. In Vitro Comparison of Marginal and Internal Fit of Press-on-Metal Ceramic (PoM) Restorations with Zirconium-Supported and Conventional Metal Ceramic Fixed Partial Dentures Before and After Veneering. *J. Prosthodont.* **2015**, *24*, 387–393. [CrossRef]

25. Bayramoğlu, E.; Özkan, Y.K.; Yildiz, C. Comparison of marginal and internal fit of press-on-metal and conventional ceramic systems for three-and four-unit implant-supported partial fixed dental prostheses: An in vitro study. *J. Prosthet. Dent.* **2015**, *114*, 52–58. [CrossRef]

26. Massignan Berejuk, H.; Hideo Shimizu, R.; Aparecida de Mattias Sartori, I.; Valgas, L.; Tiossi, R. Vertical microgap and passivity of fit of three-unit implant-supported frameworks fabricated using different techniques. *Int. J. Oral Maxillofac. Implants* **2014**, *29*, 1064–1070. [CrossRef]

27. Rapone, B.; Palmisano, C.; Ferrara, E.; Di Venere, D.; Albanese, G.; Corsalini, M. The Accuracy of Three Intraoral Scanners in the Oral Environment with and without Saliva: A Comparative Study. *Appl. Sci.* **2020**, *10*, 7762. [CrossRef]

28. Lee, S.J.; Kim, S.W.; Lee, J.J.; Cheong, C.W. Comparison of Intraoral and Extraoral Digital Scanners: Evaluation of Surface Topography and Precision. *Dent. J.* **2020**, *8*, 52. [CrossRef]

29. Russo, L.L.; Caradonna, G.; Biancardino, M.; De Lillo, A.; Troiano, G.; Guida, L. Digital versus conventional workflow for the fabrication of multiunit fixed prostheses: A systematic review and meta-analysis of vertical marginal fit in controlled in vitro studies. *J. Prosthet. Dent.* **2019**, *122*, 435–440. [CrossRef]

30. Hasanzade, M.; Shirani, M.; Afrashtehfar, K.I.; Naseri, P.; Alikhasi, M. In Vivo and In Vitro Comparison of Internal and Marginal Fit of Digital and Conventional Impressions for Full-Coverage Fixed Restorations: A Systematic Review and Meta-analysis. *J. Evid. Based Dent. Pract.* **2019**, *19*, 236–254. [CrossRef]

31. Hasanzade, M.; Aminikhah, M.; Afrashtehfar, K.I.; Alikhasi, M. Marginal and internal adaptation of single crowns and fixed dental prostheses by using digital and conventional workflows: A systematic review and meta-analysis. *J. Prosthet. Dent.* **2020**, in press. [CrossRef] [PubMed]

32. Goodacre, C.J.; Campagni, W.V.; Aquilino, S.A. Tooth preparations for complete crowns: An art form based on scientific principles. *J. Prosthet. Dent.* **2001**, *85*, 363–376. [CrossRef] [PubMed]

33. Dancey, C.; Reidy, J. *Statistics without Maths for Psychology: Pearson Higher*; Pearson: London, UK, 2014.

Publisher's Note: MDPI stays neutral with regard to jurisdictional claims in published maps and institutional affiliations.

Journal of
Clinical Medicine

Article

Congruence between Meshes and Library Files of Implant Scanbodies: An In Vitro Study Comparing Five Intraoral Scanners

Francesco Mangano [1,2,*], Henriette Lerner [3,4], Bidzina Margiani [2], Ivan Solop [2], Nadezhda Latuta [2] and Oleg Admakin [2]

1   Private Practice, Gravedona, 22015 Como, Italy
2   Department of Prevention and Communal Dentistry, Sechenov First Moscow State Medical University, 119991 Moscow, Russia; margiani.b@gmail.com (B.M.); solopivan@yandex.ru (I.S.); latuta.n@mail.ru (N.L.); admakin1966@mail.ru (O.A.)
3   Private Practice, Ludwing-Wilhelm Strasse, 76530 Baden-Baden, Germany; h.lerner@web.de
4   Academic Teaching and Research Institution of Johann Wolfgang Goethe-University, 60323 Frankfurt am Main, Germany
*   Correspondence: francescomangano1@mclink.net; Tel.: +7-(980)-0195356

Received: 17 June 2020; Accepted: 8 July 2020; Published: 9 July 2020

Abstract: Purpose. To compare the reliability of five different intraoral scanners (IOSs) in the capture of implant scanbodies (SBs) and to verify the dimensional congruence between the meshes (MEs) of the SBs and the corresponding library file (LF). Methods. A gypsum cast of a fully edentulous maxilla with six implant analogues and SBs screwed on was scanned with five different IOSs (PRIMESCAN®, CS 3700®, MEDIT i-500®, ITERO ELEMENTS 5D®, and Emerald S®). Ten scans were taken for each IOS. The resulting MEs were imported to reverse engineering software for 3D analysis, consisting of the superimposition of the SB LF onto each SB ME. Then, a quantitative and qualitative evaluation of the deviations between MEs and LF was performed. A careful statistical analysis was performed. Results. PRIMESCAN® showed the highest congruence between SB MEs and LF, with the lowest mean absolute deviation (25.5 ± 5.0 µm), immediately followed by CS 3700® (27.0 ± 4.3 µm); the difference between them was not significant ($p = 0.1235$). PRIMESCAN® showed a significantly higher congruence than MEDIT i-500® (29.8 ± 4.8 µm, $p < 0.0001$), ITERO ELEMENTS 5D® (34.2 ± 9.3 µm, $p < 0.0001$), and Emerald S® (38.3 ± 7.8 µm, $p < 0.0001$). CS 3700® had a significantly higher congruence than MEDIT i-500® ($p = 0.0004$), ITERO ELEMENTS 5D® ($p < 0.0001$), and Emerald S® ($p < 0.0001$). Significant differences were also found between MEDIT i-500® and ITERO ELEMENTS 5D® ($p < 0.0001$), MEDIT i-500® and Emerald S® ($p < 0.0001$), and ITERO ELEMENTS 5D® and Emerald S® ($p < 0.0001$). Significant differences were found among different SBs when scanned with the same IOS. The deviations of the IOSs showed different directions and patterns. With PRIMESCAN®, ITERO ELEMENTS 5D®, and Emerald S®, the MEs were included inside the LF; with CS 3700®, the LF was included in the MEs. MEDIT i-500® showed interpolation between the MEs and LF, with no clear direction for the deviation. Conclusions. Statistically different levels of congruence were found between the SB MEs and the corresponding LF when using different IOSs. Significant differences were also found between different SBs when scanned with the same IOS. Finally, the qualitative evaluation revealed different directions and patterns for the five IOSs.

Keywords: Intraoral scanner; Scanbody; Mesh; Library; Congruence; Quantitative evaluation

1. Introduction

Digital technologies are revolutionising the world of dentistry [1]. The introduction of intraoral scanners (IOSs) [2,3], cone beam computed tomography (CBCT) [4], computer-assisted design and computer-assisted manufacturing (CAD/CAM) software [5], milling machines [6], and three-dimensional (3D) printers [7], together with new highly compatible and aesthetic ceramic materials [8], are transforming workflows in dentistry.

The digital prosthetic workflow is divided into four phases: the acquisition of data with a scanner, which allows obtaining a mesh (ME), i.e., a surface reconstruction of the scanned model; the processing of the ME within CAD software, for designing the prosthetic restorations; the fabrication of the restorations by milling or 3D printing; and finally, the clinical application [1,5,9]. All these steps determine the quality of the clinical result [9,10].

In fixed implant prosthodontics, in particular, the 3D position of the implant is captured through the use of a transfer device, the scanbody (SB), which is screwed on the fixture and scanned with an IOS [11]. This scan, together with that of the master model without SB, the antagonist and the bite are sent in standard tessellation language (STL) format to the dental laboratory, which uses CAD software to model the restorations (individual abutments, temporary, and then definitive restorations) [11]. Within the CAD software, the first and fundamental step performed by the dental technician is the replacement, on the master model, of the ME of the SB with the corresponding library file (LF) [11]. This LF is aligned with all the components (titanium bonding bases with different shape and height) necessary for modelling the prosthetic restorations. Modelling on an LF is certainly preferable to modelling on an ME. The LF, originally designed using CAD software, is geometrically perfect, while the ME is a 3D surface reconstruction that derives from a scan, and is always a geometric approximation of the scanned object; when modelling on an LF, it is possible to obtain perfect marginal adaptation, without any limitation related to the visibility of subgingival structures [11]. The replacement of the ME of the SB with the corresponding LF is possible thanks to the powerful best-fit algorithm of the CAD software, and results in the integration of the LF of the SB, and therefore of the entire library, which is geometrically linked to it, in the master model [11]. From this point, the technician can model all the restorations, which will be fabricated and applied clinically.

Several clinical studies have reported how these protocols can represent a predictable solution for the fabrication of short-span implant-supported restorations (single crowns [11–13] and fixed partial prostheses supported by 4–5 implants [6,14,15]). The application of these protocols has a series of advantages, such as the elimination of the conventional impressions with trays and materials, which have always been unwelcome to patients [16], the simplification of clinical procedures, and the saving of time and money, especially when printing physical models is unnecessary [17].

However, several studies [18,19] and literature reviews [20,21] have shown that difficulties persist in fabricating long-span implant-supported restorations (particularly in the case of fixed full arches (FFAs) supported by six or more fixtures) via a full digital workflow, i.e., starting from an optical impression with IOS. These difficulties are mainly attributed to the intrinsic error of IOS, which the literature reports is not sufficiently accurate to capture the impression of multiple implants in the completely edentulous patient [18–21]. This seems to be mainly related to the mechanism by which the IOS acquires the images, 'attaching' frames to each other during the acquisition; therefore, the greater the extent of the scan, the larger the error [19,21]. The intrinsic error of the IOS, however, may not be the only factor determining the inaccuracy. At least four other factors must be considered when capturing an intraoral digital impression: the environment [22], the operator [23], the patient, and the SB [24]. In particular, the SB is still little investigated in the literature [24], but plays a fundamental role in the acquisition. Study of SBs should consider design, material, colour, and tolerances in the fabrication [24–27]. To date, only a few studies have investigated the influence of these parameters on the quality of the scan [24–27], and unfortunately, no studies have analysed in depth what happens in the very early stages of CAD modelling, i.e., when the dental technician replaces the SB ME with the corresponding LF. This phase is particularly delicate. If dimensional congruence is not exact between the LF of the SB with the corresponding ME acquired with IOS, problems can arise in the superimposition in CAD, which may result in positional errors [25,27].

Hence, the aim of this in vitro study was to assess and compare the reliability of five different IOSs in the capture of implant SBs, to verify the dimensional congruence between the MEs of the scan abutments captured during the scan of a complete arch model with six implants, and the corresponding LF. The evaluation of the deviations between the MEs of the SBs and the LF was performed using reverse engineering software able to quantitatively and qualitatively assess the incongruences. The null hypothesis was that there was no quantatitive nor qualitative difference between the MEs of the SBs and the LF, and that there were no differences between the different IOSs evaluated.

2. Materials and Methods

2.1. Study Design

In this study, a gypsum cast representative of a fully edentulous maxilla with six implant analogues and SBs screwed on was scanned with five different IOSs (PRIMESCAN®, Dentsply-Sirona, York, PA, USA; CS 3700®, Carestream Dental, Atlanta, GA, USA; MEDIT i-500®, Medit, Seoul, Korea; ITERO ELEMENTS 5D®, Align Technologies, San José, CA, USA; and Emerald S®, Planmeca, Helsinki, Finland) and with a desktop scanner (Freedom UHD®, Dof Inc., Seoul, Korea). The desktop scans were taken only as a reference and were not included in the comparison. In total, 10 scans were taken for each IOS, for a total of $10 \times 5 = 50$ MEs, plus 3 desktop scans, for a total of 53 MEs collected. These MEs were saved in specific folders, trimmed to make them uniform and imported to reverse engineering software (Studio®, Geomagics, Morrisville, NC, USA) for 3D analysis. The 3D analysis consisted of the superimposition of the SB LF onto each SB ME, using a best-fit algorithm, to replicate the scenario when prosthetic CAD modelling starts. In total, 300 superimpositions were performed for the IOSs plus 18 for the desktop scanner. Then, a quantitative and qualitative evaluation of the deviations between MEs and LF was performed. The results obtained with the different IOSs were evaluated and compared to verify the degree of reliability in the capture of the SB with the different machines. The study design is summarised in Figure 1.

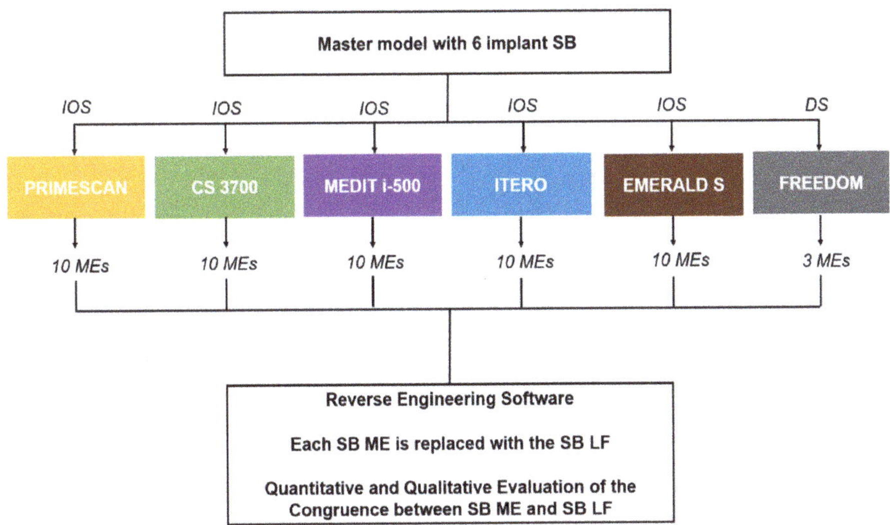

Figure 1. Schematic drawing of the design of the study.

2.2. Master Model, SB, and Scanning Procedures

The gypsum cast was made of type IV plaster with pink gingiva in the scan abutment area, and the implant analogues in positions # 16 (S1), # 14 (S2), # 11 (S3), # 21 (S4), # 24 (S5), and # 26 (S6; Figure 2A). The implant analogues were positioned specularly and not particularly inclined, so that the SBs were positioned fairly parallel to each other (Figure 2B).

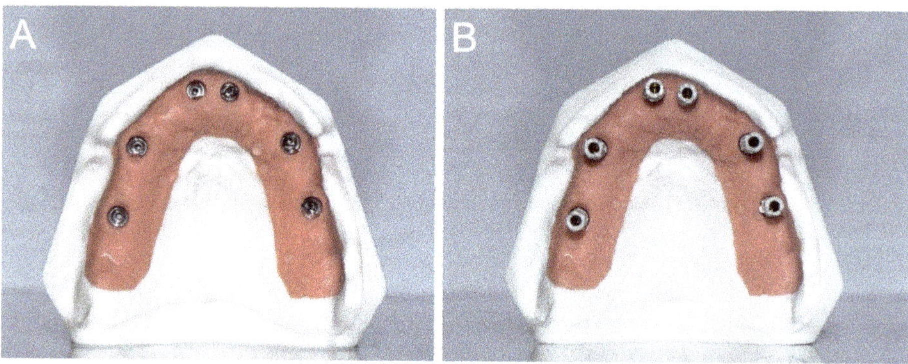

Figure 2. (**A**) A type IV gypsum cast with pink gingiva in the scanabutment area, and the implant analogues in position # 16 (S1), # 14 (S2), # 11 (S3), # 21 (S4), # 24 (S5), and # 26 (S6), respectively, was prepared for the study. (**B**) The gypsum cast with the SBs in position. The SBs were positioned fairly parallel each other.

The SBs were all identical, 13 mm in height, and fabricated by the same manufacturer (Megagen, Gyeongbuk, Korea) with the scanning area in opaque white polyether ether ketone (PEEK; Figure 3).

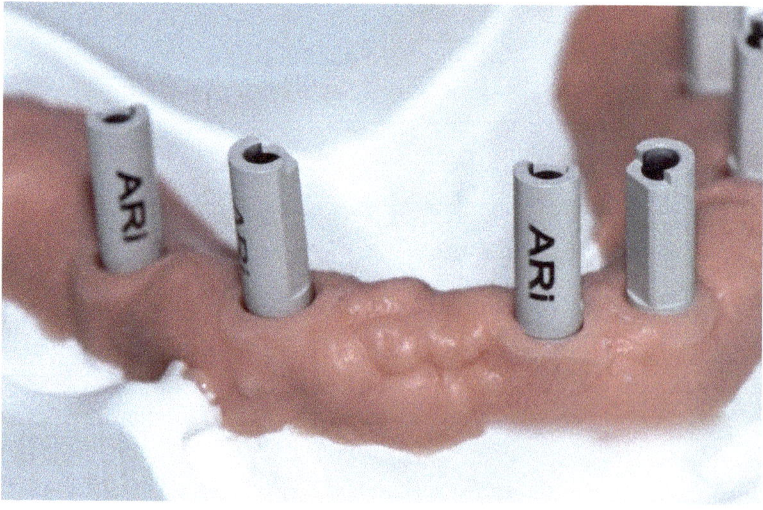

Figure 3. The SBs were all identical, 13 mm in height, and fabricated by the same manufacturer (Megagen, Gyeongbuk, Korea) with the scanning area in opaque white polyether-ether-ketone (PEEK).

The manufacturer reported a maximum tolerance of ±20 μm in the SB production phase. In total, 10 scans were taken per IOS, for a total of 50 MEs captured, by the same implant scanning expert operator (FM). The characteristics of the different IOSs used in this study are summarised in Table 1. For each IOS, the scans were taken using the latests available software version in March 2020.

Table 1. Features of the different intraoral scanners (IOSs) used in this study.

Name	Producer	Technology	Colour	Output
PRIMESCAN®	Dentsply-Sirona	High-resolution sensors and shortwave light with optical high-frequency contrast analysis for dynamic deep scan (20 mm)	yes	dxd (proprietary format) and stl (open format) with Connect
CS 3700®	Carestream Dental	Active triangulation with smart-shade matching via bidirectional reflectance distribution function	yes	dcm (proprietary format); ply and stl (open formats)
MEDIT i-500®	Medit	3D in motion video technology	yes	obj, ply and stl (open formats)
ITERO ELEMENTS 5D®	Align Technologies	Parallel confocal microscopy	yes	3ds (proprietary format); ply and stl (open formats)
Emerald S®	Planmeca	Projected pattern triangulation	yes	ply and stl (open formats)

The IOS scan was limited to the area of the pink gingiva, which was scanned in full, and included capturing the entire SB. To avoid the potential negative effects of operator fatigue, the sequence of scans with the different IOSs was randomised and a 5-min break was scheduled between scans to rest the operator and change the scanner. The scanning strategy used was the same described in a previous study [18]: the zig-zag technique. The operator started from the buccal surface of the model and precisely from the first SB of the right posterior maxilla (# 16), then moved to the occlusal and then palatal side; the operator then returned to the occlusal and then buccal side, moving slowly forward. The progress was slow and constant and the operator tried to capture all the details of the different SBs, without insisting too much on them from the same angle, to avoid excessive reflection. The movement described by the operator was therefore arched, with the scanner head moving over the SB and pink gingiva in a continuous passage from outside to inside, and through a progressive advancing movement. All scans were captured in the same environmental conditions, i.e., in a room with constant temperature (22 °C), controlled humidity (45%) and ambient light, without interference from external light sources. The MEs of the models captured with the different IOSs (10 STL files for each of the 5 IOSs, for a total of 50 MEs) were saved in dedicated folders, labelled with the name of the scanner used. Within each of these folders, the models were numbered from 1 to 10; within each ME, the SBs were numbered from 1 to 6, starting from right posterior area to left. The three desktop scans were taken with an industrial-derived desktop scanner (Freedom UHD®, Dof Inc., Seoul, Korea). The Freedom UHD® scanner is a structured light scanner (white light-emitting diode) that acquires the models through two 5.0-megapixel cameras, using patented stable scan stage technology. This technology allows the cameras to move above and around the model to be scanned. It is not the model plate that moves in different positions to facilitate the acquisition of all the details; instead, the lights and cameras move, rotating around the centre of the scan plate, while the model remains stationary. This allows the capture of all the details of the model in a relatively short time (less than 50 s). The scanner has a certified accuracy of 5 μm and generates STL files immediately usable by any CAD. The scanner weighs 15 kg, has dimensions of 330 × 495 × 430 mm, is powered at 110–240 V and 50–60 Hz, and works with Windows operating systems 7, 8 and 10 (64 bit). The three different desktop scans captured were also saved in a dedicated folder as STL files. All the MEs were then cut and trimmed using an individual template to be uniform in size and shape; when uniform, they were saved again in the respective dedicated folders and were ready for analysis.

2.3. 3D Analysis of the Congruence between ME and LF

After all scans were captured, each ME was imported into reverse engineering software (Studio®, Geomagics, Morrisville, NC, USA), and the LF of the implant SB was superimposed onto the corresponding parts. Six superimpositions were therefore made for each model, for a total of 60 superimpositions per each IOS. Each superimposition entailed two steps. First, the numbered ME, labelled with the name of the IOS used, was loaded into the software; then, six identical SB LFs were loaded, taken directly from the official library of the manufacturer of the implants. These LFs were superimposed, one by one, on the corresponding ME, i.e., on the SB captured by intraoral scanning. The first overlap was by points. The operator (FM) identified three points on each of the SBs present in the MEs acquired with IOS, considering the reference for the overlap; the same points were searched on the SB LF, and the software could thus proceed to a first rough alignment. After this first manual overlap, the operator launched the best-fit algorithm, through which the software perfected the overlaps, one by one, of the SB LFs onto the corresponding SB MEs. The parameters were set with a minimum of 100 iterations per case and the registration used a robust iterative closest point algorithm. With this algorithm, the distances between the SB from ME and library were minimised using a point-to-plane method, and it was possible to calculate the congruence between the structures, expressed quantitatively as the mean ± standard deviation (SD) of the distances between all points of the superimposed models. Finally, for a better qualitative evaluation of the distances between the files and understanding of the directionality of the deviation (i.e., to allow the correct evaluation of the inward and outward deviations), the software allowed generating a colorimetric map. This map was generated through the '3D deviation' function, which made it possible to evaluate the distances between specific points, globally and in all space planes. In this case, the SB LF was considered a reference. Therefore, the colorimetric map indicated inward deviations (defects) with different shades of blue, and outward deviations (excesses) in yellow and red. Minimal deviations were coloured green. The same setting of the colorimetric map was fixed, with the scale ranging from a maximum deviation of +50 to −50 μm, and the best results between +1 and −1 μm (green). The screenshots of the quantitative evaluation and of the colorimetric maps were saved in special folders; particular attention was devoted to capturing screenshots from different angles, to better qualitatively understand on which portion of the SB the major deviations were concentrated. The same process was repeated for the desktop scans.

2.4. Outcome Variables

Quantitative deviation between the SB ME and SB LF. This value was calculated with the reverse engineering software, after the application of the best-fit algorithm. It represented the average deviation between the two objects, expressed in mean ± SD, median, range, and 95% confidence interval (CI), in μm.

Qualitative deviation between the SB ME and SB LF. The qualitative deviation was obtained through visual inspection of all samples, using the colorimetric map generated in the software, after the application of the best-fit algorithm. To define this variable, the same experienced operator (FM) who captured all the scans and performed the superimpositions assigned a label of 'outward deviation', 'no deviation' or 'inward deviation', based on the chromatic predominance of red/yellow, green or pale/dark blue, respectively, on each of the three SB portions: flat face central, flat faces lateral, and posterior (back) cylindrical area. This information was collected in a table and expressed as a qualitative variable.

2.5. Statistical Analysis

All the data collected were included in datasheets used for statistical analysis. The sample size was determined sufficient for the analysis by a professional statistician. Data analysis and visualisation were performed using R (version 3.6.3) environment for statistical computing (R Foundation for

Statistical Computing, Vienna, Austria). For the quantitative evaluation, descriptive statistics for absolute deviations were presented as means (±SD), medians (1st and 3rd quartiles, Q1–Q3), ranges, and 95% CIs. Sample distributions of absolute deviations across different scanners and SBs were visualised using box plots. An observation obtained from S1 SB using the ITERO ELEMENTS 5D® scanner was considered an outlier and not used in the parametric estimation and hypothesis testing procedures. A linear mixed-effects model (implemented in lme4 1.1–21 package) was used to estimate and compare the mean absolute deviations between the IOSs (this model allowed accounting that data have hierarchical properties: scanner → SB). The Tukey method (implemented in emmeans 1.4.5) was used to adjust *p*-values and confidence limits. A linear model (two-way ANOVA with interaction) was used to compare the mean absolute deviations between SBs for each scanner. The Tukey method was used to adjust *p*-values and confidence limits. The Friedman rank test was used to compare models in each type of scanner (SB was considered the blocking variable) with the Holm procedure for multiple testing adjustment. Agglomerative hierarchical biclustering was used to explore scanner–SB relationships regarding averages and variability of the absolute deviations. In order to reduce the α error, the significance level for all tests was established at $p < 0.01$.

3. Results

Descriptive statistics for the absolute deviations in each scanner–SB pair are reported in Figure 4 and Table 2.

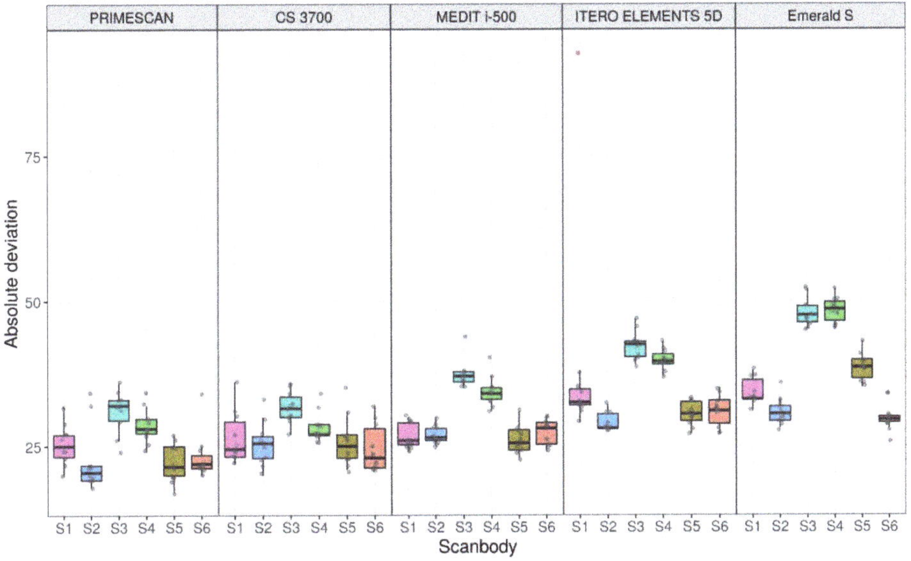

Figure 4. Sample distributions of absolute deviations across IOS and scanbody (SB), in μm.

The results of the comparison of the different scanners (estimation and testing using a linear mixed-effects model) are summarised in Figure 5 (means and 95% CIs for each type of scanner) and Table 3 (pairwise differences between means, 95% CIs for differences and *p*-values for pairwise comparisons).

Overall, PRIMESCAN® was the IOS with the lowest mean absolute deviation (25.5 ± 5.0 μm), equal to that of the desktop scanner DOF UHD® (25.5 ± 2.9 μm) used as an external reference in this study. Similar excellent results were also reported for CS 3700® (27.0 ± 4.3 μm), so that the difference between PRIMESCAN® and CS3700® was not statistically significant ($p = 0.1235$). However, the congruence between SB ME and SB LF with PRIMESCAN® was statistically higher than that with MEDIT i-500®

(29.8 ± 4.8 µm, $p < 0.0001$), ITERO ELEMENTS 5D® (34.2 ± 9.3 µm, $p < 0.0001$), and Emerald S® (38.3 ± 7.8 µm, $p < 0.0001$). Statistically significant differences were also found when comparing CS 3700® with MEDIT i-500® ($p = 0.0004$), ITERO ELEMENTS 5D® ($p < 0.0001$), and Emerald S® ($p < 0.0001$). Finally, statistically significant differences were found between MEDIT i-500® and ITERO ELEMENTS 5D® ($p < 0.0001$), MEDIT i-500® and Emerald S® ($p < 0.0001$), and ITERO ELEMENTS 5D® and Emerald S® ($p < 0.0001$).

Table 2. Descriptive statistics: mean (SD); median; (Q1–Q3), in µm.

Scanner	S1	S2	S3	S4	S5	S6
PRIMESCAN®	25.4 (3.5); 25.0; (23.2–27.0)	22.6 (5.6); 20.5; (19.2–21.8)	31.0 (3.7); 32.0; (29.5–33.0)	28.5 (3.0); 28.0; (27.2–29.8)	22.2 (3.4); 21.5; (20.0–25.0)	23.3 (4.0); 22.0; (21.2–23.5)
CS 3700®	26.5 (4.5); 24.5; (23.2–29.2)	25.5 (3.9); 25.5; (23.0–26.8)	31.6 (2.8); 31.5; (30.0–33.5)	28.3 (2.7); 27.0; (27.0–28.8)	25.9 (4.4); 25.0; (23.0–27.0)	24.7 (4.2); 23.0; (21.2–28.0)
MEDIT i-500®	27.0 (2.3); 26.0; (25.2–29.0)	27.0 (1.7); 26.5; (26.0–28.0)	37.3 (2.6); 37.0; (36.0–37.8)	34.4 (2.6); 34.0; (33.0–35.0)	26.1 (2.4); 25.5; (24.2–27.8)	27.4 (2.2); 28.0; (25.2–29.0)
ITERO ELEMENTS 5D®	38.9 (19.2); 32.5; (32.0–34.8)	29.1 (1.6); 28.0; (28.0–30.5)	42.4 (2.6); 42.5; (40.2–43.0)	39.8 (1.8); 39.5; (39.0–40.8)	30.5 (2.1); 30.5; (29.2–32.5)	30.9 (2.8); 31.0; (28.8–32.8)
Emerald S®	34.2 (2.3); 33.0; (33.0–36.2)	30.9 (2.3); 30.5; (29.2–31.8)	48.0 (2.5); 47.5; (46.2–49.0)	48.3 (2.2); 48.5; (46.5–49.8)	38.5 (2.5); 38.5; (36.5–39.8)	29.9 (2.5); 29.5; (29.0–30.0)

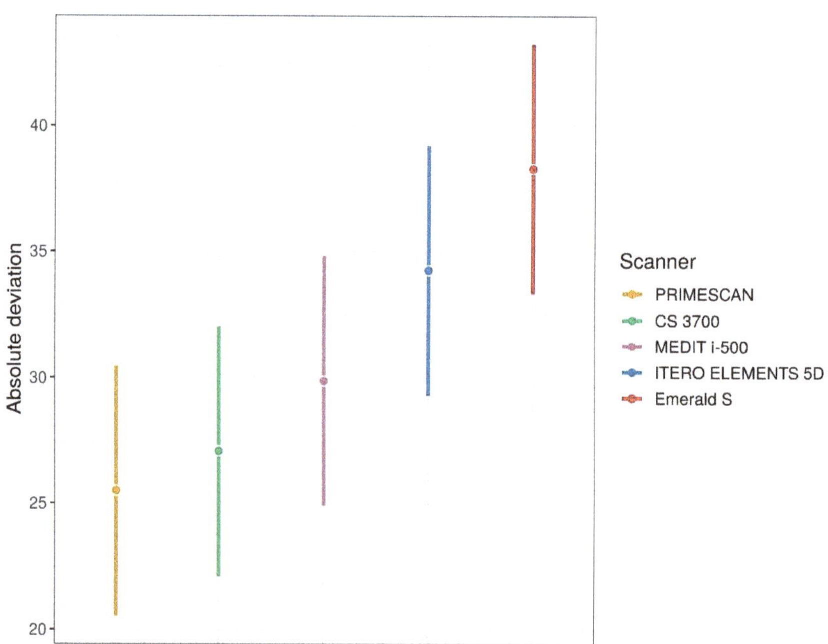

Figure 5. Mean absolute deviations estimates (with 95% confidence interval (CI)) for each type of IOS (these quantities were estimated using linear mixed effects model).

Table 3. Comparisons of mean absolute deviations between the IOS.

Contrast (Pairwise Comparisons)	Difference	95% CI for Difference	*p*
PRIMESCAN®—CS 3700®	−1.58	[−3.41; 0.24]	0.1235
PRIMESCAN®—MEDIT i-500®	−4.37	[−6.19; −2.54]	<0.0001
PRIMESCAN®—ITERO ELEMENTS 5D®	−8.76	[−10.59; −6.92]	<0.0001
PRIMESCAN®—Emerald S®	−12.80	[−14.63; −10.97]	<0.0001
CS 3700®—MEDIT i-500®	−2.78	[−4.61; −0.96]	0.0004
CS 3700®—ITERO ELEMENTS 5D®	−7.17	[−9.01; −5.34]	<0.0001
CS 3700®—Emerald S®	−11.22	[−13.04; −9.39]	<0.0001
MEDIT i-500®—ITERO ELEMENTS 5D®	−4.39	[−6.22; −2.56]	<0.0001
MEDIT i-500®—Emerald S®	−8.43	[−10.26; −6.61]	<0.0001
ITERO ELEMENTS 5D®—Emerald S®	−4.04	[−5.88; −2.21]	<0.0001

In the quantitative evaluation, among the 60 superimpositions performed in each group, the best single result obtained with PRIMESCAN® was 17 ± 19 μm (Figure 6A), with CS 3700® 20 ± 18 μm (Figure 6B), with MEDIT i-500® 23 ± 26 μm (Figure 6C), with ITERO ELEMENTS 5D® 27 ± 27 μm (Figure 6D), and with Emerald S® 26 ± 28 μm (Figure 6E). The best result obtained with the reference desktop scanner DOF UHD® was 21 ± 21 μm (Figure 6F).

The comparison of the deviations between the SBs in each group of IOSs with the results of estimation and testing using the linear model (two-way ANOVA with interaction) with a *p*-value for interaction <0.0001 is summarised in Figure 7 (means and 95% CIs for each type of scanner), Table 4 (pairwise differences between means and 95% CIs for differences), and Table 5 (*p*-values for pairwise comparisons).

With regard to the deviations between the models in each group of scanner, the results of the Friedman test are presented in Table 6. There were no statistically significant differences between the models in each scanner group.

Hierarchical biclustering results were reported when IOSs and SBs were grouped based on average absolute deviation (Figure 8) and variability (SD) of absolute deviation (Figure 9).

From these latter figures, it was evident that the best results in terms of trueness were obtained by PRIMESCAN® and CS 3700®, in correspondence with the SBs in position S2 and S6. Conversely, ITERO ELEMENTS 5D® and MEDIT i-500® revealed the highest repeatability (precision) of the scans with less SD, in correspondence with the SBs in position S4 and S3.

With regard to the qualitative evaluation, the scanners showed different features (Figure 10).

In PRIMESCAN®, the SB LF usually included the SB ME; the SB ME did not grow, and appeared to be included within the SB LF (Figure 10A). In contrast, with CS 3700®, all the SB MEs included the SB LF, with a marked tendency for the ME to grow, although with minimal quantitative deviations, and in a fairly uniform way (Figure 10B). At the qualitative evaluation, MEDIT i-500® showed a remarkable interpolation between the SB ME and the SB LF, with no clear direction for the deviation (Figure 10C). Finally, ITERO ELEMENTS 5D® (Figure 10D) and Emerald S® (Figure 10E) revealed a pattern and direction of deviation similar to that of PRIMESCAN®, with an SB ME that appeared to be included within the SB LF, although with a higher quantitative deviation. DOF UHD® (Figure 10F) showed a deviation pattern similar to that of CS3700®.

After visual inspection of all samples, the flat surfaces of almost all SBs showed the best results in terms of deviations (Table 7).

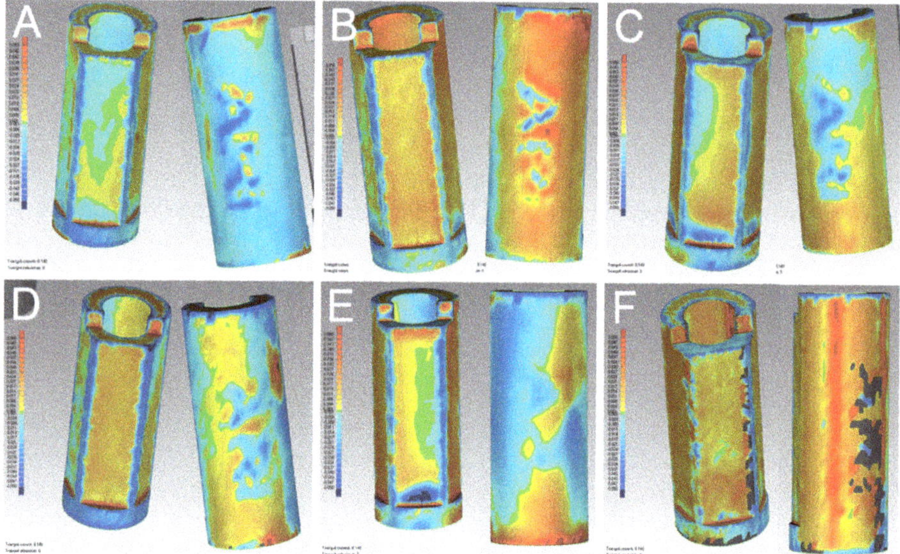

Figure 6. Quantitative evaluation with colorimetric map (frontal and back surface): best results obtained in the study. (**A**) With PRIMESCAN®, the best single result amounted to 17 ± 19 μm. (**B**) With CS 3700®, the best single result amounted to 20 ± 18 μm. (**C**) With MEDIT i-500® the best single result amounted to 23 ± 26 μm. (**D**) With ITERO ELEMENTS 5D®, the best single result amounted to 27 ± 27 μm. (**E**) With Emerald S®, the best single result amounted to 26 ± 28 μm. (**F**) With DOF UHD®, the best single result amounted to 21 ± 21 μm.

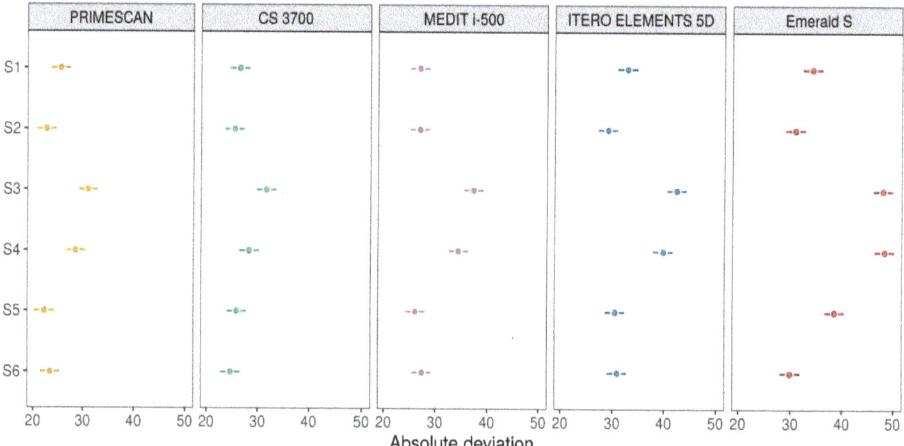

Figure 7. Mean absolute deviations estimates (with 95% confidence intervals) for SBs in each group of IOSs (these quantities were estimated using two-way ANOVA).

Table 4. Comparisons of mean absolute deviations between SBs in each group of IOSs (mean differences and 95% CIs for them).

	PRIMESCAN®	CS 3700®	MEDIT i-500®	ITERO ELEMENTS 5D®	Emerald S®
S6 − S5	1.10 [−2.81; 5.01]	−1.20 [−5.11; 2.71]	1.30 [−2.61; 5.21]	0.40 [−3.51; 4.31]	−8.60 [−12.51; −4.69]
S6 − S4	−5.20 [−9.11; −1.29]	−3.60 [−7.51; 0.31]	−7.00 [−10.91; −3.09]	−8.90 [−12.81; −4.99]	−18.40 [−22.31; −14.49]
S6 − S3	−7.70 [−11.61; −3.79]	−6.90 [−10.81; −2.99]	−9.90 [−13.81; −5.99]	−11.50 [−15.41; −7.59]	−18.10 [−22.01; −14.19]
S6 − S2	0.70 [−3.21; 4.61]	−0.80 [−4.71; 3.11]	0.40 [−3.51; 4.31]	1.80 [−2.11; 5.71]	−1.00 [−4.91; 2.91]
S6 − S1	−2.10 [−6.01; 1.81]	−1.80 [−5.71; 2.11]	0.40 [−3.51; 4.31]	−1.99 [−6.01; 2.03]	−4.30 [−8.21; −0.39]
S5 − S4	−6.30 [−10.21; −2.39]	−2.40 [−6.31; 1.51]	−8.30 [−12.21; −4.39]	−9.30 [−13.21; −5.39]	−9.80 [−13.71; −5.89]
S5 − S3	−8.80 [−12.71; −4.89]	−5.70 [−9.61; −1.79]	−11.20 [−15.11; −7.29]	−11.90 [−15.81; −7.99]	−9.50 [−13.41; −5.59]
S5 − S2	−0.40 [−4.31; 3.51]	0.40 [−3.51; 4.31]	−0.90 [−4.81; 3.01]	1.40 [−2.51; 5.31]	7.60 [3.69; 11.51]
S5 − S1	−3.20 [−7.11; 0.71]	−0.60 [−4.51; 3.31]	−0.90 [−4.81; 3.01]	−2.39 [−6.41; 1.63]	4.30 [0.39; 8.21]
S4 − S3	−2.50 [−6.41; 1.41]	−3.30 [−7.21; 0.61]	−2.90 [−6.81; 1.01]	−2.60 [−6.51; 1.31]	0.30 [−3.61; 4.21]
S4 − S2	5.90 [1.99; 9.81]	2.80 [−1.11; 6.71]	7.40 [3.49; 11.31]	10.70 [6.79; 14.61]	17.40 [13.49; 21.31]
S4 − S1	3.10 [−0.81; 7.01]	1.80 [−2.11; 5.71]	7.40 [3.49; 11.31]	6.91 [2.89; 10.93]	14.10 [10.19; 18.01]
S3 − S2	8.40 [4.49; 12.31]	6.10 [2.19; 10.01]	10.30 [6.39; 14.21]	13.30 [9.39; 17.21]	17.10 [13.19; 21.01]
S3 − S1	5.60 [1.69; 9.51]	5.10 [1.19; 9.01]	10.30 [6.39; 14.21]	9.51 [5.49; 13.53]	13.80 [9.89; 17.71]
S2 − S1	−2.80 [−6.71; 1.11]	−1.00 [−4.91; 2.91]	0.00 [−3.91; 3.91]	−3.79 [−7.81; 0.23]	−3.30 [−7.21; 0.61]

Table 5. Comparisons of mean absolute deviations between SBs in each group of scanners (*p*-values).

	PRIMESCAN®	CS 3700®	MEDIT i-500®	ITERO ELEMENTS 5D®	Emerald S®
S6 − S5	0.9661	0.9509	0.9318	0.9997	<0.0001
S6 − S4	0.0023	0.0910	<0.0001	<0.0001	<0.0001
S6 − S3	<0.0001	<0.0001	<0.0001	<0.0001	<0.0001
S6 − S2	0.9956	0.9918	0.9997	0.7734	0.9776
S6 − S1	0.6382	0.7734	0.9997	0.7147	0.0219
S5 − S4	0.0001	0.4929	<0.0001	<0.0001	<0.0001
S5 − S3	<0.0001	0.0006	<0.0001	<0.0001	<0.0001
S5 − S2	0.9997	0.9997	0.9860	0.9085	<0.0001
S5 − S1	0.1789	0.9979	0.9860	0.5290	0.0219
S4 − S3	0.4455	0.1526	0.2762	0.3997	0.9999
S4 − S2	0.0003	0.3147	<0.0001	<0.0001	<0.0001
S4 − S1	0.2082	0.7734	<0.0001	<0.0001	<0.0001
S3 − S2	<0.0001	0.0002	<0.0001	<0.0001	<0.0001
S3 − S1	0.0007	0.0030	<0.0001	<0.0001	<0.0001
S2 − S1	0.3147	0.9776	1.0000	0.0774	0.1526

Table 6. Comparisons between models (M) in each group of scanners (row p-values and adjusted for multiple comparisons).

Scanner	p	p_{adj}
PRIMESCAN®	0.0147	0.0737
CS 3700®	0.3971	1.0000
MEDIT i-500®	0.3879	1.0000
ITERO ELEMENTS 5D®	0.3452	1.0000
Emerald S®	0.2700	1.0000

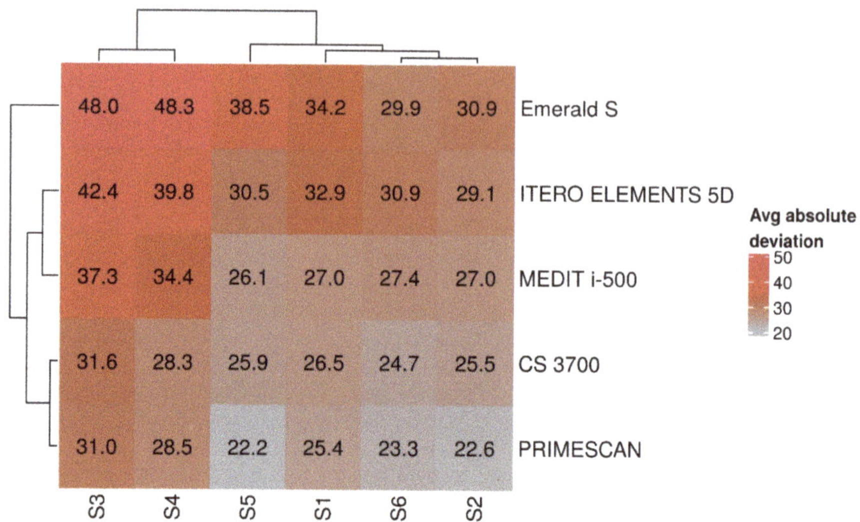

Figure 8. Results of IOSs and SBs biclustering based on average absolute deviation.

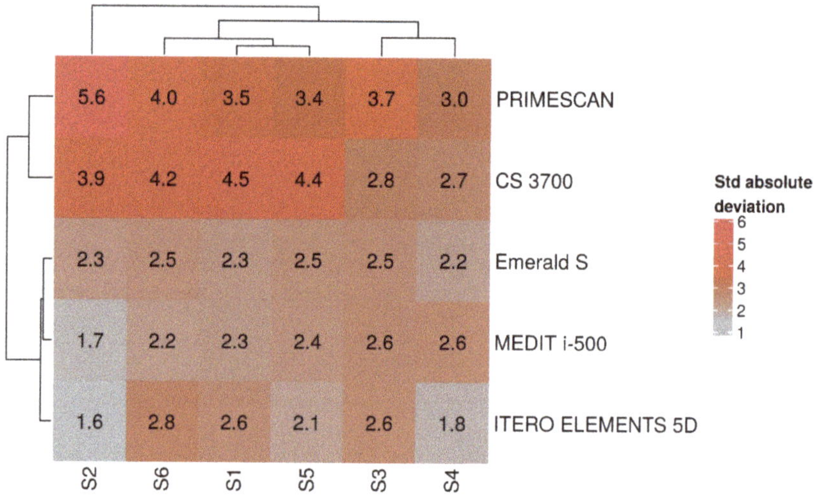

Figure 9. Results of IOSs and SBs biclustering based on standard deviation of absolute deviation.

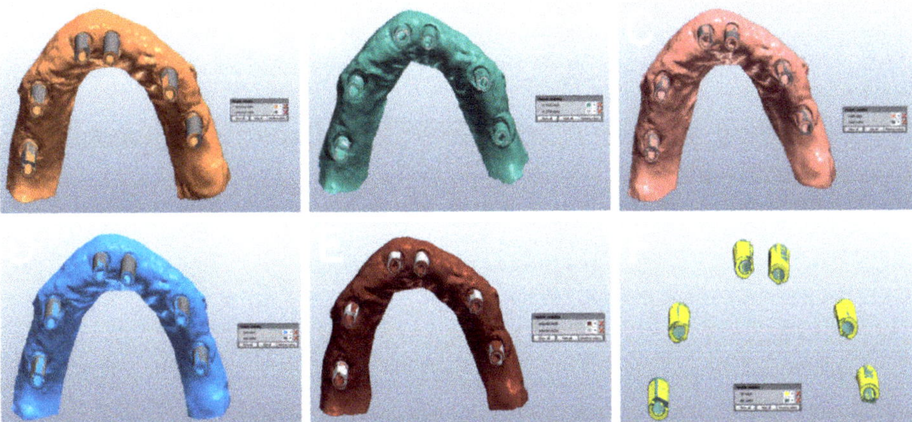

Figure 10. Qualitative evaluation of the deviation patterns (occlusal view of the model). (**A**) With PRIMESCAN®, the SB meshes (MEs) were included inside the SB library files (LFs). (**B**) With CS 3700®, the SB LFs were included inside the SB MEs. (**C**) With MEDIT i-500®, a remarkable interpolation between the SB ME and the SB LF was seen, with no clear direction for the deviation. (**D**) With ITERO ELEMENTS 5D®, the SB MEs were included inside the SB LFs. (**E**) With Emerald S®, the SB MEs were included inside the SB LFs. (**F**) With DOF UHD®, the SB LFs were included inside the SB MEs.

Table 7. Distribution of the deviations in the different surfaces of the SBs, by visual inspection.

Scanner	Deviation	Flat Central	Flat Lateral	Back
PRIMESCAN®	Outward deviation	+	+	+
	No deviation	++	++	++
	Inward deviation	++	+++	+++
CS 3700®	Outward deviation	+++	+++	+++
	No deviation	++	+	+
	Inward deviation	+	+	+
MEDIT i-500®	Outward deviation	++	++	++
	No deviation	++	+	+
	Inward deviation	++	++	++
ITERO ELEMENTS 5D®	Outward deviation	+	+	+
	No deviation	++	+	+
	Inward deviation	+++	+++	+++
Emerald S®	Outward deviation	+	+	+
	No deviation	+	+	+
	Inward deviation	+++	+++	+++

4. Discussion

Until now, most studies on the direct digital workflow in implant prosthodontics have focused on the use of IOSs and the intrinsic accuracy of these devices [3,12,18–21]. The intrinsic error generated during the progression of the intraoral scan has been considered the main reason for the insufficient accuracy of IOSs in taking impressions of completely edentulous patients for the fabrication of implant-supported FFAs [19–21]. This intrinsic error exists and certainly plays a role in determining the final inaccuracy of the process, as unequivocally demonstrated by the literature [2,3,12,19–21]; however, it is not the only source of error in the full digital workflow in implant prosthodontics.

Other elements contribute to increasing the error: environmental factors (light conditions) [22], factors related to the patient (position, depth and inclination of the implants) [28], factors related to the operator (scanning strategy [23] and experience of the clinician), and finally, the SB [24]. The SB is the transfer that allows capturing the position of the implants in the digital workflow and is therefore crucial. To date, few studies have analysed the influence of factors such as the design of the SB [25,26], the material used to build it [24,26], and the manufacturing tolerances [27] on the error in intraoral scanning. All these elements play a fundamental role and deserve to be adequately investigated by the scientific literature [24].

Even less investigated, however, are the first stages of modelling, in which the dental technician uploads in the CAD software the ME captured by the clinician through intraoral scanning and replaces portions of the SB with the corresponding LF. This moment is key, since an error in this phase can compromise the entire workflow: if a mistake is made at this stage, the individual abutment and prosthetic restoration will be modelled starting from an incorrect implant position [24]. It is therefore important to investigate this phase too, and in particular the congruence between the SB ME and the corresponding LF. Only in the presence of adequate dimensional congruence between these parts can the best-fit algorithm in the CAD software superimpose the files without difficulty, replacing the SB ME with the SB LF [24]. In contrast, in the presence of incongruence between the parts or dimensional deviations, positional errors may arise. Deviations have a detrimental effect when applying the best-fit algorithm, since they can lead to a positional error of the library components on which the dental technician models the prosthetic restorations [24,29]. This may ultimately contribute to a misfit of the prosthetic structure, especially in the case of long-span restorations such as FFAs [24,29].

The purpose of our present in vitro study was therefore to verify the dimensional congruence of MEs of SBs captured with five different IOSs, with the respective LF. This was to quantify the possible error, in micrometres, and to understand not only its presence, but also its direction. For this purpose, a completely edentulous maxilla model was used, with six analogues to which six SBs from the same manufacturer were screwed. This gypsum cast was scanned with five different IOS, and the MEs derived from these scans were loaded into reverse engineering software, where the portions of SBs were aligned to the corresponding LF using the best-fit algorithm. These scans were therefore not considered in their entirety, but the attention was concentrated only on the surface reconstruction of the SB, whose congruence with the corresponding LF was investigated. Therefore, the purpose of the study was not to investigate the general accuracy of the different scanners, which is usually determined by the correct 3D measurement of the distances between the different SBs; rather, the goal was to investigate whether inconsistencies existed between SB MEs and LF, and the presence of any deviations between the files, after the superimposition. To evaluate this, a quantitative and qualitative analysis of the congruence between the files was performed, to quantify the degree of deviation between LF and SB MEs obtained with different IOSs, and to analyse the qualitative characteristics of this deviation, where present. The qualitative analysis, in particular, aimed to establish whether the ME reconstruction of the SB occurred by excess or defect in the various positions, with respect to the reference file of the implant library. The null hypothesis was that there was no quantitative nor qualitative difference between the MEs of the SBs and the LF, and that there were no differences between the different IOSs evaluated.

At the end of the study, this null hypothesis was rejected. In fact, our present work has highlighted incongruence between SB MEs and SB LF, and that this inconsistency is quantitatively different, with the different scanners. In the present study, the best performance was obtained by PRIMESCAN®, which had an average deviation (25.5 ± 5.0 µm) equal to that of the desktop scanner used as an external reference in this study, DOF UHD® (25.5 ± 2.9 µm). The difference between PRIMESCAN® and CS 3700® (27.0 ± 4.3 µm) was minimal and not statistically significant; conversely, MEDIT i-500® (29.8 ± 4.8 µm), ITERO ELEMENTS 5D® (34.2 ± 9.3 µm), and Emerald S® (38.3 ± 7.8 µm) had higher average deviations, and significant differences were found between them, PRIMESCAN® and CS 3700®. Finally, significant differences were found also between MEDITi-500®, ITERO ELEMENTS 5D® and Emerald S®, and between ITERO ELEMENTS 5D® and Emerald S®. These differences, measured on numerous samples, can be important in determining the actual accuracy of a CAD project, since this absolute mean error or dimensional discrepancy applies to each SB of the model, and could lead to positional errors.

The qualitative data that emerge from our present in vitro work are also interesting. The error seemed to have a peculiar directionality based on the scanner used. In the case of the desktop scanner and CS 3700®, although the average error was quantitatively small, the ME always tended to grow; therefore, the SB LF was always contained within the SB ME, which was larger. With PRIMESCAN®, Emerald S®, and ITERO ELEMENTS 5D®, the opposite happened: the SB ME was contained within the SB LF, at the end of the overlap. This could be linked to post-processing or 'smoothing' of the images, through the removal of triangles that exceed the surface reconstruction, by the reconstruction software. The relationship between ME and LF in MEDIT i-500® seems to be more balanced, since there is a sort of interpenetration between the files. More generally, on inspection, the error appears to be less marked on the flat face of the SB; however, this is only qualitative data, because the software used in this study does not allow separately calculating the error present in each of the different faces or geometric parts of the SB. Notably, it is extremely difficult to establish how much these discrepancies can contribute to determining a positional error on the master model that the dental technician uses for modelling. However, having defined their existence, quantified them, and studied their direction are the greatest advantage of our present scientific work.

The CAD best-fit algorithm searches for the congruence between the surfaces of the two STL files of the SB: ME and LF. If congruence is found, superimposition can proceed without errors. If, instead, congruence is not found, i.e., in the case of dimensional differences, the algorithm proceeds using the flat face of the SB as the main reference (if they have more than one, usually the more extended). The flat face undoubtedly represents a valid reference for the best-fit algorithm within the CAD software, but it may also represent an element of dangerous attraction [30]. In case of dimensional differences between the STL files, this face 'drives' the superimposition between the files. The result of this process is that the SB LF is 'dragged' towards the flat surface of the ME, used as the main reference for the superimposition [30]. This movement can result in a positional error, with a relative shift of the centroids of the objects [30]. This shift, when added up to the intrinsic error given by the intraoral scan, and to the error in milling, may determine the failure or misfit of long-span implant-supported restorations such as FFAs [31].

Although our study is the first to address this issue and is based on the evaluation of a fair number of scans obtained with six different scanners (five IOSs and a desktop scanner), it has limitations. First, it is an in vitro study. Scanning a gypsum cast is certainly easier than in vivo intraoral scanning, which presents technical difficulties due to space limitations, the presence of saliva and possible patient movements [32]. Furthermore, in vitro scanning takes place in light conditions which, although controlled (same environmental light for all IOSs), do not replicate in any way the light conditions of the oral cavity [22]. Since SB reflectance and related ME reconstruction errors vary with light conditions, the data reported in this study need to be critically evaluated. A further limitation of the present study is that the SBs used (six identical SBs produced by the same manufacturer) had not been preliminarily probed with a coordinate measuring machine (CMM) to assess their exact physical dimensions, thus the data relating to manufacturing tolerances were not known. The library is different from the actual SB. In order to make an actual SB, it is produced by an injection method or milling based on the library. These fabrication errors can affect the trueness of the scan: therefore, for the evaluation of the trueness and in order to produce a reference model, reference data for each SB should be obtained using a high-precision CMM machine or industrial 3D optical scanner. Manufacturing tolerances, in fact, play a potentially important role that deserves to be properly investigated, as reported in previous studies [27,33]. However, our present study aimed to investigate the congruence between the SB MEs and the LF, and not the trueness of each SB scan. Not surprisingly, in our study, a statistically significant difference was found between different SBs, in all scanner groups. This could be determined by dimensional differences between the pieces, due to dimensional tolerances given by the production phase. Finally, the strategy used may have favoured some scanners in the correct reconstruction of

ME, penalising others; certainly, there is a link between the scanner acquisition technology and the scanning strategy, although the literature has not adequately clarified this [34]. Further in vivo studies are therefore necessary to obtain more information, to better define the influence of the first CAD phases on the final error in the full digital workflow in implant prosthodontics. These studies also need to be extended to other implant systems.

5. Conclusions

Our present in vitro study aimed to compare the reliability of five different IOSs (PRIMESCAN®, CS 3700®, MEDIT i-500®, ITERO ELEMENTS 5D®, and Emerald S®) in the capture of implant SBs and to verify the dimensional congruence between the MEs of the SBs and the corresponding LF. At the end of the quantitative evaluation, statistically different levels of congruence were found between the SB MEs captured with the different IOSs and the corresponding LF. PRIMESCAN® and CS 3700® showed the highest congruence between SB MEs and LF, with the lowest mean absolute deviations (25.5 ± 5.0 µm and 27.0 ± 4.3 µm, respectively); the difference between these two scanners and the other three was statistically significant. Significant differences were also found between MEDIT i-500® and ITERO ELEMENTS 5D®, MEDIT i-500® and Emerald S®, and ITERO ELEMENTS 5D® and Emerald S®. Based on these results, the null hypothesis for this study was rejected, since deviations were found between the MEs of the SBs and the LF, and significant differences were found among the different IOSs evaluated. Significant differences were also found among different SBs, but no differences were found between the models scanned with the same IOS. Finally, the qualitative evaluation revealed different directions and patterns for the five IOSs investigated. Further studies are needed to confirm these preliminary results.

Author Contributions: Conceptualization, F.M. and H.L.; methodology, F.M. and B.M.; software: F.M.; validation, F.M., B.M., and N.L.; formal analysis, I.S.; investigation, F.M. and N.L.; resources, H.L.; data curation, F.M., I.S., and N.L.; writing—original draft preparation, F.M.; writing—review and editing, F.M. and O.A.; visualization, F.M. and O.A.; supervision, O.A.; project administration, B.M.; funding acquisition, F.M. and H.L. All authors have read and agreed to the published version of the manuscript.

Funding: This research received no external funding. This study was self-funded.

Acknowledgments: The authors are grateful to Uli Hauschild, Master Dental Technician, and Federico Manes, CAD designer, for help with the preparation of the master model used in this study.

Conflicts of Interest: The authors declare no conflict of interest for the present study.

References

1. Joda, T.; Ferrari, M.; Gallucci, G.O.; Wittneben, J.G.; Brägger, U. Digital technology in fixed implant prosthodontics. *Periodontology 2000* **2017**, *73*, 178–192. [CrossRef] [PubMed]

2. Mangano, F.; Gandolfi, A.; Luongo, G.; Logozzo, S. Intraoral scanners in dentistry: A review of the current literature. *BMC Oral Health* **2017**, *17*, 149. [CrossRef] [PubMed]

3. Roig, E.; Garza, L.C.; Álvarez-Maldonado, N.; Maia, P.; Costa, S.; Roig, M.; Espona, J. In vitro comparison of the accuracy of four intraoral scanners and three conventional impression methods for two neighboring implants. *PLoS ONE* **2020**, *15*, e0228266. [CrossRef] [PubMed]

4. Jacobs, R.; Salmon, B.; Codari, M.; Hassan, B.; Bornstein, M.M. Cone beam computed tomography in implant dentistry: Recommendations for clinical use. *BMC Oral Health* **2018**, *18*, 88. [CrossRef]

5. Joda, T.; Zarone, F.; Ferrari, M. The complete digital workflow in fixed prosthodontics: A systematic review. *BMC Oral Health* **2017**, *17*, 124. [CrossRef]

6. Mühlemann, S.; Benic, G.I.; Fehmer, V.; Hämmerle, C.H.F.; Sailer, I. Randomized controlled clinical trial of digital and conventional workflows for the fabrication of zirconia-ceramic posterior fixed partial dentures. Part II: Time efficiency of CAD-CAM versus conventional laboratory procedures. *J. Prosthet. Dent.* **2019**, *121*, 252–257. [CrossRef]

7. Barazanchi, A.; Li, K.C.; Al-Amleh, B.; Lyons, K.; Waddell, J.N. Additive Technology: Update on Current Materials and Applications in Dentistry. *J. Prosthodont.* **2017**, *26*, 156–163. [CrossRef]

8. Zarone, F.; Di Mauro, M.I.; Ausiello, P.; Ruggiero, G.; Sorrentino, R. Current status on lithium disilicate and zirconia: A narrative review. *BMC Oral Health* **2019**, *19*, 134. [CrossRef]

9. Cagidiaco, E.F.; Grandini, S.; Goracci, C.; Joda, T. A pilot trial on lithium disilicate partial crowns using a novel prosthodontic functional index for teeth (FIT). *BMC Oral Health* **2019**, *19*, 276. [CrossRef]

10. Rutkunas, V.; Larsson, C.; Vult von Steyern, P.; Mangano, F.; Gedrimiene, A. Clinical and laboratory passive fit assessment of implant-supported zirconia restorations fabricated using conventional and digital workflow. *Clin. Implant Dent. Relat. Res.* **2020**, *22*, 237–245. [CrossRef]

11. Lerner, H.; Mouhyi, J.; Admakin, O.; Mangano, F. Artificial intelligence in fixed implant prosthodontics: A retrospective study of 106 implant-supported monolithic zirconia crowns inserted in the posterior jaws of 90 patients. *BMC Oral Health* **2020**, *20*, 80. [CrossRef]

12. Joda, T.; Bragger, U.; Zitzmann, N.U. CAD/CAM implant crowns in a digital workflow: Five-year follow-up of a prospective clinical trial. *Clin. Implant Dent. Relat. Res.* **2019**, *21*, 169–174. [CrossRef] [PubMed]

13. Mangano, F.; Veronesi, G. Digital versus Analog Procedures for the Prosthetic Restoration of Single Implants: A Randomized Controlled Trial with 1 Year of Follow-Up. *Biomed Res. Int.* **2018**, *2018*, 5325032. [CrossRef] [PubMed]

14. Benic, G.I.; Sailer, I.; Zeltner, M.; Gütermann, J.N.; Özcan, M.; Mühlemann, S. Randomized controlled clinical trial of digital and conventional workflows for the fabrication of zirconia-ceramic fixed partial dentures. Part III: Marginal and internal fit. *J. Prosthet. Dent.* **2019**, *121*, 426–431. [CrossRef] [PubMed]

15. Sailer, I.; Mühlemann, S.; Fehmer, V.; Hämmerle, C.H.F.; Benic, G.I. Randomized controlled clinical trial of digital and conventional workflows for the fabrication of zirconia-ceramic fixed partial dentures. Part I: Time efficiency of complete-arch digital scans versus conventional impressions. *J. Prosthet. Dent.* **2019**, *121*, 69–75. [CrossRef] [PubMed]

16. Joda, T.; Ferrari, M.; Bragger, U.; Zitzmann, N.U. Patient Reported Outcome Measures (PROMs) of posterior single-implant crowns using digital workflows: A randomized controlled trial with a three-year follow-up. *Clin. Oral Implants Res.* **2018**, *29*, 954–961. [CrossRef]

17. Zitzmann, N.U.; Kovaltschuk, I.; Lenherr, P.; Dedem, P.; Joda, T. Dental Students' Perceptions of Digital and Conventional Impression Techniques: A Randomized Controlled Trial. *J. Dent. Educ.* **2017**, *81*, 1227–1232. [CrossRef]

18. Mangano, F.G.; Hauschild, U.; Veronesi, G.; Imburgia, M.; Mangano, C.; Admakin, O. Trueness and precision of 5 intraoral scanners in the impressions of single and multiple implants: A comparative in vitro study. *BMC Oral Health* **2019**, *19*, 101. [CrossRef]

19. Di Fiore, A.; Meneghello, R.; Graiff, L.; Savio, G.; Vigolo, P.; Monaco, C.; Stellini, E. Full arch digital scanning systems performances for implant-supported fixed dental prostheses: A comparative study of 8 intraoral scanners. *J. Prosthodont. Res.* **2019**, *63*, 396–403. [CrossRef]

20. Kihara, H.; Hatakeyama, W.; Komine, F.; Takafuji, K.; Takahashi, T.; Yokota, J.; Oriso, K.; Kondo, H. Accuracy and practicality of intraoral scanner in dentistry: A literature review. *J. Prosthodont. Res.* **2020**, *64*, 109–113. [CrossRef]

21. Wulfman, C.; Naveau, A.; Rignon-Bret, C. Digital scanning for complete-arch implant-supported restorations: A systematic review. *J. Prosthet. Dent.* **2019**. [CrossRef] [PubMed]

22. Revilla-León, M.; Jiang, P.; Sadeghpour, M.; Piedra-Cascón, W.; Zandinejad, A.; Özcan, M.; Krishnamurthy, V.R. Intraoral digital scans-Part 1: Influence of ambient scanning light conditions on the accuracy (trueness and precision) of different intraoral scanners. *J. Prosthet. Dent.* **2019**. [CrossRef] [PubMed]

23. Reich, S.; Yatmaz, B.; Raith, S. Do "cut out-rescan" procedures have an impact on the accuracy of intraoral digital scans? *J. Prosthet. Dent.* **2020**. [CrossRef] [PubMed]

24. Mizumoto, R.M.; Yilmaz, B. Intraoral scan bodies in implant dentistry: A systematic review. *J. Prosthet. Dent.* **2018**, *120*, 343–352. [CrossRef] [PubMed]

25. Motel, C.; Kirchner, E.; Adler, W.; Wichmann, M.; Matta, R.E. Impact of Different Scan Bodies and Scan Strategies on the Accuracy of Digital Implant Impressions Assessed with an Intraoral Scanner: An In Vitro Study. *J. Prosthodont.* **2020**, *29*, 309–314. [CrossRef] [PubMed]

26. Moslemion, M.; Payaminia, L.; Jalali, H.; Alikhasi, M. Do Type and Shape of Scan Bodies Affect Accuracy and Time of Digital Implant Impressions? *Eur. J. Prosthodont. Restor. Dent.* **2020**, *28*, 18–27.

27. Schmidt, A.; Billig, J.W.; Schlenz, M.A.; Rehmann, P.; Wöstmann, B. Influence of the Accuracy of Intraoral Scanbodies on Implant Position: Differences in Manufacturing Tolerances. *Int. J. Prosthodont.* **2019**, *32*, 430–432. [CrossRef]

28. Tan, M.Y.; Yee, S.H.X.; Wong, K.M.; Tan, Y.H.; Tan, K.B.C. Comparison of Three-Dimensional Accuracy of Digital and Conventional Implant Impressions: Effect of Interimplant Distance in an Edentulous Arch. *Int. J. Oral Maxillofac. Implants* **2019**, *34*, 366–380. [CrossRef]

29. Flügge, T.; van der Meer, W.J.; Gonzalez, B.G.; Vach, K.; Wismeijer, D.; Wang, P. The accuracy of different dental impression techniques for implant-supported dental prostheses: A systematic review and meta-analysis. *Clin. Oral Implants Res.* **2018**, *29* (Suppl. 16), 374–392. [CrossRef]

30. Arcuri, L.; Pozzi, A.; Lio, F.; Rompen, E.; Zechner, W.; Nardi, A. Influence of implant scanbody material, position and operator on the accuracy of digital impression for complete-arch: A randomized in vitro trial. *J. Prosthodont. Res.* **2020**, *64*, 128–136. [CrossRef]

31. Abduo, J.; Elseyoufi, M. Accuracy of Intraoral Scanners: A Systematic Review of Influencing Factors. *Eur. J. Prosthodont. Restor. Dent.* **2018**, *26*, 101–121. [PubMed]

32. Keul, C.; Güth, J.F. Accuracy of full-arch digital impressions: An in vitro and in vivo comparison. *Clin. Oral Investig.* **2020**, *24*, 735–745. [CrossRef] [PubMed]

33. Huang, R.; Liu, Y.; Huang, B.; Zhang, C.; Chen, Z.; Li, Z. Improved scanning accuracy with newly designed scan bodies: An in vitro study comparing digital versus conventional impression techniques for complete-arch implant rehabilitation. *Clin. Oral Implant Res.* **2020**, *31*, 625–633. [CrossRef] [PubMed]

34. Latham, J.; Ludlow, M.; Mennito, A.; Kelly, A.; Evans, Z.; Renne, W. Effect of scan pattern on complete-arch scans with 4 digital scanners. *J. Prosthet. Dent.* **2020**, *123*, 85–95. [CrossRef]

Journal of
Clinical Medicine

Article

A Double-Blind Crossover RCT Analyzing Technical and Clinical Performance of Monolithic ZrO₂ Implant Fixed Dental Prostheses (iFDP) in Three Different Digital Workflows

Aiste Gintaute [1,†], Karin Weber [2,†], Nicola U. Zitzmann [1], Urs Brägger [3], Marco Ferrari [4] and Tim Joda [1,*]

1   Department of Reconstructive Dentistry, University Center for Dental Medicine Basel, University of Basel, 4058 Basel, Switzerland; aiste.gintaute@unibas.ch (A.G.); n.zitzmann@unibas.ch (N.U.Z.)
2   Private Dental Office, 4314 Zeiningen, Switzerland; karinwe17@hotmail.com
3   Department of Reconstructive Dentistry and Gerodontology, School of Dental Medicine, University of Bern, 3010 Bern, Switzerland; urs.braegger@zmk.unibe.ch
4   Department of Prosthodontics and Dental Materials, University of Siena, 53100 Siena, Italy; ferrarm@gmail.com
*   Correspondence: tim.joda@unibas.ch; Tel.: +41-61-267-26-30
†   These authors contributed equally to the manuscript.

check for **updates**

Citation: Gintaute, A.; Weber, K.; Zitzmann, N.U.; Brägger, U.; Ferrari, M.; Joda, T. A Double-Blind Crossover RCT Analyzing Technical and Clinical Performance of Monolithic ZrO₂ Implant Fixed Dental Prostheses (iFDP) in Three Different Digital Workflows. *J. Clin. Med.* **2021**, *10*, 2661. https:// doi.org/10.3390/jcm10122661

Academic Editor: James Kit-hon Tsoi

Received: 20 April 2021
Accepted: 14 June 2021
Published: 16 June 2021

Publisher's Note: MDPI stays neutral with regard to jurisdictional claims in published maps and institutional affiliations.

Abstract: This double-blind randomized controlled trial with a crossover design analyzed the technical and clinical performance of three-unit monolithic ZrO₂ implant-fixed dental prostheses (iFDPs), prepared using two complete digital workflows (Test-1, Test-2) and one mixed analog–digital workflow (Control). Each of the 20 study patients received three iFDPs, resulting in 60 restorations for analysis. The quality of the restorations was assessed by analyzing laboratory cross-mounting and calculating the chairside adjustment time required during fitting. All iFDPs could be produced successfully with all three workflows. The highest cross-mounting success rate was observed for the original pairing iFDP/model of the Control group. Overall, 60% of iFDPs prepared with Test-1 workflow did not require chairside adjustment compared with 50% for Test-2 and 30% for Controls. The mean total chairside adjustment time, as the sum of interproximal, pontic, and occlusal corrections was 2.59 ± 2.51 min (Control), 2.88 ± 2.86 min (Test-1), and 3.87 ± 3.02 min (Test-2). All tested workflows were feasible for treatment with iFDPs in posterior sites on a soft tissue level type implant system. For clinical routine, it has to be considered that chairside adjustments may be necessary, at least in every second patient, independent on the workflow used.

Keywords: dental implant; fixed dental prosthesis (FDP); monolithic; zirconia; zirconium-dioxide (ZrO₂); digital workflow; accuracy; precision; clinical trial

1. Introduction

Digitization has significantly influenced dentistry in recent years and continues to enable new options in clinical routines [1]. Translating established, conventional dental procedures into digitized protocols requires a detailed understanding of digital dental processes [2]. Conventional procedures cannot always be transferred identically to digital workflows. Continuous progress inevitably leads to an adaptation of the established workflows in the clinic and in dental technology. Consequently, the digital processes must be critically analyzed and recalibrated [3].

Intraoral optical scanning (IOS) plays a crucial role in digital dentistry, especially in the field of prosthodontics [4,5]. The contactless transfer of the individual patient situation to the virtual dental laboratory, without the use of any physical models, is central to completely digital workflows [6]. The continuous and incremental improvement in the quality of IOS systems, both hardware and software, is enabling the production of more and more accurate and precise final restorations [7]. In this context, the level of digital experience of the dental team (operating the different IOS and CAD/CAM systems) has a significant impact on the quality of the final outcome [8]. While the interfaces of available digital systems have been opened, enabling free data transfer between systems from different manufacturers, it is still unclear whether there is a loss of data quality and, subsequently, a fluctuating quality of reconstructions when non-proprietary data is transferred between different CAD/CAM systems [9].

Among fixed dental prostheses (FDP) supported by dental implants, digital pathways have demonstrated clear superiority compared with conventional workflows for the production of single implant crowns. Implant crowns fabricated digitally were superior in terms of clinical fitting [10] and economic parameters, such as time-efficiency [11] and costs [12] compared with classical impressions and gypsum casts. The use of standardized scan bodies makes the IOS technology predestined for therapy with fixed implant reconstructions. Whereas in single restorations, secure occlusion is usually ensured by the adjacent teeth, in multi-unit implant-supported fixed dental prostheses (iFDP), the lack of stable occlusal support poses the challenge for IOS systems during computerized bite registration [13]. It is not the optical resolution of the IOS devices or the power of the software that are the limiting features, rather that the missing occlusal units become the bottleneck in digital impression taking for iFDPs [14]. The question arises to what extent dental models are required for fabricating iFDPs, or whether a completely digital approach is possible, similar to what can be done for single implant crowns, i.e., IOS > Design and Mill > Delivery [15].

While the dental technology industry may promise that their IOS systems will deliver smooth processes with reproducible results in prosthodontic indications, clinical evidence is lacking. In particular, evidence for the feasibility of fully digital workflows for iFDPs, starting with IOS in the clinic, to CAD/CAM-processing in the dental lab, and back to clinical delivery. Therefore, the aim of the present study was to analyze the technical and clinical performance of monolithic zirconium-dioxide (ZrO_2) iFDPs prepared according to two complete digital workflows and one mixed analog–digital workflow. The null hypothesis was that the process quality of iFDPs in these three digital workflows is similar for the two investigated outcomes, namely (i) laboratory cross-mounting [16]; and [1] clinical performance indicator (CPI) defined by restoration fit, chairside adjustment time, and consecutive density changes of the modified iFDPs.

2. Materials and Methods

2.1. Trial Setting

The study was designed as a prospective, double-blind, triple-armed randomized controlled trial (RCT) with a crossover approach in a university-based setting. Neither the clinician nor the patient was aware of the three different treatment groups. The protocol was officially approved by the Ethics Committee Basel, Switzerland (EKNZ-ID 2019-00706) and registered at ClinTrials.gov (NCT 04029025). This RCT was conducted in compliance with the study protocol, the current version of the Declaration of Helsinki, the ICH-GCP, as well as all national legal and regulatory requirements. Patients provided an informed consent to participate in the trial. No changes were made to methods after trial commencement. The RCT followed the CONSORT 2010 statements (http://www.consort-statement.org/consort-2010, accessed on 19 April 2021).

At the time of development of the study protocol, no data from trials investigating the clinical and technical performance of three-unit monolithic ZrO_2 iFDPs were available. Therefore, the power analysis for the present study was based on our own preliminary findings for treatment with implant-supported single units considering time efficiency in terms of clinical adjustment time: with a coefficient of determination of $R^2 = 0.5$, a statistical power of 0.8, and a significance level of $\alpha = 0.05$, a sample size of $n = 16$ would be required. Therefore, 20 study patients presenting with a three-unit edentulous space or free-end situation in one posterior quadrant qualifying for a three-unit FDP supported by two dental implants (Tissue Level Implant System RN/WN, Institut Straumann AG, Basel, Switzerland) were considered sufficient. Inclusion criteria were periodontal health or successfully treated [17], consumption of no more than 10 cigarettes per day, the presence of antagonistic contacts, and at least one adjacent tooth.

Each of the 20 study patients received three iFDPs prepared, according to three different workflows, resulting in a total of 60 restorations. All iFDPs were designed and produced out of monolithic ZrO_2 (VITA YZ ST Super Translucent Multicolor, Bad Säckingen, Germany) as three-unit screw-retained, full-contour reconstructions (Figure 1). Two of the workflows were completely digital and the third combined analog and digital steps:

- **Test-1 "Complete Digital Workflow"** (3Shape, Copenhagen, Denmark) IOS Trios 3 + Dental System Lab-Software;
- **Test-2 "Complete Digital Workflow"** (Dental Wings Inc., Montreal, Canada) IOS Virtuo Vivo + DWOS Lab-Software;
- **Control "Analog–Digital Workflow"** Polyether Impression/Gypsum Cast/Lab-Scan + EXOCAD Lab-Software.

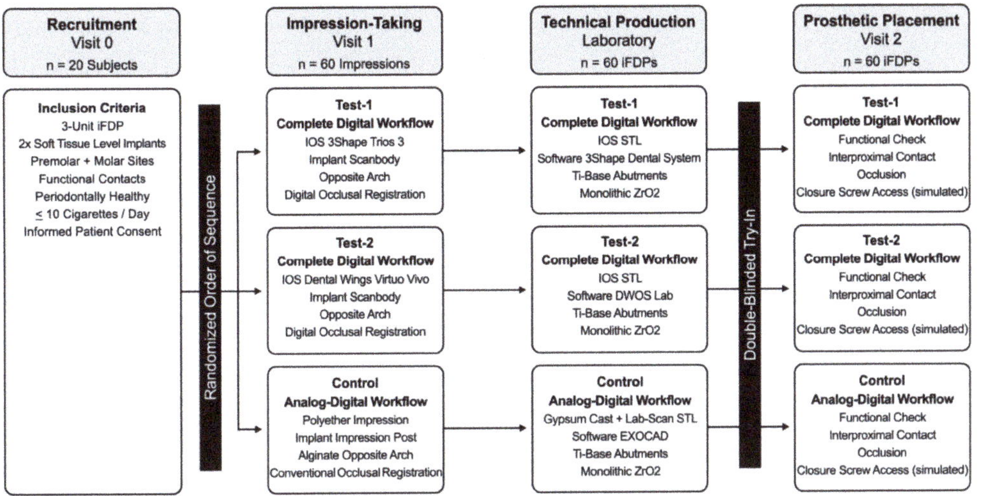

Figure 1. Study flow-chart for clinical and technical work steps.

The baseline of the study was the start of the prosthetic therapy. The study involved three key stages: (i) impression-taking for clinical registration of the implants' 3D-positioning, including antagonists and occlusal relation [16]; (ii) CAD/CAM fabrication of iFDPs [16]; and (iii) clinical try-in/delivery (Figure 2). Distribution of the work steps, whether beginning with Test-1, Test-2, or Control, as well as the order of sequence during iFDP try-in, were randomly chosen per study patient using the envelope-technique. The principal investigator (T.J.) performed the random allocation sequence and the enrollment of all study patients. Blinding persisted at the time of try-in/delivery of the final iFDPs to the study patients and to the clinical operator. For each of the three workflows, outcomes were evaluated according to performance of the produced reconstructions based on (i) the feasibility of laboratory cross-mounting of each iFDP [16]; and (ii) clinical fit and assessment of adaptation time for clinical adjustments of interproximal surfaces, pontic areas and occlusal surfaces—if required.

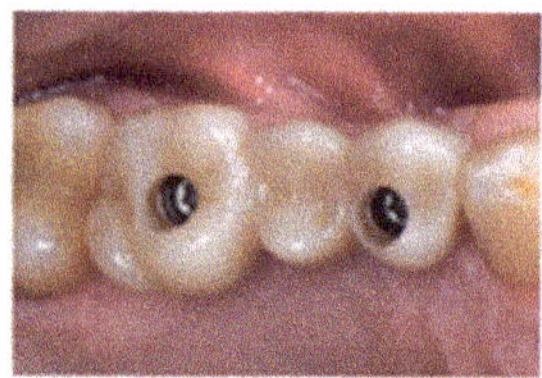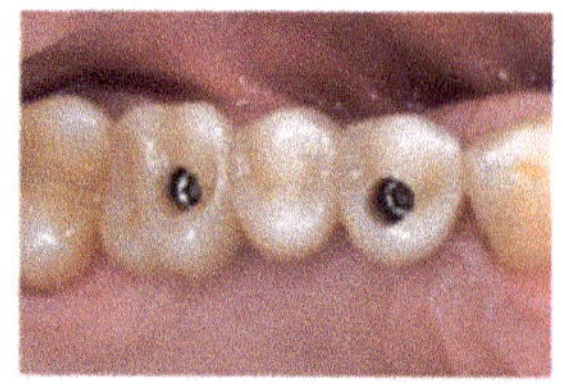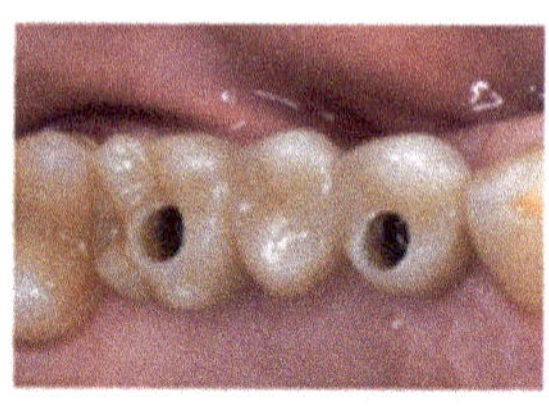

Figure 2. Clinical try-in/delivery of the iFDPs for Test-1 (**a**), Test-2 (**b**), and Control (**c**).

2.2. Clinical and Technical Work Steps

Clinical work steps were performed by one experienced dentist (K.W.) and observed by one spectator (A.G.). For the complete digital workflows (Test-1, Test-2), the intraorally obtained STL-files were directly transferred to the relevant laboratory CAD/CAM software specific for the IOS system used: Trios 3 > Dental System Lab-Software (3Shape, Copenhagen, Denmark); and Virtuo Vivo > DWOS Lab-Software (Dental Wings Inc., Montreal, Canada). Full-arch digital impressions were taken and scanned according to the manufacturer's recommendations.

For the analog–digital workflow (Control), the pick-up impression technique with individualized open trays and polyether impression material (Impregum, 3M ESPE, Neuss, Germany) was used. In addition, a high-viscosity alginate impression was taken from the antagonistic arch (Palgat Plus Quick, 3M Espe GmbH, Neuss, Germany) as well as occlusal registration with fast-setting vinyl polysiloxane Blu-Mousse (Parkell Inc., Edgewood, USA). Gypsum implant master casts were manmade, conventionally, with low-expansion die stone ISO Type 4 (Silky Rock, Whip Mix, Louisville, KY, USA) under consideration of the company's recommendations. The gypsum implant master casts were scanned in the laboratory (Ceramill Map 400+, Amann Girrbach, Koblach, Austria) and the STL-files further processed with EXOCAD Lab-Software (EXOCAD, Darmstadt, Germany).

Finally, all 60 monolithic ZrO_2 iFDPs were milled with a five-axis unit (Ceramill 2 Motion, Amann Girrbach, Koblach, Austria) and bonded to pre-fabricated titanium base abutments (Variobase RN/WN, Institut Straumann AG, Basel, Switzerland). For both complete digital workflows, the iFDPs were finalized without using physical models.

2.3. Laboratory Cross-Mounting

Only for testing cross-mounting of the restorations, implant models for Test-1 and Test-2 were additionally made out of a high-precision photopolymer dental resin (Form 3B, Formlabs Inc., Somerville, MA, USA) matched to the 3D printer system (Model Resin 1L, Formlabs Inc., Somerville, USA). For each study patient, iFDPs of one group were placed onto corresponding model situations of the other groups and vice versa. Success was defined as a fit to the implant analogues, interproximal plus occlusal fit in terms of dichotomic feasibility testing for all parameters (yes/no).

Calibration among the evaluators was completed in advance of the RCT. Evaluators were trained using pairs of models/reconstructions that represented mis-fitting and well-fitting situations. Analysis was done by the clinical operator (K.W.) and independently verified by the observer (A.G.). Evaluation was performed separately by the clinical operator and the observer including repetition after one week with an inter-examiner Kappa-Score of 1.0 (Figure 3).

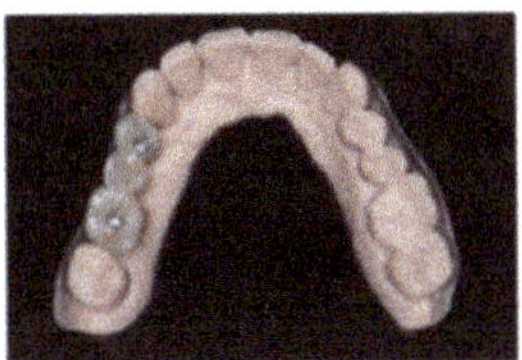 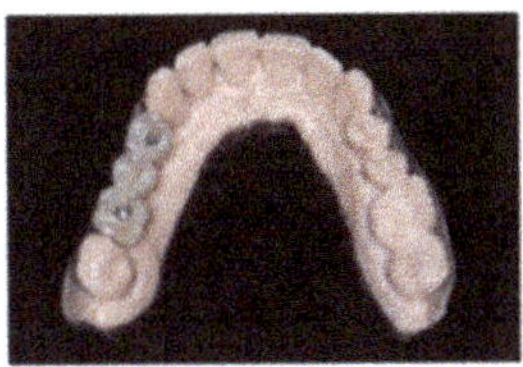 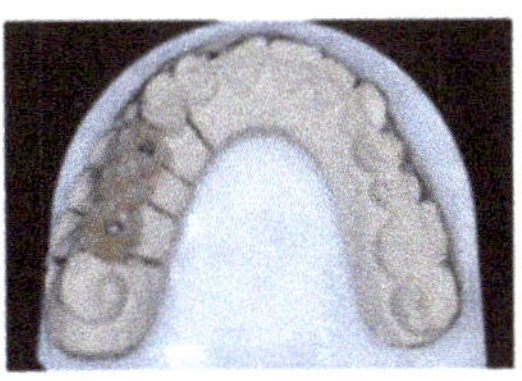

Figure 3. Set-up for cross-mounting displaying dental model situations with inserted iFDPs representing on study patient with 3D-printed implant models for Test-1 (**a**) and Test-2 (**b**), as well as gypsum implant master cast for Control (**c**).

2.4. Clinical Performance Indicators (CPI)

The CPIs comprised the clinical fit of the iFDPs, the chairside adjustment time required to fit the iFDPs, and the subsequent density changes of the adjusted iFDPs (if applicable). For clinical assessment, the healing abutments were removed and the iFDPs were mounted onto the implants with respect to the randomization process for the order of the sequence and respecting the prerequisite of blinding. Only the dental technician and the clinical observer were informed about the traceability from the iFDPs and their related workflows.

First, the interproximal fit and the seating of the iFDPs were analyzed, striving for continuity with waxed dental floss at neighboring sites. If necessary, corrections were made with diamond burs and silicone polishers to create satisfactory interproximal surfaces. Secondary, the pontic site was evaluated, aiming at Superfloss (Oral-B, Procter & Gamble, Cincinnati, OH, USA) passing through with some hindrance. Finally, the occlusal scheme was checked, aiming at occlusal contacts with 12 μm articulation foil in maximum intercuspation without contacts during mandibular movements. Again, if necessary, adjustments were made as described to achieve light occlusal contacts without dynamic interference. The observer documented the treatment time needed for each clinical step to ensure that the treatment time was accurately recorded.

Afterwards, all iFDPs in need of adjustments were analyzed for deviation compared to the original design. The clinically modified iFDPs were digitized by the same lab-side scanner, which was used previously for scanning of the gypsum implant master casts the Controls. STL-files were then imported into a 3D analysis software (PreForm, Formlabs Inc., Somerville, MA, USA) and matched with the original virtual reconstructive design related to the workflows of Test-1, Test-2, and Control. A best-fit algorithm of the 3D analysis software was applied for volumetric deviation analysis in terms of density changes in g/mm^3 and visualization of the superimposed STL-files (Figure 4).

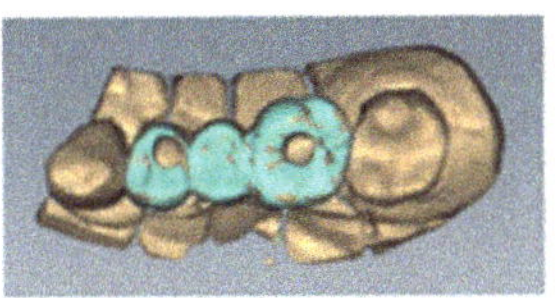 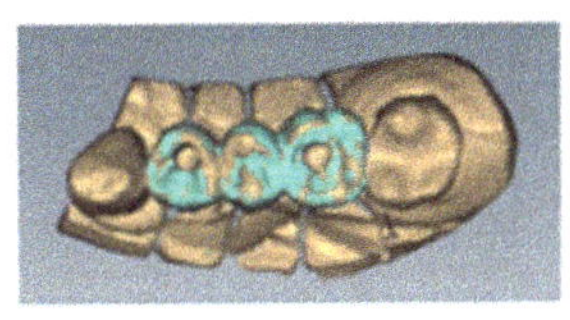 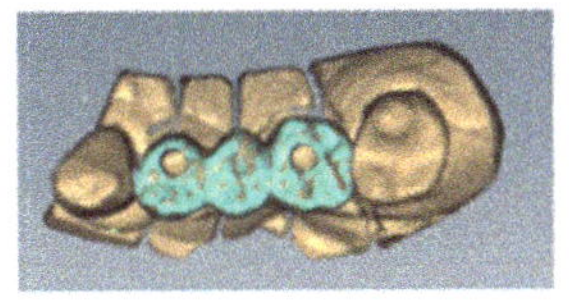

Figure 4. Deviation analysis comparing the initial virtual design and the digitized iFDPs after clinical adjustments for Test-1 (**a**), Test-2 (**b**), and Control (**c**) mounted on the gypsum implant cast guaranteeing a standardized evaluation.

2.5. Statistical Analysis

Statistical analysis was conducted with the program "Software R" (version 4.0.4). Kruskal–Wallis Tests were used for all comparisons in this RCT with crossover design. Since no carry-over effects were expected, the data of investigated measurement-rounds were used for analysis of Test-1, Test-2, and Control, respectively. A *p*-value of < 0.05 was considered as statistically significant.

3. Results

The mean age of the patients was 63 years (range 30 to 76 years) and 55% were female. The location of the iFDPs were equally distributed between the maxilla and mandible, and most implants had a regular neck configuration (n = 35 RN and n = 5 WN). A total of 17 iFDPs were examined in free-end situations and three iFDPs in edentulous spaces with two adjacent teeth. All analyses were performed on originally assigned groups. Recruitment started in January 2020 and no study patient was excluded after enrollment.

3.1. Laboratory Cross-Mounting

Based on the three workflows, all 60 iFDPs were tried in onto the conventionally manufactured gypsum implant casts (Control) and onto the 3D printed implant models (Test-1 and Test-2), resulting in nine possible pairings for laboratory cross-mounting (Table 1). Correlation analysis of crossover investigation revealed that none of the iFDP could be successfully mounted on all three corresponding model situations. The highest transfer success rate was the original pairing iFDP Control/model Control; and the lowest was iFDP Control/model Test-2. Per definition, failure modes of laboratory cross-mounting were characterized by interproximal, and subsequently, occlusal misfit of the iFDPs.

Table 1. Feasibility testing (yes/no) for laboratory cross-mounting summarizing success rates of 3 × 20 iFDPs and 3 × 20 model situations in nine possible pairings.

		Model Situation		
		Test-1	**Test-2**	**Control**
iFDP	**Test-1**	12/20 (60%)	6/20 (30%)	13/20 (65%)
	Test-2	3/20 (15%)	5/20 (25%)	12/20 (60%)
	Control	3/20 (15%)	2/20 (10%)	15/20 (75%)

3.2. Clinical Performance Indicators (CPI)

All iFDPs could be produced successfully. For Test-1, IOS had to be redone for one study patient due to an unusable STL-dataset. The need for clinical adjustment was highest for iFDPs of Controls, followed by Test-2 and Test-1. Overall, 14 iFDPs of the mixed analog–digital workflow required clinical corrections to achieve sufficient fit in terms of interproximal contact and occlusion; whereby 10 (Test-2) and 12 (Test-1) iFDPs of the complete digital workflows could be inserted without any modifications, representing success rates for adjustment-free delivery of 30% (Control), 50% (Test-2), and 60% (Test-1), respectively. In three (15%) of the 20 study patients, all three iFDPs could be immediately inserted without any corrections. Results for the mean total adjustment time, as the sum of interproximal plus occlusal corrections plus pontic-area modifications, were 2.59 min (SD ± 2.51) for Control, 2.88 min (SD ± 2.86) for Test-1, and 3.87 min (SD ± 3.02) for Test-2. The mean values for density changes of the clinically modified iFDPs were 4.92 g/mm^3 (SD ± 9.64) for Test-1, 35.83 g/mm^3 (SD ± 114.14) for Test-2, and 14.13 g/mm^3 (SD ± 40.51) for Control.

4. Discussion

This double-blind, crossover RCT investigated digital and conventional workflows for producing monolithic ZrO$_2$ iFDPs in posterior sites. The results revealed differences in the technical and clinical performance of the three different treatment workflows. Therefore, the tested hypothesis for equivalence for the two complete digital workflows plus one mixed analog–digital workflow had to be rejected.

Oral rehabilitation of partially edentulous patients with implant-retained multi-span fixed reconstructions is a comprehensive treatment requiring multiple work steps in an interdisciplinary approach with knowledge and skills in prosthodontics, implant surgery, and dental technology [18]. The team of clinicians and technicians has to rely on the therapy protocols to achieve long-term success. For the implementation of new workflows in clinical routine use, these must withstand the comparison with the established approach [19]. In general, the feasibility of a new therapy's protocol is the minimal prerequisite; predictable and reproducible treatment outcomes are the key to become a serious alternative to the gold standard–or even to be disruptive enough to replace it [20]. The specific trial setting of the current study with a crossover design allowed direct comparisons of three different workflows for each study patient.

In the present study, laboratory cross-mounting was conducted to provide a simple quality-check of the three iFDP workflows prior to the clinical try-in. Since the three iFDPs derived from the different workflows for each patient were fabricated for the same clinical situation, it should be expected that the implant reconstructions could be exchanged among each other on the corresponding dental models for Test-1, Test-2, and Control, respectively. However, no iFDP could be successfully mounted on all three-model pairings indicating that either the impression technique, the CAD/CAM process, the model fabrication, or a combination of these factors affected this outcome. The completely digitally produced iFDPs had worse laboratory transfer rates than the analog–digital workflow, with Test-1 workflow performing slightly better than Test-2. In this context, it has to be emphasized that iFDPs from the Test-1 and Test-2 workflows were virtually designed and produced without the 3D printed models. These models were used only for cross-mounting analysis. No other investigation has reported on cross-mounting of multi-unit iFDPs based on a prospective clinical trial; however, the present findings are consistent with a previously published investigation of in vitro cross-mounting comparing implant-supported single crowns produced with digital and conventional workflows [10].

In the present study, CPIs comprised clinical fit, the time needed for chairside adjustments, and the subsequent density changes of the adjusted iFDPs. Both the dentist and the patient were blinded to the workflow that was used to generate the individual iFDPs to minimize bias. While the success rate for adjustment-free delivery of iFDPs of Test-1 was minimally better than that of Test-2, the success rate of Test-1 was twice as high as that of Controls. The shortest chair-time was required for correction of Control iFDPs (relevant for cases requiring corrections); however, the overall clinical adjustment time for Test-1, Test-2, and Controls differed only slightly within a narrow range of 2.59 min to 3.87 min without statistical significance. Therefore, the general need for clinical adaptation of iFDPs (yes/no) must be rated higher than the amount of time required for adjustment. For treatment with single-unit implant crowns in posterior sites, clinical trials also reported on complete digital workflows in favor of conventional workflows in terms of time-efficiency [21,22].

The number of adjustments required in the present study was quantified by comparing the initial iFDP design with the modified restoration, and the volumetric deviation was expressed as density changes. Applying this novel method, Test-1 performed more than seven-fold better than Test-2, and almost three-fold better than Controls. Adjustment time and density changes can only represent surrogate parameters as CPIs for the evaluation of the process quality. The best performance indicator, however, is the successful insertion of the iFDP–ideally without corrections to the prosthetic reconstruction [23]. Not only do clinical step-by-step modifications require valuable chairside time, the adjusted surfaces of the iFDPs represent predilection areas and must be re-polished to a high gloss prior to final insertion [24,25].

Overall, the results of the present RCT investigating three-unit monolithic ZrO_2 iFDPs revealed lower performance with complete digital treatment protocols compared with previously published data for single-unit implant crowns [15]. One possible explanation for this may be that multi-span implant-retained reconstructions lack unilaterally antagonistic pairs for a secure positional relationship of the jaws to each other, depending on the individual patient situation. Based on the present findings, it can only be speculated whether the reason for the need for clinical modifications of the iFDPs is the technical process itself or the patient-specific situation affecting the bite registration [13]. The typical indication for an iFDP is the rehabilitation of posterior free-end situations to reestablish masticatory function without the need for removable dental prostheses. Therefore, it is not surprising that 85% of the iFDPs in the present study were located in free-end situations. This classic treatment "premolar-pontic-molar", unlike single-unit implant crowns, makes bite registration more complex and less predictable. In the complete digital workflows, the systems work with the data recorded in a single moment, whereas in the mixed analog–digital workflow, the dental technician can incorporate experience in designing the occlusion in the physical articulator. Therefore, digital bite registration still seems to be the critical point today [14].

To interpret the findings from the digital and complete digital workflows in the present study, the specific systems used should be considered. Test-1 and Test-2 exemplified synchronized workflows of IOS and corresponding laboratory CAD-software from the same company, while the Control workflow used lab-side scanning of the dental implant master casts and EXOCAD as laboratory CAD-software. For further production following the CAM-process, all three workflows used the same milling unit and the same ZrO_2 blanks in order to harmonize the final production and to focus on the initial work steps of data generation and technical processing. In this RCT, only proprietary data transfer within system-specific solutions of STL generation and processing was analyzed. Therefore, future research should also consider whether different data flows involving mixing systems might create better clinical results. Moreover, prospective follow-up studies need to investigate the aspect of digital bite registration in order to define a system-specific minimal number of antagonistic units as threshold indicating if a complete digital workflow is feasible.

Based on the results of laboratory cross-mounting and clinical adjustment time from the present study it can be concluded that all three workflows worked successfully. Conventional impression-taking plus lab-side scanning of dental gypsum casts demonstrated high precision for analysis of the iFDP/model pairing, but not necessarily in the clinical situation. Conversely, the complete digital workflows with IOS and model-free processing achieved higher rates for adjustment-free delivery of the iFDPs in the clinic, while fitting on the 3D printed models fabricated from the STL data was worse. It cannot be generalized that digital workflows using different IOS and CAD/CAM system might produce identical results per se. The important factors seem to be: (i) what system is used for which indication? [16] (ii) Is a proprietary data transfer used? (iii) Are the hardware and software up to date [7]?

Complete digital workflows represent the future of (fixed) prosthodontics. Reducing the absolute number of laboratory steps has the advantage of lower susceptibility to technical errors. Therefore, the potential advantage of digital workflows should not be diminished by leaving the pathway of model-free fabrication. It is feasible to use all three digital workflows for treatment with three-unit monolithic ZrO_2 iFDPs in posterior sites on a soft tissue level type implant system (with given emergence profile). For a clinical routine, chairside adjustments may be necessary, at least in every second patient, independent on the workflow used.

Author Contributions: Conceptualization, T.J.; methodology, T.J.; software, A.G. and K.W.; validation, T.J., A.G. and K.W.; formal analysis, T.J.; investigation, A.G. and K.W.; resources, T.J.; data curation, A.G. and T.J.; writing—original draft preparation, T.J., A.G. and K.W.; writing—review and editing, N.U.Z., U.B. and M.F.; visualization, T.J., A.G. and K.W.; supervision, T.J.; project administration, T.J. and A.G.; funding acquisition, T.J. and U.B. All authors have read and agreed to the published version of the manuscript.

Funding: This research was funded by the International Team for Implantology (ITI); grant number "ITI 1214_2017". The APC was funded by the Department of Reconstructive Dentistry, University Center for Dental Medicine Basel, University of Basel, Switzerland.

Institutional Review Board Statement: The study was conducted according to the guidelines of the Declaration of Helsinki, and approved by the Ethics Committee Basel, Switzerland (EKNZ-ID. 2019-00706, 26 June 2019).

Informed Consent Statement: Informed consent was obtained from all subjects involved in the study. Written informed consent has been obtained from the patient(s) to publish this paper.

Data Availability Statement: The data are not publicly available due to privacy restrictions.

Acknowledgments: The authors express their gratitude to Dental Technician Markus Link (LABOR-LINK AG Zahntechnik, Basel, Switzerland) for the fabrication of the implant-supported reconstructions.

Conflicts of Interest: The authors declare no conflict of interest. The funders had no role in the design of the study; in the collection, analyses, or interpretation of data; in the writing of the manuscript, or in the decision to publish the results.

References

1. Blatz, M.B.; Conejo, J. The Current State of Chairside Digital Dentistry and Materials. *Dent. Clin. N. Am.* **2019**, *63*, 175–197. [CrossRef] [PubMed]
2. Joda, T.; Bornstein, M.M.; Jung, R.E.; Ferrari, M.; Waltimo, T.; Zitzmann, N.U. Recent Trends and Future Direction of Dental Research in the Digital Era. *Int. J. Environ. Res. Public Health* **2020**, *17*, 1987. [CrossRef]
3. Joda, T.; Ferrari, M.; Bragger, U.; Zitzmann, N.U. Patient Reported Outcome Measures (PROMs) of posterior single-implant crowns using digital workflows: A randomized controlled trial with a three-year follow-up. *Clin. Oral Implant. Res.* **2018**, *29*, 954–961. [CrossRef] [PubMed]
4. Ahlholm, P.; Sipilä, K.; Vallittu, P.; Jakonen, M.; Kotiranta, U. Digital Versus Conventional Impressions in Fixed Prosthodontics: A Review. *J. Prosthodont.* **2018**, *27*, 35–41. [CrossRef] [PubMed]
5. Joda, T.; Bragger, U.; Zitzmann, N.U. CAD/CAM implant crowns in a digital workflow: Five-year follow-up of a prospective clinical trial. *Clin. Implant. Dent. Relat. Res.* **2019**, *21*, 169–174. [CrossRef] [PubMed]
6. Schlenz, M.A.; Schubert, V.; Schmidt, A.; Wöstmann, B.; Ruf, S.; Klaus, K. Digital versus Conventional Impression Taking Focusing on Interdental Areas: A Clinical Trial. *Int. J. Environ. Res. Public Health* **2020**, *17*, 4725. [CrossRef]
7. Schmidt, A.; Klussmann, L.; Wöstmann, B.; Schlenz, M.A. Accuracy of Digital and Conventional Full-Arch Impressions in Patients: An Update. *J. Clin. Med.* **2020**, *9*, 688. [CrossRef]
8. Greenberg, A.M. Digital Technologies for Dental Implant Treatment Planning and Guided Surgery. *Oral Maxillofac. Surg. Clin. N. Am.* **2015**, *27*, 319–340. [CrossRef]
9. Joda, T.; Ferrari, M.; Gallucci, G.O.; Wittneben, J.; Brägger, U. Digital technology in fixed implant prosthodontics. *Periodontology* **2016**, *73*, 178–192. [CrossRef]
10. Joda, T.; Katsoulis, J.; Brägger, U. Clinical Fitting and Adjustment Time for Implant-Supported Crowns Comparing Digital and Conventional Workflows. *Clin. Implant. Dent. Relat. Res.* **2015**, *18*, 946–954. [CrossRef]
11. Joda, T.; Brägger, U. Time-Efficiency Analysis Comparing Digital and Conventional Workflows for Implant Crowns: A Prospective Clinical Crossover Trial. *Int. J. Oral Maxillofac. Implant.* **2015**, *30*, 1047–1053. [CrossRef]
12. Joda, T.; Brägger, U. Digital vs. conventional implant prosthetic workflows: A cost/time analysis. *Clin. Oral Implant. Res.* **2015**, *26*, 1430–1435. [CrossRef] [PubMed]
13. Pol, C.W.; Raghoebar, G.M.; Cune, M.S.; Meijer, H.J. Implant-Supported Three-Unit Fixed Dental Prosthesis Using Coded Healing Abutments and Fabricated Using a Digital Workflow: A 1-Year Prospective Case Series Study. *Int. J. Prosthodont.* **2020**, *33*, 609–619. [CrossRef] [PubMed]
14. Gintaute, A.; Keeling, A.; Osnes, C.; Zitzmann, N.; Ferrari, M.; Joda, T. Precision of maxillo-mandibular registration with intraoral scanners in vitro. *J. Prosthodont. Res.* **2020**, *64*, 114–119. [CrossRef]
15. Mühlemann, S.; Kraus, R.D.; Hämmerle, C.H.F.; Thoma, D.S. Is the use of digital technologies for the fabrication of implant-supported reconstructions more efficient and/or more effective than conventional techniques: A systematic review. *Clin. Oral Implant. Res.* **2018**, *29*, 184–195. [CrossRef]
16. Miyazaki, T.; Hotta, Y.; Kunii, J.; Kuriyama, S.; Tamaki, Y. A review of dental CAD/CAM: Current status and future perspectives from 20 years of experience. *Dent. Mater. J.* **2009**, *28*, 44–56. [CrossRef]
17. Wennstrom, J.L.; Tomasi, C.; Bertelle, A.; Dellasega, E. Full-mouth ultrasonic debridement versus quadrant scaling and root planing as an initial approach in the treatment of chronic periodontitis. *J. Clin. Periodontol.* **2005**, *32*, 851–859. [CrossRef]
18. Zitzmann, N.U.; Krastl, G.; Hecker, H.; Walter, C.; Weiger, R. Endodontics or implants? A review of decisive criteria and guidelines for single tooth restorations and full arch reconstructions. *Int. Endod. J.* **2009**, *42*, 757–774. [CrossRef] [PubMed]
19. Cardoso, J.R.; Pereira, L.M.; Iversen, M.D.; Ramos, A.L. What is gold standard and what is ground truth? *Dent. Press J. Orthod.* **2014**, *19*, 27–30. [CrossRef] [PubMed]
20. Joda, T.; Yeung, A.; Hung, K.; Zitzmann, N.; Bornstein, M. Disruptive Innovation in Dentistry: What It Is and What Could Be Next. *J. Dent. Res.* **2021**, *100*, 448–453. [CrossRef] [PubMed]
21. Joda, T.; Bragger, U. Time-efficiency analysis of the treatment with monolithic implant crowns in a digital workflow: A randomized controlled trial. *Clin. Oral Implant. Res.* **2016**, *27*, 1401–1406. [CrossRef]
22. Joda, T.; Ferrari, M.; Brägger, U. Monolithic implant-supported lithium disilicate (LS2) crowns in a complete digital workflow: A prospective clinical trial with a 2-year follow-up. *Clin. Implant. Dent. Relat. Res.* **2017**, *19*, 505–511. [CrossRef] [PubMed]
23. Joda, T.; Zarone, F.; Ferrari, M. The complete digital workflow in fixed prosthodontics: A systematic review. *BMC Oral Health* **2017**, *17*, 124. [CrossRef] [PubMed]

24. Matzinger, M.; Hahnel, S.; Preis, V.; Rosentritt, M. Polishing effects and wear performance of chairside CAD/CAM materials. *Clin. Oral Investig.* **2019**, *23*, 725–737. [CrossRef]
25. Scherrer, D.; Bragger, U.; Ferrari, M.; Mocker, A.; Joda, T. In-vitro polishing of CAD/CAM ceramic restorations: An evaluation with SEM and confocal profilometry. *J. Mech. Behav. Biomed. Mater.* **2020**, *107*, 103761. [CrossRef] [PubMed]

Journal of
Clinical Medicine

Review

3D Printing in Digital Prosthetic Dentistry: An Overview of Recent Developments in Additive Manufacturing

Josef Schweiger [1,*], Daniel Edelhoff [1] and Jan-Frederik Güth [2,*]

1 Department of Prosthetic Dentistry, University Hospital, Ludwig-Maximilians University Munich, 80336 Munich, Germany; Daniel.Edelhoff@med.uni-muenchen.de
2 Poliklinik für Zahnärztliche Prothetik, Center for Dentistry and Oral Medicine (Carolinum), Goethe-University, 60596 Frankfurt am Main, Germany
* Correspondence: Josef.Schweiger@med.uni-muenchen.de (J.S.); gueth@med.uni-frankfurt.de (J.-F.G.); Tel.: +49-89-440-059-520 (J.S.); +49-89-4400-5903 (J.-F.G.)

Abstract: Popular media now often present 3D printing as a widely employed technology for the production of dental prostheses. This article aims to show, based on factual information, to what extent 3D printing can be used in dental laboratories and dental practices at present. It attempts to present a rational evaluation of todays´ applications of 3D printing technology in the context of dental restorations. In addition, the article discusses future perspectives and examines the ongoing viability of traditional dental laboratory services and manufacturing processes. It also shows which expertise is needed for the digital additive manufacturing of dental restorations.

Keywords: 3D printing; digital one-piece casting; multi-material 3D printing; graphic 3D models; 3D printing using composite resin; digital pressing technology; 3D printing using zirconia; hybrid production

Citation: Schweiger, J.; Edelhoff, D.; Güth, J.-F. 3D Printing in Digital Prosthetic Dentistry: An Overview of Recent Developments in Additive Manufacturing. *J. Clin. Med.* **2021**, 10, 2010. https://doi.org/10.3390/jcm10092010

Academic Editor: Tim Joda

Received: 24 March 2021
Accepted: 30 April 2021
Published: 7 May 2021

Publisher's Note: MDPI stays neutral with regard to jurisdictional claims in published maps and institutional affiliations.

1. Introduction

The pace of development in digital dental manufacturing has become impressive. High levels of productivity and accuracy of fit have been achieved by subtractive processes, while additive processes (3D printing) are increasingly coming to the fore. Combinations of different manufacturing methods—such as laser sintering plus CNC machining or digital design and 3D printing plus analog ceramic pressing—display the enormous potential [1,2].

2. Current State of Technology

2.1. A Rationale for Digital Manufacturing and 3D Printing in Dentistry

Fundamental changes in society are also affecting dental technology, like any other area. One of these changes is the shortage of skilled workers; the number of trainees in dental technology is continuously decreasing [3] even as the demand for dental prostheses remains high due to changing demographics [4,5]. In addition, patients are increasingly subject to time constraints created by rising expectations in the workplace, limiting their ability to undergo dental procedures. The digital transformation can help us meet these challenges, as digital processes are often characterized by their efficiency. Digital processes in the dental laboratory provide for greater accuracy and reproducibility (precision) as well as improved material properties and user comfort.

The interesting combination of a digital working environment and an analog craft makes dental technology attractive to young people looking for a varied and diverse work experience. Many dental laboratories are already managing the balancing act between craftsmanship and the digital world, tradition and disruption, and existing values and necessary changes. 3D printing as a digital manufacturing process is an important aspect of this development. In simplified terms, the process can be described as follows: The dental technician creates a digital data set on the computer (computer-aided design, CAD) and then designs a three-dimensional object whose data are transferred to the 3D printer, where it is converted into a physical object.

A major advantage of all additive processes is that three-dimensional objects can be designed and realized on screen to allow for an almost unlimited variety of shapes and levels of complexity. One aspect that has received little attention is that the mechanical and esthetic properties of the object to be printed can still be influenced during the 3D building process. This is not possible with subtractive manufacturing, where the material properties are defined by the manufacturer of the prefabricated blank. This customization option and the fact that digitally designed objects are available more quickly and easily, or even at lower cost, makes additive manufacturing a cornerstone of digital dentistry (Dentistry 4.0) [6–11].

2.2. History of 3D Printing

The first industrial-level units for additive manufacturing (commonly termed 3D printing) appeared on the market in the early 1980s. Pioneers of 3D printing include Charles W. Hull (founder of 3D Systems), S. Scott Crump (founder of Stratasys), and Hans J. Langer and Hans Steinbichler (founders of EOS). The first 3D printer was patented by Charles W. Hull in 1986 [12]. At the time, 3D printers were mainly used for rapid prototyping.

However, the technology advanced rapidly in the ensuing years. Following the expiration of the patent for the fused deposition modeling (FDM) process [13] in 2009, the 3D printers began to make enormous inroads into the consumer sector. This dynamic was ultimately carried over to the dental sector. Printing units became smaller and cheaper, and their fields of application changed. The range of printable materials expanded to include plastics, metal, ceramics, and even human tissue. Rapid-prototyping processes can be categorized by the type of materials used (plastics, metals, or powder).

2.3. Nomenclature and Classification of Additive CAD/CAM-Based Manufacturing

In additive manufacturing (AM) processes, objects are produced layer by layer on the basis of three-dimensional models. The term used in common parlance as a synonym for all additive processes is *3D printing* [14].

According to the EN ISO/ASTM 52,900 terminology standard, an AM process is the "process of joining materials to make objects from 3D model data, usually layer by layer, as opposed to subtractive manufacturing methods" [15].

EN ISO 17296-2 describes the process fundamentals of additive manufacturing. It also provides an overview of the existing process categories, although such an overview can never be comprehensive, given the dynamic development of innovative technologies (Figure 1).

The following seven process categories can be distinguished within additive manufacturing [16]:

- Vat photopolymerization (**VPP**)
- Material extrusion (**MEX**)
- Material jetting (**MJT**)
- Binder jetting (**BJT**)
- Powder-bed fusion (**PBF**)
- Directed energy deposition (**DED**)
- Sheet lamination (**SHL**)

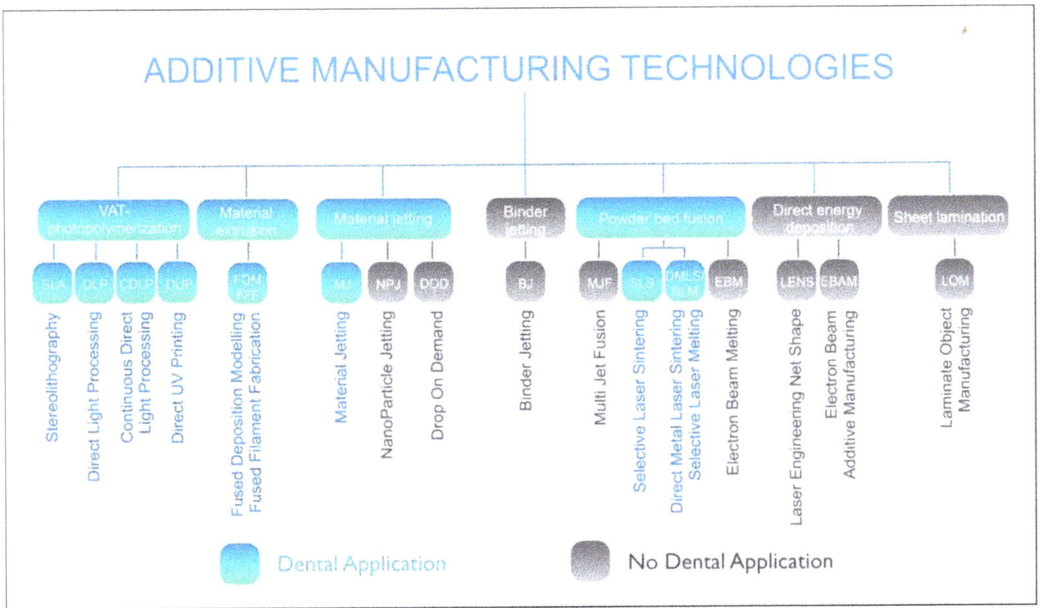

Figure 1. Overview of the existing process categories in additive manufacturing. (According to EN ISO 17296-2).

3. The Status Quo of Dental 3D Printing

Additive manufacturing has raised high expectations. Its market potential is thought to be considerable. The Gartner Hype Cycle [17], providing a powerful snapshot of current trends, reviews public attention to a specific technology (such as 3D printing) in the context of its development over time. The Hype Cycle is divided into five parts. For the innovation trigger, a potential technology breakthrough and media interest trigger significant publicity as commercial viability is unproven. At the peak of inflated expectations, the topic is hot, and unrealistic expectations are fueled by excessive enthusiasm. This is followed by the trough of disillusionment and the slope of enlightenment, in which public interest has decreased, but the technology is being improved. At the plateau of productivity, the technology is recognized and maturing.

Dental 3D printing follows this hype cycle (Figure 2). The Gartner analysis predicted in 2014 that 3D printing would take about 10 to 15 years to full adoption. This could be roughly true for the dental sector, if probably not as much as marketing claims would suggest. Neutral institutions should be tasked with attenuating inflated forecasts and supporting the continuous establishment of the technology as a function of the state of research and development, and to modulate expectations. Yet, the potential is, in fact, immense. Dental technicians and dentists should familiarize themselves with 3D printing technology and objectively assess possible areas of application.

Common Processes in Dental 3D Printing

The technology is not entirely new. Additive manufacturing has been established in the dental sector for almost 20 years, represented, for example, by the laser sintering (selective laser melting, SLM) processes of Bego Medical (Bremen, Germany) and EOS (Krailing, Germany). When presented for the first time in November 2002, this technology for printing metals caused a sensation. Experts recognized the enormous potential of this technology. Moreover, SLM enjoys worldwide acceptance as the basis for manufacturing metallic structures (such as crowns, bridges, or clasp-retained cast-metal frameworks). Stereolithography (SL), too, has been used in the dental industry for many years, for example, in the production of surgical templates (drilling guides). Stereolithography is based on point-by-point solidification within a resin vat (epoxy resins, acrylates) by means of a laser beam or with the aid of blue-light LEDs (digital light processing, DLP). Until a few years ago, 3D printers for dental applications were the preserve of industry or large manufacturing centers, given the considerable capital outlays required, but for some time now, many printers have come down to within reach of "regular" dental laboratories. Moreover, industry outsiders are entering the dental market and offering additive manufacturing technologies. Using comparatively inexpensive equipment, dental laboratories can now realize objects made of acrylics or composite resins to be used in the preparatory stages of a workflow, such as jaw models or surgical templates.

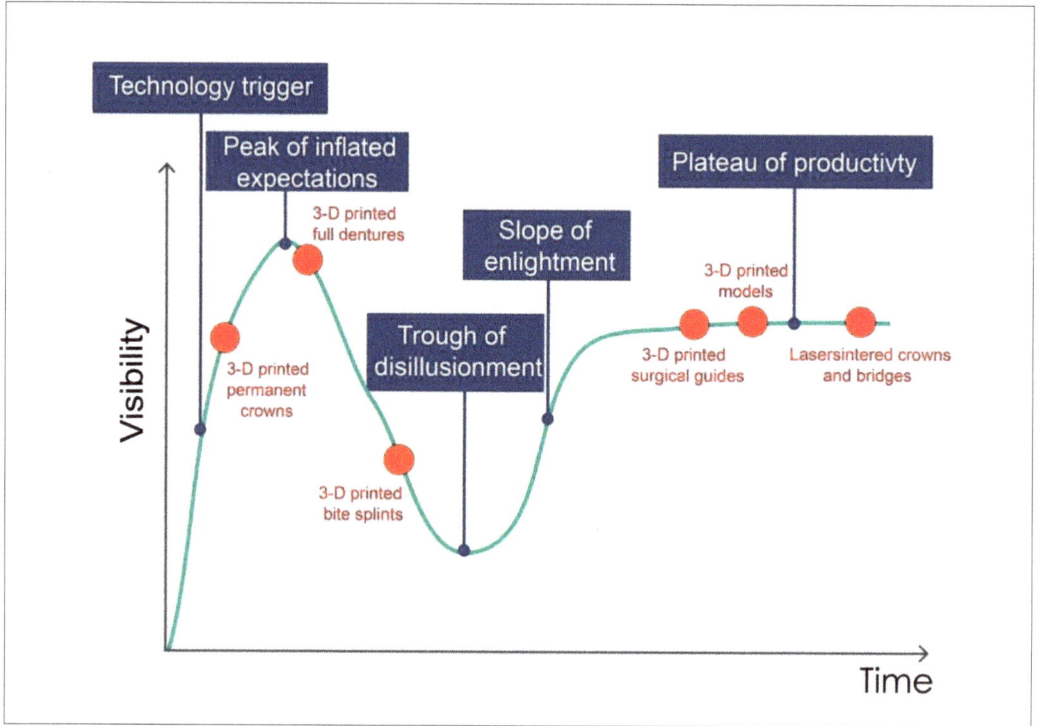

Figure 2. Dental 3D printing follows the characteristics of the Gartner hype cycle.

4. Dental Indications and Applications of 3D Printing

Not all additive technologies are suitable for use in the dental laboratory or practice. The following sections will discuss indications and applications for 3D printing that are sensible and economical to use in dental technology or else have great future potential. These will be differentiated not on the basis of technologies but on the basis of the materials used, i.e., metals, plastics, and ceramics.

4.1. Additive Manufacturing and Metals

Additive manufacturing using metal alloys has been successfully used in the dental sector since 2002. The use of laser sintering in the dental field represented a revolution in the processing of non-precious alloys at the time [18].

4.1.1. Laser Sintering of Crowns and Bridges Made from Non-Precious Alloys

Laser sintering has now become a standard process for the production of CoCr crowns and bridges [19]. By optimizing post-processing after the actual building process, it is now possible to manufacture absolutely stress-free and accurately fitting non-precious alloy frameworks even for larger bridge spans. The large number of units that can be positioned on a single platform has reduced the production time per unit to a few minutes (Figure 3). The procedure is extremely cost-effective and is well established when it comes to fixed restorations made of non-precious alloys.

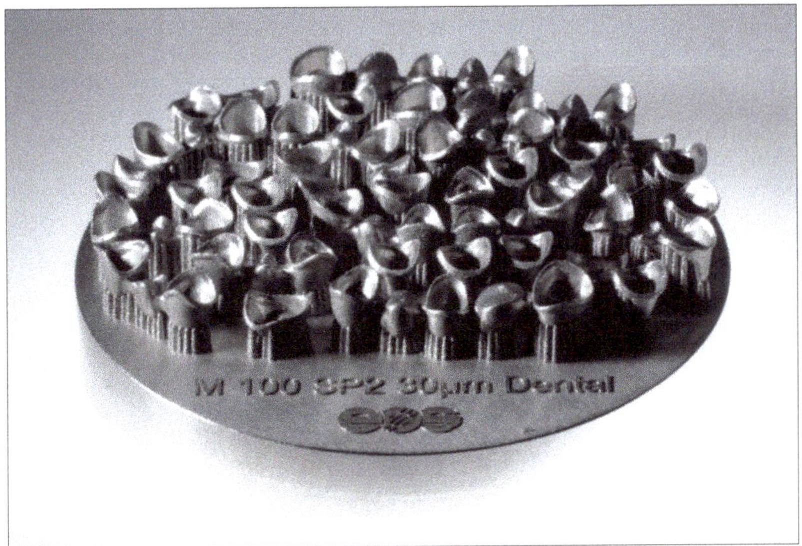

Figure 3. Lasersintered CoCr Crowns and Bridges.

To generate stress-free restorative frameworks, the build platforms are summarily subjected to a thermal post-treatment in a downstream processing step before the individual restorations are separated from the build platform. Most production centers automate this step. The support structures are then removed by manual finishing.

The physical and mechanical properties of laser-sintered non-precious alloy crown and bridge frameworks are comparable to cast restorations [20,21]. The rougher surface compared to cast or milled restorations actually has a positive effect on the cementation of laser-sintered crowns and bridges. Inside the crown and at the crown margins, laser-sintered restorations exhibit small but macroscopically visible ledges parallel to the z-axis of the building process. Nevertheless, the fit of laser-sintered crowns is within the clinically acceptable range [22]. Other studies have found that laser-sintered CoCr-alloy crowns have even a better marginal fit than casted CoCr-alloy crowns [23–25]. Ceramic veneers are very easily applied to laser-sintered frameworks as their rougher surface makes them highly wettable by the opaquer.

4.1.2. Laser Sintering of Clasp-Retained Cast-Metal Frameworks

Clasps are one of the oldest forms of denture retention [26]. Clasp-retained dentures, also referred to as one-piece cast dentures, are a simple form of restoration and allow a wide range of variations, making them universally applicable [27]. For more than 100 years, clasps have been a proven means of retaining removable dentures in the presence of withdrawing forces—for example, when speaking or chewing—and of distributing occlusal forces as evenly as possible to the residual teeth and soft tissue. In 1930, Dr. F. E. Roach wrote in the Journal of the American Dental Association [28]: "The clasp is the oldest and still is and probably will continue to be the most practical and popular means of anchoring partial dentures."

The introduction of digital techniques for the production of dentures, such as computer-aided design/computer-aided manufacturing (CAD/CAM) and additive manufacturing techniques, allows one-piece prostheses to be planned digitally and manufactured subtractively using CNC milling units, or additively, using 3D printing [29]. Here, we can distinguish between indirect and direct fabrication methods. In the indirect method, the frameworks are printed in wax or plastics and then produced by casting using the lost-wax technique. In the direct method, the CAD data set is directly converted into a Co-Cr alloy object by laser sintering [30–32] (Figure 4). This method is currently still in the prototype stage. Recent publications have claimed advantages for laser sintering in digital manufacturing in terms of standardization, reduced production times, and easy transfer of digital data [33]. However, its economic viability is still critically assessed [34]. Additional research is required before this method can be definitely recommended. Particular attention must be paid to the retaining elements (clasps), as these are permanently exposed to high mechanical loads as they serve in their retaining and supporting function.

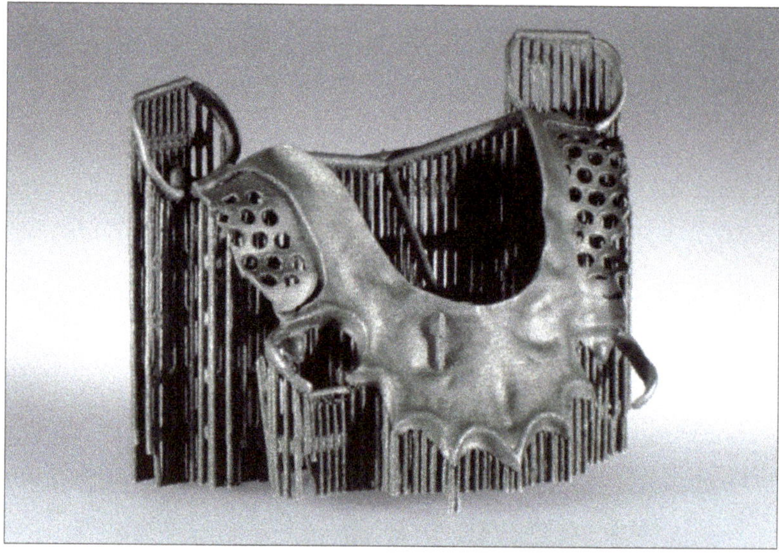

Figure 4. Laser sintered removable partial denture with support structures.

An in-vitro study by the authors, conducted at the Department for Dental Prosthetics University Hospital, Ludwig-Maximilians University Munich, examined the mechanical quality of cast versus laser-sintered clasps for cast-metal frameworks. The results of the study are very promising and show the high mechanical potential of laser-sintered clasps. The following key statements can be made on the basis of this study [35]:

- The required initial clasp withdrawal forces were attained by the cast and laser-sintered clasps alike. After artificial aging, the laser-sintered clasps exhibited no decrease in retention force.
- Pores and flaws were smaller and more evenly distributed overall in the laser-sintered clasps compared to the cast clasps.
- Laser-sintered clasps performed significantly better in the long term than cast clasps, with more than twice the latter's survival rates. One reason could be the superior structural quality of the laser-sintered clasps.

4.1.3. Hybrid Manufacturing

In digital dental technology, hybrid production is the term for a combination of additive and subtractive steps with a view to combining the efficiency of additive manufacturing with the precision of CNC milling [36,37]. Objects made using hybrid processes are characterized by improved surface structures, greater accuracy of fit, and lower cost (Figure 5). The company Datron (Mühltal, Germany) has been working on the implementation of dental manufacturing using hybrid technology for more than 8 years. A collaboration project by Datron, Concept Laser (Lichtenfels, Germany), and the Follow Me Technology Group (Munich, Germany) is working on mapping the hybrid workflow to standard milling machines through smart networking. An essential part of hybrid manufacturing is the transfer of the zero point (origo) from the additive process to the CNC milling unit. For this purpose, three measuring pins are built on the build platform during sintering. These pins are detected by the Datron D5 milling unit by means of an infrared touch probe developed especially for the hybrid manufacturing process, allowing the unit to determine the exact positions of the laser-sintered objects. The correction values are calculated directly by the unit, thus that no new CAM calculations are required. Since the objects remain firmly attached to the platform for post-processing (no pick-up via the grid structure), maximum positioning accuracy and precision are ensured. For implant superstructures, the screw hole is machined from the basal side via the screw access canal using special form cutters. Manufacturing costs can be reduced in the range of 30% to 50%, depending on the production volume.

Figure 5. Hybrid manufacturing combines additive manufacturing. With CNC-milling (Source: Datron AG, Mühltal, Germany).

4.2. Additive Manufacturing and Polymers

Several 3D printing technologies exist for the additive manufacturing of plastic objects [38–40] and exhibit, when compared to each other, different characteristics regarding speed, resolution, size, and process reliability depending on the underlying technology (Table 1). Currently, stereolithographic processes predominate in the dental sector, including classic stereolithography using a laser source (stereolithography, SLA) and the so-called mask exposure processes (digital light processing, DLP). In both processes, the object is solidified by the action of light in a vat of photopolymer.

Table 1. Characteristics of 3D printing technologies for plastics used in the dental sector.

	Filament-Based 3D Printing	Light-Based 3D Printing			Material Jetting
	FDM/FFF	**SLA**	**DUP**	**DLP**	**MJT**
Speed	medium	medium	medium	high	high
Resolution	low	high	medium	high	high
Size	scalable	scalable	scalable	scalable	scalable
Process reliability	medium	medium	low	high	high
Cost	low	medium	low	medium to high	high

For about 3 years now, 3D printers have been available that use low-cost liquid crystal displays (LCD). The technology is called direct ultraviolet printing (DUP); it uses the LCD displays for pixel-by-pixel exposure of the build platform. UV LEDs with a wavelength range of 395 to 405 nm are usually used for background lighting.

Direct 3D printing processes (material jetting, MJT) are also used in dental applications. A special process worth mentioning is multi-material 3D printing by Stratasys, which allows different colors and materials with different properties to be processed simultaneously in a single build. Material extrusion (MEX) processes such as fused-filament fabrication (FFF) or fused deposition modeling (FDM) are currently of lesser relevance on the dental market since they require long printing times and are restricted to lower resolutions. Of the technologies mentioned for the plastics sector, SLA, DLP, and MJT appear to be the most interesting from a technical and economic point of view [41–46].

4.2.1. Stereolithography Using a Laser Source (SLA)

Stereolithographic systems, which use laser beams to solidify liquids, were the first 3D printing systems to appear on the market. Charles Hull had applied for a patent for the first stereolithography printer as early as in the 1980s. The first devices were very extensive—and expensive. The latest generation of stereolithographic printers, by contrast, has become quite economical. Formlabs (Sommerville, MA, USA) has been offering a 3D printer for dental applications for about five years now. This very affordable system is an ideal entry-level system for 3D printing technology, even if building takes much longer than with to DLP printers.

4.2.2. Digital Light Processing (DLP)

Along with stereolithography, digital light processing is probably one of the most popular additive manufacturing processes in the dental sector right now. The design of a DLP printer is similar to that of an SLA printer, the main difference being the light source used. In the SLA printer, the photopolymer is cured with the help of a laser beam. DLP printers use projection technology from Texas Instruments instead, where short-wave light (currently used wavelengths: 380 nm and 405 nm) is guided through a digital micromirror device (DMD) that constitutes the core of the DLP technology. The system uses controlled square micromirrors with an edge length of approximately 16 μm. The light is guided optically either onto the build platform, which resides in a translucent vat of photopolymer (photopolymer bath) or onto a diffuse surface (absorber). This is made possible by tilting the individual micromirrors in the unit, which are triggered by forces exerted by electrostatic fields [47,48]. The exposure mask is projected onto the build platform through an optical lens, causing the photopolymer to cure at the exposed areas. After each exposed mask, the build platform moves along the z-axis, and new material flows into the space beneath the object and can be exposed with the next mask. When using DLP technology, the building time is, therefore, almost independent of the objects produced, the decisive factor being the dimension of the object along the z-axis.

Resolution of DLP Printers

One micromirror corresponds to one image point (pixel). Since a DMD has a limited number of these micromirrors, when the build platform is increased in size, edge lengths along the x and y axes also increase, resulting in lower precision. There are currently three ways to, nevertheless, realize larger build platforms, although the first one will not find its way into standard lab printers for the time being on account of the cost involved:

- Using a DMD chip with higher resolution (e.g., 4K resolution)

Less expensive DLP printers use DMD chips with lower resolution (e.g., 1280 × 720 pixels) and a correspondingly smaller footprint. If DMD chips with high resolution (e.g., HD 1920 × 1080 pixels) are used, greater object accuracy can be achieved with the same footprint. When using 4K DMD chips (3840 × 2160 pixels), it is possible to achieve high resolutions while maintaining a large build area (e.g., Rapid Shape D70+; Rapid Shape, Heimsheim, Germany) [49]. However, the prices for 4K DMD chips are still very high.

- Two DLP projectors with HD resolution connected in parallel

This approach creates a "joint" on the build platform caused by the use of two light sources. As a result, no objects can be printed that are positioned across the projection field. Example: Rapid Shape D40 II (Rapid Shape) [50]

- Moving DLP projectors (W2P Engineering, Vienna, Austria)

DLP projectors whose optical subsystem moves below the material vat are able to expose a larger area [51]. One advantage of the Moving DLP is that the object will feature no joint line and that, consequently, the entire extent of the build platform can be used at full resolution. This makes for a higher resolution, greater printing accuracy, and better utilization of the capacity of the device.

- Prodways MovingLight technology (Prodways Group, Paris, France)

The MovingLight technology was developed and patented by the French company Prodways [52]. This AM technology is based on the DLP process. It differs from its competitors' approaches in that the projector is not rigidly fixed in one location within the printer but moves around across the complete working area in several steps, achieving high resolutions (42 μm) and high accuracy despite the extensive build platform [53]. Examples include Prodways' ProMaker LD10 Dental Plus, LD10 Dental Models, LD20 Dental Plus, and LD20 Dental Models. The latter two have two movable projector heads, reducing build times by another 40%. For example, it takes about 1 hour to print 55 dental arches.

DLP Printer Build Process Optimization

The DLP printers also use various techniques to detach objects from the material vat during the build process. This detachment occurs after each exposure cycle when the build platform is lifted along the z-axis. Four different techniques are applied:

- Fixed intervals

The build platform covers a defined path in a defined time after the exposure cycle. The path/time ratio here remains the same within a build process, even if the object could be removed sooner in the process (e.g., when using fewer support structures). The fixed-interval principle is very simple, but the duration of the building processes is not altered.

- Force Feedback technology (Rapid Shape, Heimsheim, Germany)

The force needed for detachment can be measured via force sensors. The smart control technology is then used to calculate an optimum path/time ratio, which speeds up the building process [54]. A particular advantage is that the separation process is controlled and gentle. The patented Force Feedback technology is used, for example, by the Rapid Shape D30.

- Vat deflection feedback system (VDFS; W2P, Vienna, Austria)

The patented vat deflection feedback system uses an additional sensor to speed up the building process. In addition, the material tray can be deformed (FlexVat), allowing the detachment force to be minimized and resulting in increased printing speed and quality [55,56].

- Continuous direct light processing (CDLP; Carbon3D, Redwood City, CA, USA)

In 2015, Carbon3D first released information on its patented continuous liquid interface production (CLIP) technology, which is classified as a CDLP process [57,58]. Unlike the incremental build-up of objects in DLP printers, the CLIP process involves a continuous build process without the steps normally required to detach objects from the build platform in DLP printing. This process is enabled by the fact that there is an oxygen-rich zone ("dead zone") immediately above the build platform where no curing of the photopolymer takes place. Oxygen is conducted into the "dead zone" through a window that is permeable to oxygen. Since there is no adhesion of the object to the build platform, a continuous build process is possible. The result is extremely high build speeds with high object precision and continuous object geometries along the z-axis. Examples of dental applications include additively manufactured Lucitone Digital Print denture bases from DentsplySirona (York, PA, USA) or bite splints made of KeyPrint or KeySplint Soft Clear, both additively manufactured using a Carbon3D printer.

4.2.3. Material Jetting (MJT)

In material jetting, the material is applied directly to the build platform via the print head (similar to the 2D printing process) and then cured in an intermediate exposure step, building up the object layer by layer. The best-known representative of this technology is the Polyjet method (Stratasys, Eden Prairie, MN, USA), characterized by an extremely fast build process and high precision [41–43]. A special feature is multi-material 3D printing, where five different grades of materials can be printed in more than 500,000 colors [59–61]. The Stratasys product portfolio includes, for example, the J720 Dental or J750 Digital Anatomy printers that operate in multi-material multicolor mode.

4.2.4. Useful Indications for AM of Polymers

- Model fabrication based on intraoral scan data

Due to the high efficiency of DLP printers in combination with high precision, the digital fabrication of master models and segmented models is one of the primary domains of DLP printers [62]. In particular, the additive manufacturing of models for oral implantology would appear to be an interesting application for these systems (Figure 6). Precise positioning of the laboratory analogs in the printed model is crucial, as it has a decisive influence on the proximal and occlusal fit of the restorations.

- Templates (drilling stents) for guided implant surgery

Software developments in recent years have made it possible to overlay (match) volume data sets from radiology (DICOM) with surface data sets (STL) from the laboratory or from intraoral scanners. This allows optimizing the implant position, taking into account anatomical, surgical, and prosthetic aspects. The planned positions are then realized with the help of a surgical template inserted into the patient's mouth. DLP printing technology offers particular advantages here, as it allows very quick production at low costs (Figure 7). Unlike subtractive methods, there are no restrictions on the design of the three-dimensional geometry [63].

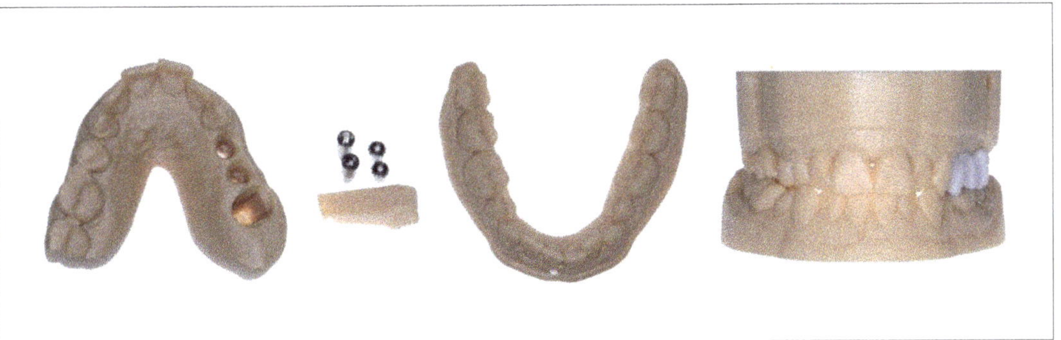

Figure 6. Additive manufactured models for implantology.

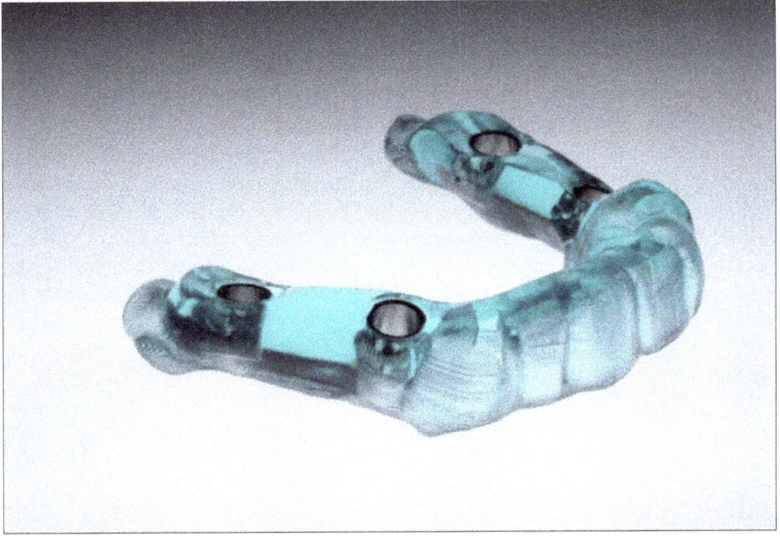

Figure 7. 3D printed surgical guide with drilling sleeves.

- Custom impression trays

The production of custom impression trays is made particularly enticing by DLP printing technology due to the sped of this technology. The CAD software solutions available on the market allow custom impression trays to be designed with optimum fit parameters in just a few steps, saving considerable time, especially when undercuts are blocked out virtually and can be dimensioned more precisely. It is important to avoid irreversible deformation of the impression during removal [64]. Despite their technical advantages, it should be pointed out that the materials currently intended for the fabrication of functional impression trays are expensive, making them viable only for use in implant impression trays (Figure 8). It would appear advisable to combine them with digital implant planning, as digital models will already be available in this context, and the position of the planned implants can be used as a basis for tray fabrication.

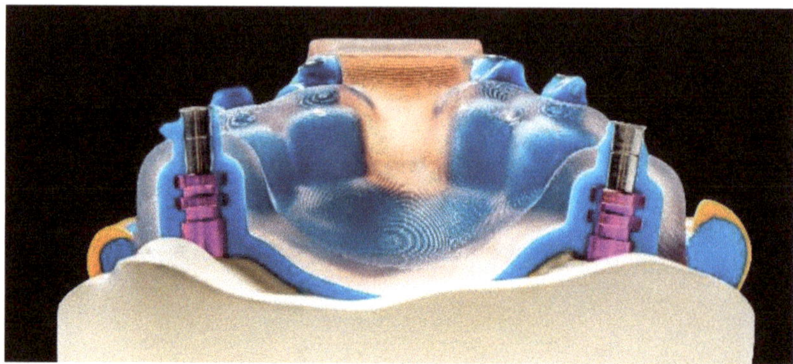

Figure 8. Cross-section through an implant impression tray on the implant model (Source: Shera Werkstofftechnologie, Lemförde, Germany).

- Production of occlusal splints

In addition to production using the scatter-and-press method or subtractive milling, it is also possible to produce precision-fit occlusal splints by 3D printing. However, in addition to the overall production accuracy, the quality of the material and the associated long-term stability and biocompatibility are determining factors. No long-term clinical experience with additively manufactured occlusal splints has as yet been reported. At the same time, it is necessary to investigate the elution behavior of additively manufactured occlusal splints under laboratory and oral conditions [65]. Comparisons with current procedures would be desirable to decide which manufacturing process yields the best long-term results. The bar has generally been set very high for homogeneity and biocompatibility as achieved by high-performance polymers machined in the subtractive CAD/CAM process (e.g., milled splints). Factors such as the positioning and alignment of the objects and their influence on accuracy, stability, and durability must also be investigated. The working angle on the build platform and, hence, the direction of the layers seem to be of particular importance here (Figure 9). Initial studies have shown that 3D-printed occlusal splints are similarly accurate as CAD/CAM-milled splints but exhibit higher material wear and less favorable material properties [66–68].

- Production of realistic training models

Realistic patient models for training and continuing education courses have been developed by the Department for Dental Prosthetics of the University of Munich. These models can be fixed on standard phantom heads (Figure 10). Their design is based on scanned models, with connection geometries (threading, anti-rotational features) added in the CAD software. In order to save on weight and material, the models are hollow on the inside and possess a reinforcing grid. After adding the support structures and subsequent slicing, the models were printed on the SheraPrint D30. The SheraPrint-model was used as the material for the models, a material that is also excellently suited for the preparation of various restorative shapes with irrigated rotary instruments. These models can be used to easily simulate the cementation of various restorations in the phantom head [61].

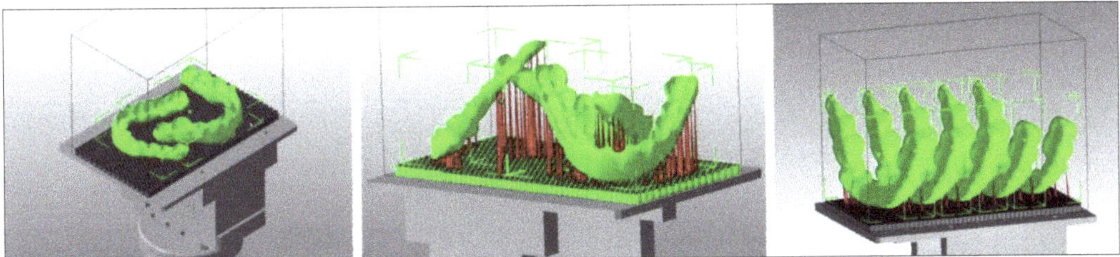

Figure 9. Different orientations for occlusal splints on the build platform.

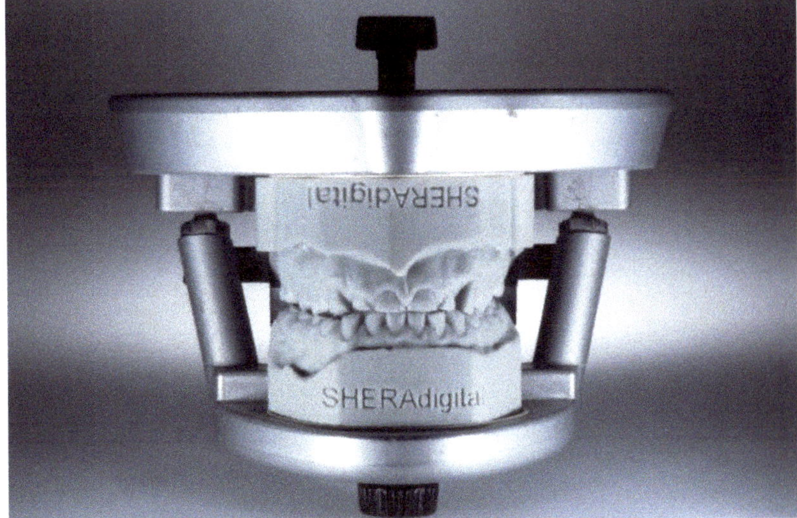

Figure 10. 3D printed realistic training models in a standard phantom head.

The next stage of development for the production of training models are multi-layer models. This multi-layering can refer to tooth structures as well as to layered structures of the entire jaw. Such models are extremely versatile to use; they may cover nature-identical simulation teeth for endodontic exercises [69] (Figure 11) to multi-layered models of the complete jaw for surgical simulations and trainings [70] (Figure 12).

- Production of graphic 3D models (3D Medical Print, Lenzing, Austria)

Several intraoral 3D scanners now allow the digital capture of shade information in addition to surface data. Available file formats include PLY, OBJ, and VRML. Using Polyjet technology, it is possible to convert these data into physical models. The pertinent shade information is geometry-related, i.e., the two-dimensional shade information is uniquely assigned to 3D surfaces. Model builder software is used to generate a virtual shade model, which is then converted into a physical shade model using multi-material 3D printing (Polyjet technology; Stratasys, Rheinmünster, Germany) (Figure 13). Since the transfer of shade information is not possible with analog impressions, graphic 3D models are a veritable "killer application." Data generation and model production mandate the use of a digital workflow. In the future, new possibilities will emerge here that will be associated with enormous improvements and simplified procedures, especially for highly esthetic dental restorations [2,71].

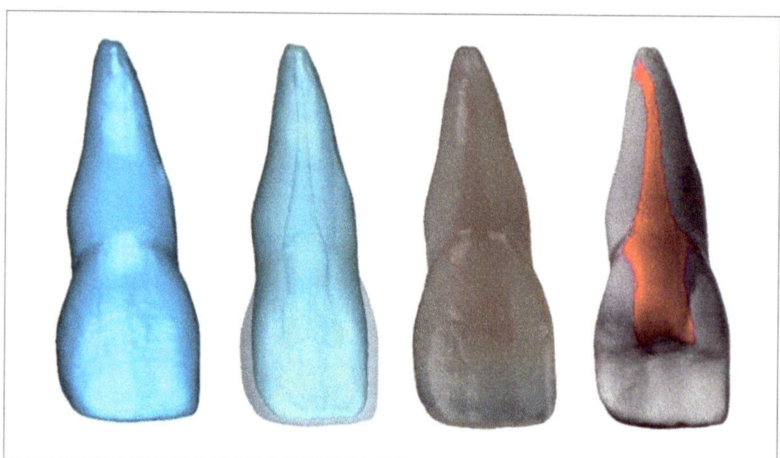

Figure 11. Nature-identical simulation teeth.

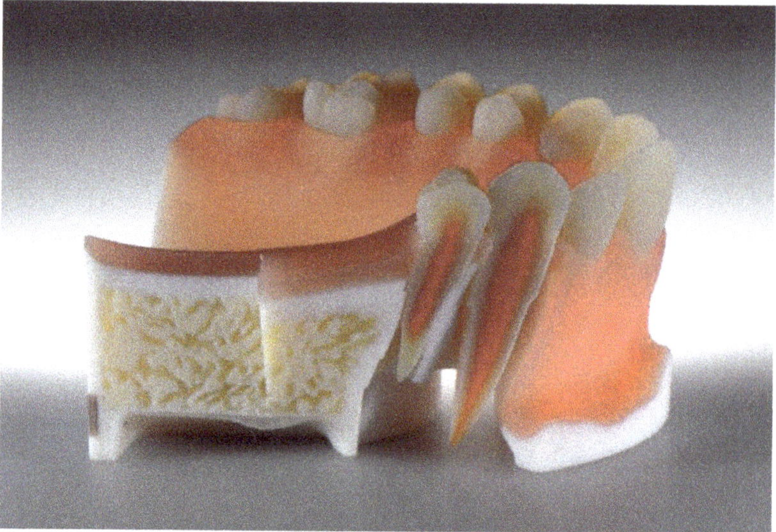

Figure 12. Multi-layered model of the complete jaw for surgical simulations and trainings.

Another possible application of multi-material 3D printing in the dental field could be the fabrication of multi-layered dentures made from different materials. With regard to the identical reproduction of natural teeth by crowns or bridges, this technology is currently in the prototype phase. It is based on the tooth-structure database, according to Schweiger [72–74], which allows the multilayer structure of natural teeth to be copied and the data thus generated to be used in an additive manufacturing process (Figure 14). The ultimate aim is to produce biomimetic dental restorations that reflect the multi-layered three-dimensionality as well as the complex mechanical and optical properties of natural teeth. Taking into account the light-optical properties of the different tooth layers (pulp, dentin, enamel), an identical esthetic reproduction of natural teeth can be achieved.

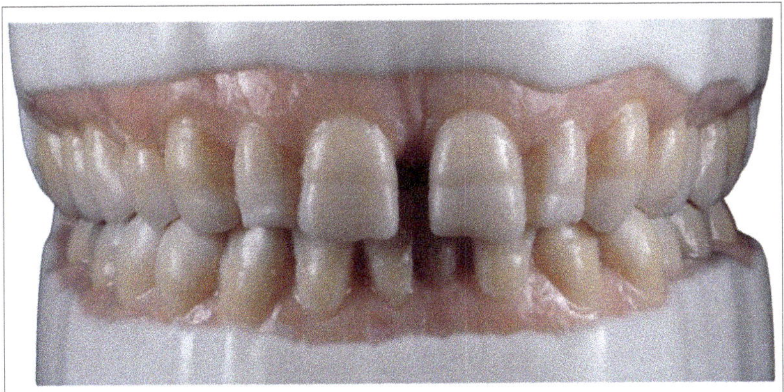

Figure 13. 3D printed graphic 3D model based on 3D data from an intraoral 3D scanners.

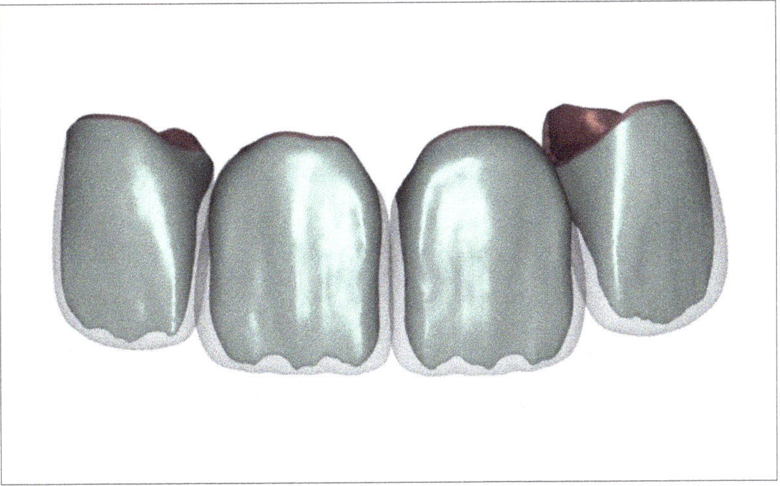

Figure 14. CAD–construction of 4 upper incisal crowns using the tooth-structure database.

Current research at the Department for Dental Prosthetics at the University of Munich uses data from the tooth-structure database in a Stratasys Polyjet process to implement the concept. At present, the process allows the fabrication of esthetic try-in crowns or bridges from light-polymerizing resins (Figure 15). The materials used are approved for use in the mouth for up to 24 h, permitting the evaluation of functional and not least esthetic criteria.

Since the layering process includes no analogous steps, the result is not influenced by manual imponderables. The composition of the printing materials in combination with the three-dimensional multi-layer structure of the denture are the only determinants of the structural and esthetic results. Fine-tuning the material composition in the multi-material 3D printing process is likely to permit fine-tuning of optical properties in the future. For example, various mixtures for the enamel compound are currently being tested in in-vitro studies to replicate the light transmission behavior of natural tooth enamel as closely as possible. Likewise, the shade and translucency of different dentin qualities can be adjusted by mixing. The layered 3D design of the restorations is reproducible thanks to the digital design process. After the try-in, the layered structure can be transferred to the final ceramic restoration, for example, by way of subtractive manufacturing [61].

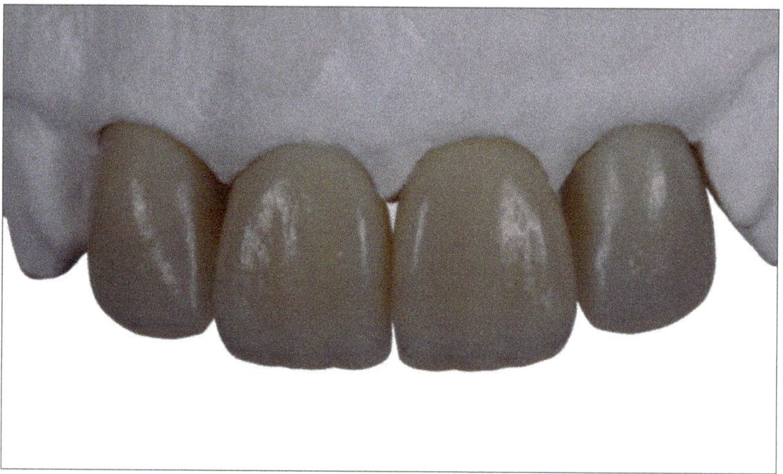

Figure 15. 3D printed multi-layered upper incisal crowns.

- VarseoSmile Crown plus–3D printing of permanent single-tooth restorations (Bego, Bremen, Germany)

The question of whether definitive dental restorations can be produced by a 3D printer can be answered by looking at the findings of materials science regarding 3D printing materials. The requirements of materials for dental prostheses permanently installed in the mouth are by necessity high. Definitive restorations require the use of materials that can withstand both high mechanical stress and the various chemical processes present in the oral cavity. No harmful substances must be released during the wearing period, and the materials must have a smooth surface to forestall bacterial deposits (plaque). In addition, a practical and economical manufacturing process must be available that can ensure precision in the micrometer range. Since February 2020, Bego has been offering the world's first method for manufacturing single-tooth restorations using 3D printing and a ceramically reinforced hybrid material. The Bego VarseoSmile Crown ^{plus} can be used to produce single-tooth crowns, inlays, onlays, and veneers using an additive process. The material has been extensively studied in scientific testing and has yielded excellent results. In particular, its fracture load (at baseline and after artificial aging), abrasion resistance, long-term stability of the cementing agent, solubility, and cytotoxicity were investigated [75]. Production in the Bego Varseo XS, which is a low-cost, high-resolution DLP 3D printer with excellent detail resolution, appears particularly interesting. Up to 20 individual restorations can be printed simultaneously on the build platform. The printer is network-compatible, facilitating fast and uncomplicated data exchange with CAD PCs. After the printing process, cleaning is carried out using ethanol and air-abrading using gloss beads (e.g., Perlablast micro; Bego). The restorations are then post-polymerized in the Bego Otoflash light-curing unit. As the surface of the printed restorations is smooth and homogeneous, the finishing step can be limited to smoothing the surface and subsequent polishing. Alternatively, the polymerized restorations can be customized using commercially available composite-resin stains (Figure 16). Bego VarseoSmile Crown plus restorations are cemented with self-adhesive luting materials (e.g., RelyX Unicem; 3M, Seefeld, Germany) or with luting composites with a separate primer (e.g., Variolink Esthetic DC and Monobond Plus; Ivoclar Vivadent, Schaan, Liechtenstein). The VarseoSmile Crown plus hybrid material is available in 7 shades (A1, A2, A3, B1, B3, C2, D3).

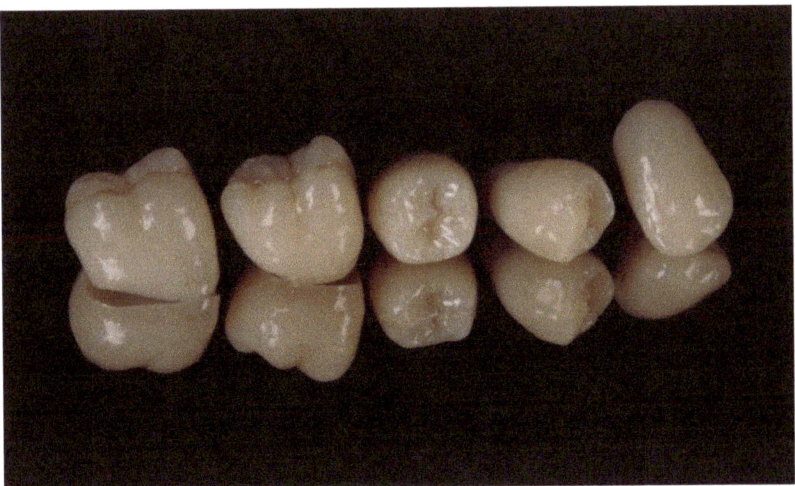

Figure 16. Single-tooth crowns printed with VarseoSmile Crown ^plus (Bego).

4.3. Additive Manufacturing and Ceramics

A number of different build-up methods now allow even ceramic materials to be processed using either indirect or direct techniques.

Indirect technique

- Trix print process by Dekema (Freilassing, Germany)
- IPS e.max Digital Press Design–Wax Tree by Ivoclar Vivadent (Schaan, Liechtenstein)

Direct technique

- SLA process, e.g., 3DCeram (Limoges, France)
- DLP process, e.g., LCM (lithography-based ceramic manufacturing, LCM) by Lithoz (Vienna, Austria)
- Material extrusion (fused-filament fabrication, FFF; paste-extrusion modeling, PEM)
- Material jetting/nanoparticle jetting, e.g., XJET (Rehovot, Israel)
- Binder jetting, e.g., 3D Systems (Rock Hill, SC, USA)
- SLS process (research project at the Department for Dental Prosthetics of the University of Munich, the Friedrich Baur Institute for Biomaterials at Bayreuth, Germany, and Concept Laser at Lichtenfels, Germany)
- LOM process (laminated object layering)

4.3.1. Indirect 3D Printing of Ceramics Example: Dekema Trix Print

Dekema (Freilassing, Germany) uses a novel approach to pressable ceramics with its innovative Trix system. It combines the advantages of digital design with the unbeatable efficiency of proven ceramic pressing technology. The system maps the entire pressing workflow digitally, from wax-up to the pressing itself. The individual steps are explained below using the example of several partial crown restorations.

Scanning and CAD design

The oral situation can be digitally acquired directly, by means of an intraoral scanner, or indirectly, by scanning a master cast after taking an analog impression. The digital pressing technology is suitable for both acquisition methods. The partial crowns can be efficiently using standard CAD software tools.

Automatic addition of sprues and placeholders for up to three pressing plungers

After selecting the objects to be pressed from the respective CAD system, Trix CAD automatically designs the complete wax-up, including the placeholders for up to three pressing plungers, in order to press up to three pressing pellets (which can be of different shades) in one process. Trix CAM determines the required layer pattern and sends it to the Dekema Trix print 3D printer.

3D printing using the Dekema Trix print 3D printer

The sliced layer data are printed on the build platform of the Trixpress muffle system. The associated Trix cast printable burnout material is also made by Dekema.

Investing and pressing

3D printing is followed by cleaning and curing the objects and investing them in the Trixpress muffle. After heating in the preheating furnace and residue-free calcination, the pressing ceramic is inserted into the muffle and usually pressed with the Trixpress punches (Figure 17). The project-specific pressing program has already been streamed from the Trix CAM to the Austromat 654i for this purpose. Alternatively, the data can also be transferred by way of a USB stick.

Figure 17. Comparison of 3D printed and pressed ceramic inlays.

Finishing and glazing

After pressing, the partial crowns are finalized following standard procedure; there is no difference between this workflow and the analog workflow. When working with a completely digital workflow, it is recommended to record the scan data of the dentition by means of a 3D-printed model thus that it is possible to check the fit along with the proximal and occlusal contacts. Staining and glaze firing then completes the fabrication of the partial crowns.

4.3.2. Direct 3D Printing of Ceramics Example: LCM Technology

No market-ready applications are yet available for direct 3D printing in the dental realm. The most advanced approach is probably the patented LCM process by Lithoz (Vienna, Austria) [76,77]. We will illustrate the current state of the art in dental zirconia 3D printing using the example of a mandibular molar crown. After scanning the jaws and CAD-designing the restoration, the fully contoured crown is fabricated using Lithoz' lithography-based ceramic manufacturing (LCM) technology. The LCM process is based on digital light processing (DLP). Here a photosensitive ceramic slurry is selectively cured, achieving a high filler content and a dense packing of the ceramic particles in the pre-sintered blank. This is necessary to produce defect-free and dense ceramic objects. The polymer network connects the ceramic particles. For dental applications, Lithoz has developed the CeraFab 7500 Dental 3D printer. The fully contoured posterior crown was made from LithaCon 3Y 230 (zirconia stabilized with 3 mol-% yttria-stabilized zirconia, 3Y-TZP). The printing process took approximately 7 h for 20 crowns, for a printing time of 21 min per crown.

After the additive manufacturing process, the crowns are available as "green bodies" that still contain the organic binder material, which must be removed in the next step—thermal debinding at 1000 °C over a period of several hours. This creates the so-called "white body," which no longer contains any binder and will already have formed solid sintering bridges that prevent the object from disintegrating. At this point, individual staining is performed using staining solutions, with three variants being available:

- Immersing the crown in the staining solution
- Custom painting of the crown using a brush and staining solution
- A combination of the two

The combination variant has shown itself to be the preferred variant. Here, a basic stain is achieved by immersion, followed by individual characterization using various intensive staining solutions, particularly at the crown margin and in the incisal/occlusal area. After staining, it is important that the crowns are dried before the final sintering step, ideally using infrared light. The sintering process is carried out at 1600 °C, at a heating rate of 8 °C/min and a holding time at the final temperature of 2 h. The cooling rate was also 8 °C/min down to 500 °C with subsequent ambient cooling to room temperature. The crowns are finalized with a stain firing and a glaze firing at 770 °C; IPS e.max Ceram Stains were used for this purpose in the case illustrated (Figure 18).

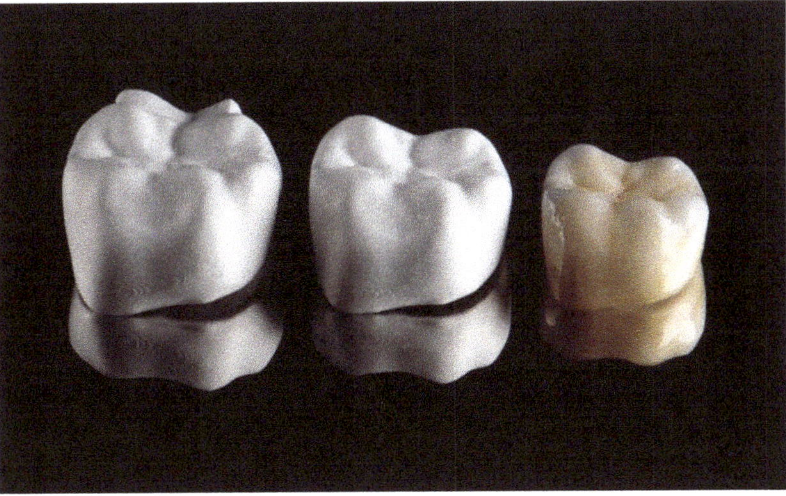

Figure 18. 3D printed Zirconia crown in the green and the white state and finally sintered.

Assessment of the final result

The crown was made from 3Y-TZP using Lithoz' LCM process. This classic zirconia was originally designed intended for the fabrication of crown or bridge frameworks, which were manually veneered using a ceramic material made of silicate ceramics. The translucency of the frameworks was, therefore, low. Nevertheless, the LCM process can achieve pleasing esthetics even with fully contoured crowns. The excellent reproduction of the sharp-edged crown margins and the exact reproduction of the occlusal surfaces with a well-defined and natural representation of the fissures were particularly striking. Since subtractive machining invariably requires crown margins to be reinforced and the occlusal fissures will always be rounded due to the finite diameter of the burs, additive manufacturing proves advantageous here.

4.3.3. Multi-Material 3D Printing of Ceramics

The most interesting development in the field of additive manufacturing using ceramics is multi-material 3D printing. The first prototypes were presented by the WZR company in 2014 [15], which combined two processes, namely binder jetting (BJT) and material jetting (MJT). Here, particle-filled inks are applied directly to the powder bed via the print head. If, for example, a different material is selected for the ink it is possible to alter the structural composition of the workpiece. Inks filled with metal particles can also be injected into a ceramic powder bed, thus that, for example, an object made of silicate ceramics can be built that integrates electrical conductor paths in silver.

The latest development in this field was presented by Lithoz in mid-2020. A specially developed LCM printer (CeraFab Multi 2M30) makes it possible to produce objects from different materials in a single printing process. It is possible not only to combine different ceramics but also to create ceramo-metal and ceramo-polymer objects. Material combinations currently include four variants [77]:

- Two materials in a single layer
- A denser material combined with a second porous material
- Two-phase or multi-phase materials with gradual variations in composition
- Gradual variations in both density and composition

These currently available options presage the enormous potential of this technology. It is likely that this will also affect additive manufacturing for dental applications.

5. Limitations of 3D-Printing

Basically, you can distinguish between 3D-printers for hobby use and for professional use. Practical application has shown that the relatively inexpensive printers for hobby users often show poor printing results, especially noticeable gradations in the FFF printers due to the filament fibers. For this reason, the devices available for dental use are mostly expensive but show good final results. However, even the printers for professional use generate a more or less pronounced gradation in the Z-direction. This is largely dependent on the thickness of the individual layers. The thinner the building layer, the lower the graduation, but also the longer the processing time. A further limitation is the maximum achievable build speed and the size of the build space. Newly developed technologies in the field of component detachment (see Section 4.2.2) in particular can solve the speed problem and lead to extremely high construction speeds. There are also limitations in the materials that can be used for 3D-printing. Especially in the field of polymers, printers based on photopolymers are predominantly used in dental technology, especially in the field of VAT polymerization (SLA, DLP, DUP). This greatly reduces the range of resins that can be used, resulting in significant disadvantages here compared to standard manufacturing processes (e.g., CNC technologies, analog manufacturing techniques). A possible solution to this problem could be the so-called "drop-on-demand" technology in which thermoplastics approved for medical technology are melted from a granulate and applied dropwise in a plastic state to the build platform. The achievable surface quality of this technique differs substantially from the results from filament printers. Further on, there is only little data regarding the behavior of 3D printed devices or restaurations in the oral cavity. Data on plaque formation, elution behavior, and general biocompatibility of 3D printed polymer materials are scarce [65,78], and further data on specific materials are urgently required.

6. Outlook

Additive processes have the significant advantage that an object´s properties can be individually influenced during the construction process. This applies to mechanical and esthetic properties alike. In subtractive processes, by contrast, these characteristics are predetermined by the manufactured milling blank. 3D-printing thus gives users an enormous range of choices as early as during the design process. On the other hand, the precision and efficiency of subtractive machining are extremely high, thus that the combination of both manufacturing techniques seems to make eminent sense.

In addition to the production of auxiliary systems (surgical guides, models, individual impression trays) and fixed dentures, there is a trend towards 3D-printing in the field of removable dentures. RPD's made of CoCr using additive technologies have already found their way into dental laboratories and practices. Currently, more and more publications on additive manufacturing of complete dentures are being published [79–82]. The results regarding mechanical strength, fit, and surface quality are promising. Since the denture bases have large area contact with the oral mucosa, biocompatibility must be critically examined. In particular, elution behavior and cytotoxicity must be investigated before a final assessment is made [65,78].

Finally, there are areas in which the classic analog processes are unbeatable in terms of economy, for example, ceramic pressing. However, here, too, integrating digital steps can be useful. With further advances in the additive manufacturing of ceramic restorations, innovative approaches to the production of natural-looking dental restorations will soon arise. Digital acquisition of three-dimensional tooth layering using NIRI technology—a likely future achievement—could be a foundation of this technique, together with tooth-structure databases [72–74]. Additive technologies such as the Lithoz LCM process are the ideal manufacturing routes to achieving this goal. Gradient technologies can be individually adapted to restoration geometries and offer unimagined design freedom in three-dimensional space, impossible to achieve with conventional technologies—all within the scope of patient-focused, individualized, and personalized dentistry.

Author Contributions: Conceptualization, J.S., D.E., J.-F.G.; writing—original draft preparation, J.S.; writing—review and editing, D.E., J.-F.G.; resources, J.S.; supervision, J.-F.G.; project administration, D.E.; visualization, J.S. All authors have read and agreed to the published version of the manuscript.

Funding: This research received no external funding.

Conflicts of Interest: The authors declare no conflict of interest.

References

1. Schweiger, J.; Edelhoff, D.; Güth, J.F. Update digitale Zahnheilkunde 2020—Aktuelle Möglichkeiten und Limitationen. *Bayerisches Zahnärzte Blatt* **2020**, *57*, 42–52.
2. Schweiger, J.; Güth, J.F. Neue Entwicklungen in der additiven und subtraktiven Fertigung Teamwork. *J. Cont. Dent. Educ.* **2020**, *23*, 82–90.
3. Betriebe—Beschäftigte—Auszubildende im Zahntechniker-Handwerk. Available online: https://www.vdzi.de/statistik.html. (accessed on 10 December 2020).
4. Campbell, S.D.; Cooper, L.; Craddock, H.; Hyde, T.P.; Nattress, B.; Pavitt, S.H.; Seymour, D.W. Removable partial dentures: The clinical need for innovation. *J. Prosthet. Dent.* **2017**, *118*, 273–280. [CrossRef] [PubMed]
5. Jordan, R.A.; Micheelis, W. *Fünfte Deutsche Mundgesundheitsstudie (DMS V)*; Deutscher Zahnärzte: Köln, Germany, 2016.
6. Van Noort, R. The future of dental devices is digital. *Dent. Mater.* **2012**, *28*, 3–12. [CrossRef] [PubMed]
7. Horn, T.J.; Harryson, O.L.A. Overview of current additive manufacturing technologies and selected applications. *Sci. Prog.* **2012**, *95*, 255–282. [CrossRef] [PubMed]
8. Dawood, A.; Marti, B.; Sauret Jackson, V.; Darwood, A. 3D printing in dentistry. *Br. Dent. J.* **2015**, *219*, 521–529. [CrossRef] [PubMed]

9. Caviezel, C.; Grünwald, R.; Ehrenberg-Silies, S.; Kind, S.; Jetzke, T.; Bovenschulte, M. *Additive Fertigungsverfahren (3D-Druck)—Innovationsanalyse*; TAB Arbeitsbereicht: Berlin, Germany, 2017.
10. Kessler, A.; Hickel, R.; Reymus, M. 3D Printing in Dentistry—State of the Art. *Oper. Dent.* **2020**, *45*, 30–40. [CrossRef] [PubMed]
11. Kieschnick, A.; Schweiger, J.; Edelhoff, D.; Güth, J.F. Status Präsens 2020: Additive CAD/CAM-Gestützte Fertigungstechnologien im Zahntechnischen Labor. Available online: https://www.researchgate.net/publication/341990852_Status_Prasens_20 20_Additive_CADCAM-gestutzte_Fertigungstechnologien_im_zahntechnischen_Labor#fullTextFileContent. (accessed on 10 December 2020).
12. Hull, C.W. Apparatus for Production of Three-Dimensional Objects by Stereolithography. U.S. Patent 4,575,330, 8 August 1984.
13. Crump, S. Apparatus and Method for Creating Three-Dimensional Objects. U.S. Patent 5,121,329, 5 September 1989.
14. Kollenberg, W. Keramik und Multimaterial 3D-Druck. *Keram. Z.* **2014**, *66*, 233–236. [CrossRef]
15. ISO/ASTM. Additive Manufacturing—General Principles—Terminology. *Beuth* **2017**, *52900*. [CrossRef]
16. ISO/ASTM. Additive Manufacturing—General Principles—Part 2: Overview of Process Categories and Feedstock. *Beuth* **2016**, *17296-2*. [CrossRef]
17. Gartner Hype Cycle. Available online: https://www.gartner.com/en/research/methodologies/gartner-hype-cycle (accessed on 10 December 2020).
18. Dolabdjian, H.; Strietzel, R. Verfahren zur Herstellung von Zahnersatz und dentalen Hilfsteilen. European Patent Application 1 021 997 B2, 26 July 2000.
19. Revilla-León, M.; Meyer, M.J.; Özcan, M. Metal additive manufacturing technologies. *Int. J. Comput. Dent.* **2019**, *22*, 55–67.
20. Fischer, J.; Stawarczyk, B.; Trottmann, A.; Hämmerle, C.H.F. Festigkeit lasergesinterter Brückengerüste aus einer CoCr-legierung. *Quintessenz Zahntech.* **2008**, *34*, 140–149.
21. Rudolph, M.; Setz, J. Ein CAD/CAM-System mit aufbauender Lasertechnologie. *Quintessenz Zahntech.* **2007**, *33*, 582–587.
22. Quante, K.; Ludwig, K.; Kern, M. Marginal and internal fit of metal-ceramic crowns abricated with a new laser melting technology. *Dent. Mater.* **2008**, *24*, 1311–1355. [CrossRef]
23. Xu, D.; Xiang, N.; Wie, B. The marginal fit of selective laser melting-fabricated metal crowns: An in vitro study. *J. Prosth. Dent.* **2014**, *112*, 1437–1440. [CrossRef]
24. Huang, Z.; Zhang, L.; Zhu, J.; Zhang, X. Clinical marginal and internal fitt of metal ceramic crowns fabricated with a selective laser melting technolog. *J. Prosth. Dent.* **2015**, *113*, 623–627. [CrossRef]
25. Lövgren, N.; Roxner, R.; Klemendz, S.; Larsson, C. Effect of production method on surface roughness, marginal and internal fit, and retention of cobalt-chromium single crowns. *J Prosth Dent* **2017**, *118*, 95–101. [CrossRef]
26. Lehmann, K.M.; Hellwig, E.; Wenz, H.J. *Zahnärztliche Propädeutik*; Deutscher Zahnärzte: Köln, Germany, 2015.
27. Stark, H. Ist die Modellgussprothese adäquater Zahnersatz für den älteren Menschen? *Quintessenz* **2005**, *56*, 367–373.
28. Roach, F.E. Principles and essentials of bar clasp partial dentures. *J. Am. Dent. Assoc.* **1930**, *17*, 124–138.
29. Schweiger, J.; Kieschnick, A. *CAD/CAM in der digitalen Zahnheilkunde*; Teamwork Media: Fuchstal, Germany, 2017.
30. Alifui-Segbaya, F.; Williams, R.J.; George, R. Additive manufacturing: A novel method for fabricating cobalt-chromium removable partial denture frameworks. *Eur. J. Prosthodont Restor. Dent.* **2017**, *25*, 73–78.
31. Laverty, D.P.; Thomas, M.B.M.; Clark, P.; Addy, L.D. The use of 3D metal printing (direct metal laser sintering) in removable prosthodontics. *Dent. Update* **2016**, *43*, 826–835. [CrossRef] [PubMed]
32. Lima, J.M.; Anami, L.C.; Araujo, R.M.; Pavanelli, C.A. Removable partial dentures: Use of rapid prototyping. *J. Prosthodont.* **2014**, *23*, 588–591. [CrossRef] [PubMed]
33. Tregermann, I.; Renne, W.; Kelly, A.; Wilson, D. Evaluation of removable partial denture frameworks fabricated using 3 differnet techniques. *J. Prosthet. Dent.* **2019**, *122*, 390–395. [CrossRef] [PubMed]
34. Van Zeghbroeck, L.; Boons, E. Evaluation of technicians working time in the fabrication of removable partial dentures: Cad/Cam versus tradition. In Proceedings of the 14th Biennial Meeting of the International College of Prosthodontics, Waikoloa Village, HI, USA, 7–12 September 2011; p. 63.
35. Schweiger, J.; Güth, J.F.; Erdelt, K.J.; Edelhoff, D.; Schubert, O. Internal porosities, retentive force, and survival of cobalt-chromium alloy clasps fabricated by selective laser sintering. *J. Prosthodont. Res.* **2019**, *64*, 210–216. [CrossRef] [PubMed]
36. Torii, M.; Nakata, T.; Takahashi, K.; Kawamura, N.; Shimpo, H.; Ohkubo, C. Fitness and retentive force of cobalt-chromium alloy clasps fabricates with repeated laser sintering and milling. *J. Prosthodont. Res.* **2018**, *62*, 342–346. [CrossRef]
37. Nakata, T.; Shimpo, H.; Okhubo, C. Clasp fabrication using one-process molding by repeated laser sintering and high-speed milling. *J. Prosth. Research* **2017**, *61*, 276–282. [CrossRef]
38. Revilla-León, M.; Öczan, M. Additive manufacturing technologies used for processing polymers: Current status and potential application in prosthetic dentistry. *J Prosthodont.* **2019**, *28*, 146–158. [CrossRef]
39. Jokusch, J.; Öczan, M. Additive manufacturing of dental polymers: An overview on processes, materials and applications. *Dent. Mater. J.* **2020**, *39*, 345–354. [CrossRef]
40. Quan, H.; Zhang, T.; Xu, H.; Luo, S.; Nie, J.; Zhu, X. Photo-curing 3D-Printing technique and its challenges. *Bioact. Mater.* **2020**, *22*, 110–115. [CrossRef]
41. Dietrich, C.A.; Ender, A.; Baumgartner, S.; Mehl, A. A validation study of reconstructes rapid prototyping models produces by two technologies. *Angle Orthod.* **2017**, *87*, 782–787. [CrossRef]

42. Brown, G.B.; Currier, G.F.; Kadioglu, O.; Kierl, J.P. Accuracy of 3-dimensional printed dental models reconstructed from digital intraoral impressions. *Am. J. Orthod. Dentofac. Orthop.* **2018**, *154*, 733–739. [CrossRef]

43. Kim, S.Y.; Shin, Y.S.; Jund, H.D.; Hwang, C.J.; Baik, H.S.; Cha, J.Y. Precision and trueness of dental models manufactured with different 3-dimensional printing technologies. *Am. J. Orthod. Dentofac. Orthop.* **2018**, *153*, 144–153. [CrossRef]

44. Emir, F.; Ayyildiz, S. Accuracy evaluation of complete-arch models manufactured by three different 3D printing technologies: A three-dimensional analysis. *J. Prosthodont. Res.* **2021**. [CrossRef]

45. Rungrojwittayakul, O.; Kann, J.Y.; Shiozaki, K.; Swamidass, R.S.; Goodacre, B.J.; Goodacre, C.J.; Lozada, J.L. Accuracy of 3D-printed models created by two technolgies of printers with different designs of model base. *J. Prosthodont.* **2020**, *29*, 124–128. [CrossRef]

46. Etemad-Shahidi, Y.; Qallandar, O.B.; Evenden, J.; Alifui-Segbaya, F.; Ahmed, K.E. Accuracy of 3-Dimensionally printed full-arch dental models: A systematic review. *J. Clin. Med.* **2020**, *9*, 3357. [CrossRef]

47. Kallweit, D.; Mönch, W.; Zappe, H. Kontrolliert kippen: Silizium-Mikrospiegel mit integriertem optischen Feedback. *Photonik* **2006**, *4*, 62–65.

48. Viereck, V.; Li, Q.; Jäkel, A.; Hillmer, H. Großflächige Anwendung von optischen MEMS: Mikospiegel-Arrays zur Tageslichtlenkung. *Photonik* **2009**, *2*, 28–29.

49. DLP®0.47-inch 4K UHD HSSI Digital Micromirror Device (DMD). Available online: https://www.ti.com/product/DLP471TP (accessed on 10 December 2020).

50. The 3D Printing Standard in Speed, Reliability and Workflow Integration. Available online: https://www.rapidshape.de/images/kataloge/Dental_Katalog_EN.pdf#page=11 (accessed on 4 May 2021).

51. Professionelle Desktop 3D-Drucker. Available online: https://www.way2production.at/produkte (accessed on 10 December 2020).

52. Allanic, A.L. Production of a Volume Object by Lithography, Having Improved Spatial Resolution. European Patent Application 2 943 329 B1, 8 November 2015.

53. ProMaker LD20 Dental Plus. Compact High Precision 3d Printer. Available online: https://www.prodways.com/en/industrial-3d-printers/promaker-ld20-dental-plus/ (accessed on 10 December 2020).

54. Schultheiss, A. Mehr Qualität, mehr Produktivität: Professioneller 3D-Druck im Dentallabor. Zahntechnik Magazin 2018. Available online: https://www.ztm-aktuell.de/marktplatz/industrie-report/story/mehr-Qualitaet-mehr-produktivitaet-professioneller-3d-druck-im-dentallabor_670.html (accessed on 10 December 2020).

55. Stadlmann, K. Anlage zum schichtweisen Aufbau eines Körpers und Entformvorrichtung hierfür. *AT 51 4496 B1*, 2015.

56. Stadlmann, K. System for Layered Construction of a Body and Tray therefore. U.S. Patent US 10,414,091 B2, 2017.

57. DeSimone, J.M.; Ermoshkin, A.; Ermoshkin, N.; Samulski, E.T. Continuous Liquid Interphase Printing. U.S. Patent US 9,205,601 B2, 8 December 2015.

58. Tumbletone, J.R.; Shirvanyants, D.; Ermoshkin, N.; Janusziewicz, R.; Johnson, A.R.; Kelly, D.; Chen, K.; Pinschmidt, R.; Rolland, J.P.; Ermoshkin, A.; et al. Additive manufacturing. Continuous liquid interface production of 3D objects. *Science* **2015**, *347*, 1349–1352. [CrossRef]

59. Schweiger, J.; Edelhoff, D.; Stimmelmayr, M.; Güth, J.F.; Beuer, F. Automatisierte Fertigung von mehrschichtigem Frontzahnersatz mithilfe digitaler Dentinkerne. *Quintessenz Zahntech.* **2014**, *40*, 1248–1266.

60. Schweiger, J.; Edelhoff, D.; Stimmelmayr, M.; Güth, J.F.; Beuer, F. Automated production of multilayer anterior restorations with digitally produced dentin cores. *Quintessence Dent. Tech.* **2015**, *38*, 207–220.

61. Schweiger, J.; Trimpl, J.; Schwerin, C.; Güth, J.F.; Edelhoff, D. Biomaterials update—Additive manufacturing: Applications in dentistry based on materials selection. *Quintessence Dent. Tech.* **2019**, *42*, 50–69.

62. Schweiger, J.; Edelhoff, D.; Schubert, O.; Trimpl, J.; Erdelt, K.J.; Güth, J.F. Digitale Modellherstellung—eine Übersicht. *Quintessenz Zahntech.* **2019**, *45*, 41–61.

63. Güth, J.F.; Schubert, O.; Nold, E.; Trimpl, J.; Schweiger, J. Teamdisziplin-3D-Planung und Navigation in der Implantologie. *Bayerischer Zahnärzte Blatt* **2018**, *55*, 50–56.

64. Wöstmann, B.; Powers, M. *Präzisionsabformungen—Ein Leitfaden für Theorie und Praxis*; 3M ESPE: Seefeld, Germany, 2016.

65. Wedekind, L.; Güth, J.-F.; Schweiger, J.; Kollmuss, M.; Reichl, F.-X.; Edelhoff, D.; Högg, C. Elution behavior of a 3D-printed, milled and conventional resin-based occlusal splint material. *Dent. Mater.* **2021**, *37*, 701–710. [CrossRef]

66. Lutz, A.-M.; Hampe, R.; Roos, M.; Lümkemann, N.; Eichberger, M.; Stawarczyk, B. Fracture resistance and 2-body wear of 3-dimensional–printed occlusal devices. *J. Prosthet. Dent.* **2019**, *121*, 166–172. [CrossRef]

67. Berli, C.; Thieringer, F.M.; Sharma, N.; Müller, J.A.; Dedem, P.; Fischer, J.; Rohr, N. Comparing the mechanical properties of pressed, milled, and 3D-printed resins for occlusal devices. *J. Prosthet. Dent.* **2020**, *124*, 780–786. [CrossRef]

68. Reymus, M.; Hickel, R.; Kunz, A. Accuracy of CAD/CAM-fabricated bite splints: Milling vs. 3D printing. *Clin. Oral Investig.* **2020**, *24*, 4607–4615.

69. Reymus, M.; Fotiadou, C.; Kessler, A.; Heck, K.; Hickel, R.; Diegritz, C. 3D printed replicas for endodontic education. *Int. Endod. J.* **2019**, *52*, 123–130. [CrossRef] [PubMed]

70. Meglioli, M.; Naveau, A.; Macaluso, G.M.; Catros, S. 3D printed bone models in oral and cranio-maxillofacial surgery: A systematic review. *3D Print. Med.* **2020**, *6*, 1–19. [CrossRef] [PubMed]

71. Schweiger, J.; Edelhoff, D.; Güth, J.F. Update digitale Fertigung 2020—neueste Entwicklungen in der additiven und subtraktiven Fertigung. *Dent. Dialogue* **2020**, *21*, 36–51.
72. Schweiger, J.; Beuer, F.; Stimmelmayr, M.; Edelhoff, D.; Magne, P.; Güth, J.F. Histo-anatomic 3D printing of dental structures. *Br. Dent. J.* **2016**, *221*, 555–560. [CrossRef]
73. Schweiger, J. Method, Apparatur and Computer Program for Producing a Dental Prosthesis. U.S. Patent US8,775,131,B2, 8 July 2014.
74. Schweiger, J. Method, Apparatur and Computer Program for Producing a Dental Prosthesis. European Patent Application 2 363 094 B1, 10 July 2013.
75. Wissenschaftliche Untersuchungen zu VarseoSmile Crown Plus. Available online: http://www.bego.com/de/3d-druck/materialien/varseosmile-crown-plus/wissenschaftliche-untersuchungen (accessed on 10 December 2020).
76. Schweiger, J.; Bomze, D.; Schwentenwein, M. 3D-Printing of Zirconia—What is the future? *Curr. Oral Health Rep.* **2019**, *6*, 339–343. [CrossRef]
77. Geier, S.; Potestio, I. 3D-printing: From multi-material to functionally-graded ceramic. *Ceram. Appl.* **2020**, *8*, 32–35.
78. Kessler, A.; Reichl, F.-X.; Folwaczny, M.; Högg, C. Monomer release from surgical guide resins manufactured with different 3D printing devices. *Dent. Mater.* **2020**, *36*, 1486–1492. [CrossRef]
79. Prpić, V.; Schauperl, Z.; Ćatić, A.; Dulčić, N.; Čimić, S. Comparison of mechanical properties of 3D-printed, CAD/CAM, and conventional denture base materials. *J. Prosthodont.* **2020**, *29*, 524–528. [CrossRef]
80. Wemken, G.; Burkhardt, F.; Spies, B.C.; Kleinvogel, L.; Adali, U.; Sterzenbach, G.; Beuer, F.; Wesemann, C. Bond strength of conventional, subtractive, and additive manufactured denture bases to soft and hard relining materials. *Dent. Mater.* **2021**, *37*, 928–938. [CrossRef]
81. Unkovskiy, A.; Schmidt, F.; Beuer, F.; Li, P.; Spintzyk, S.; Fernandez, P.K. Stereolithography vs. direct light processing for rapid manufacturing of complete denture bases: An in vitro accuracy analysis. *J. Clin. Med.* **2021**, *10*, 1070. [CrossRef]
82. Anadioti, E.; Musharbash, L.; Blatz, M.B.; Papavasiliou, G.; Kamposiora, P. 3D printed complete removable dental prostheses: A narrative review. *BMC Oral Health* **2020**, *20*, 1–9. [CrossRef]

Journal of
Clinical Medicine

Article

Stereolithography vs. Direct Light Processing for Rapid Manufacturing of Complete Denture Bases: An In Vitro Accuracy Analysis

Alexey Unkovskiy [1,2,*], Franziska Schmidt [1], Florian Beuer [1], Ping Li [3], Sebastian Spintzyk [3] and Pablo Kraemer Fernandez [4]

[1] Department of Prosthodontics, Geriatric Dentistry and Craniomandibular Disorders, Charité-Universitätsmedizin Berlin, Corporate Member of Freie Universität Berlin, Humboldt-Universität zu Berlin, Aßmannshauser Str. 4-6, 14197 Berlin, Germany; franziska.schmidt2@charite.de (F.S.); florian.beuer@charite.de (F.B.)

[2] Department of Dental Surgery, Sechenov First Moscow State Medical University, Bolshaya Pirogovskaya Street, 19c1, 119146 Moscow, Russia

[3] Section Medical Materials Science and Technology, Tuebingen University Hospital, Osianderstr. 2-8, 72076 Tuebingen, Germany; ping.li@med.uni-tuebingen.de (P.L.); sebastian.spintzyk@med.uni-tuebingen.de (S.S.)

[4] Department of Prosthodontics at the Centre of Dentistry, Oral Medicine and Maxillofacial Surgery with Dental School, Tuebingen University Hospital, Osianderstr. 2-8, 72076 Tübingen, Germany; pablo.kraemer-fernandez@med.uni-tuebingen.de

* Correspondence: dr.unkovskiy@gmail.com; Tel.: +49-030-450-662517

Citation: Unkovskiy, A.; Schmidt, F.; Beuer, F.; Li, P.; Spintzyk, S.; Kraemer Fernandez, P. Stereolithography vs. Direct Light Processing for Rapid Manufacturing of Complete Denture Bases: An In Vitro Accuracy Analysis. *J. Clin. Med.* **2021**, *10*, 1070. https://doi.org/10.3390/jcm10051070

Academic Editor: Tim Joda

Received: 9 February 2021
Accepted: 25 February 2021
Published: 4 March 2021

Publisher's Note: MDPI stays neutral with regard to jurisdictional claims in published maps and institutional affiliations.

Abstract: The topical literature lacks any comparison between stereolithography (SLA) and direct light processing (DLP) printing methods with regard to the accuracy of complete denture base fabrication, thereby utilizing materials certified for this purpose. In order to investigate this aspect, 15 denture bases were printed with SLA and DLP methods using three build angles: 0°, 45° and 90°. The dentures were digitalized using a laboratory scanner (D2000, 3Shape) and analyzed in analyzing software (Geomagic Control X, 3D systems). Differences between 3D datasets were measured using the root mean square (RMS) value for trueness and precision and mean and maximum deviations were obtained for each denture base. The data were statistically analyzed using two-way ANOVA and Tukey's multiple comparison test. A heat map was generated to display the locations of the deviations within the intaglio surface. The overall tendency indicated that SLA denture bases had significantly higher trueness for most build angles compared to DLP ($p < 0.001$). The 90° build angle may provide the best trueness for both SLA and DLP. With regard to precision, statistically significant differences were found in the build angles only. Higher precision was revealed in the DLP angle of 0° in comparison to the 45° and 90° angles.

Keywords: stereolithography; direct light processing; complete denture; edentulism; rapid prototyping; 3D printing; additive manufacturing; rapid manufacturing

1. Introduction

With extensive applications of computer-aided design and computer-aided manufacturing (CAD/CAM) in modern clinical dentistry, various additive manufacturing (AM) methods have become available for the fabrication of surgical guides, dental models, provisional crowns and complete dentures [1–3]. In the past decade, a large number of clinical and technical protocols has been introduced to fabricate a complete denture in a fully digital workflow. Many of them utilize additive manufacturing for printing either a try-in or a definitive denture [4–6]. In a clinical study by Cristache et al., patients' high levels of satisfaction with digitally produced dentures were recorded in a follow-up after 18 months [7]. The fabrication of complete dentures using AM can be considered as a promising technique with regard to its clinical and technical performance [8,9]. Furthermore, it was reported that additively produced denture bases show a comparable tissue adaptation to milled ones [10].

Direct light processing (DLP) is the most widely applied AM method for 3D printing of denture bases [4,10–13]. It utilizes a micro-mirror device and ultraviolet light for a layer-wise build-up of photopolymerizable resin [14]. Stereolithography (SLA) is an alternative vat polymerization method based on a laser beam raster scanning of the surface within a tank with photosensitive liquid, generally also a photopolymerizable resin [15,16].

Different principles in the printing process between SLA and DLP methods may potentially cause anisotropy with regard to the dimensional accuracy of the printed parts [17]. In case of SLA, the laser beam travels across the layer surface, causing localized polymerization of the photosensitive resin in the area of the illuminated field, whereas in case of DLP, the whole material portion in the x/y space is cured simultaneously by a one-time projection of the whole layer through the light projector [14]. Besides the type of resin and the light intensity, the accuracy of the mentioned AM methods may also be influenced by the build angle of the printing process [18–20]. This aspect has been investigated for DLP with regard to complete denture manufacturing and yielded no significant differences among the various build angles [12]. Another study highlighted potential differences in accuracy between DLP and SLA in the maxillofacial field [17]. Choi et al. reported that SLA may produce more accurate dental models than DLP [21]. The influence of build angle on the accuracy of SLA-produced surgical guides has also been widely investigated, reporting the 90° angle to yield the best clinical outcome [22]. Moreover, the layer thickness may influence the final result, whereby 50 μm layer printing provides a better dimensional accuracy [23].

However, no clinical report can be found regarding the utilization of SLA for manufacturing complete dentures in a digital workflow. The study of Hada et al. investigated the influence of the build angle on the accuracy of stereolithographically printed bases [24]. It must be emphasized that the material used in this study was transparent and not specified by the manufacturer for denture base fabrication.

However, there are various studies devoted to accuracy investigations of various SLA- and DLP-printed objects. However, the authors are unaware of any comparison between these AM methods with regard to their dimensional accuracy for a direct denture base fabrication in a fully digital workflow from certified denture base materials. Recently, You et al. compared SLA and DLP methods with regard to denture metal base fabrication, though within a semi-analog production chain [25].

Thus, the aim of the present study was to find out which of these vat polymerization techniques may produce the most accurate denture base for a fully digital workflow, in both cases using a certified denture base material and a uniform layer thickness. The study should also provide a recommendation for the printing preferences with regard to the build angle for this new application of SLA denture material. The potential finding of the study should aid a better understanding of how printing process parameters may influence the clinical performance of additively manufactured denture bases.

The first null hypothesis is that there will be no significant dimensional differences in SLA- and DLP-printed bases. Furthermore, it is hypothesized that the build angle of the SLA printing processes using the certified denture base material will not show any significant influence on the final dimensional accuracy.

2. Experimental Section

2.1. Specimen Design and Fabrication

A maxillary complete denture was designed in CAD software (DentalCAD 2.3 Matera, exocad GmbH, Darmstadt, Germany) and exported in a surface tessellation language (STL) format. For the SLA group, the denture base data was imported into slicing software (PreForm, Formlabs, Somerville, MA, USA) and nested on the build platform in 0°, 45°, and 90° orientations (Figure 1).

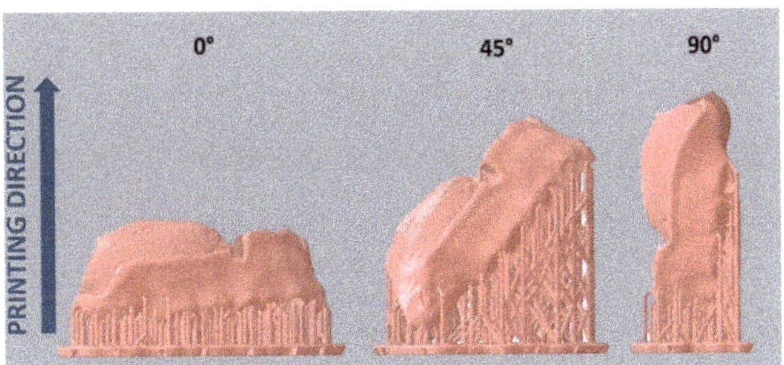

Figure 1. Nesting with three alternative build angles used in the study.

The supporting structures were generated automatically using the software script and then it was checked that none of them were connected to the intaglio surface. The denture bases were printed, $n = 5$ for each printing direction, with a liquid resin (Denture Base OP Resin, Formlabs, Somerville, USA) using an SLA printer (Form 3B, Formlabs, Somerville, MA, USA) (Figure 2). Afterwards, the bases were washed in isopropanol in a specific machine with the help of a stirrer to circulate the liquid (FormWash, Formlabs, Somerville, MA, USA) and photopolymerized in natural glycerin preheated to 80 °C for 60 min in a light chamber (FormCure, Formlabs, Somerville, MA, USA) according to the manufacturer's specifications.

Figure 2. Exemplary 3D printed denture bases (here: stereolithography (SLA) with 45° build angle).

For the DLP group, the STL file was imported into slicing software (Netfabb Premium 2021, Autodesk, San Rafael, CA, USA). The supporting structures were also generated automatically using a software script for DLP printers and a base grid. Care was taken that none of them touched the intaglio surface. The denture bases were printed, $n = 5$ for each printing direction, with liquid resin (V-Print dentbase, VOCO GmbH, Cuxhaven, Germany) using a DLP printer (Solflex 350 PLUS, W2P Engineering GmbH, Vienna, Austria) with a flexible silicone vat (FlexVat, W2P Engineering GmbH, Vienna, Austria). Afterwards, the support structures were removed; the bases were washed out for 5 min in total in an ultrasonic cleaner with isopropanol, dried for 15 min and photopolymerized for 30 min in a light chamber (LC-3DPrint Box, 3D Systems Inc., Rock Hill, SC, USA) according to the manufacturer's specifications.

2.2. Accuracy Analysis

The accuracy investigation encompassed trueness and precision analysis as per ISO 5725-1. The SLA- and DLP-printed denture bases were digitalized using a laboratory scanner (D2000, 3Shape, Copenhagen, Denmark). The gathered scans were exported in STL format and used for the accuracy test. For analysis of trueness, the obtained STL file of each printed denture was aligned with the reference CAD model using first the three-point-fit and then best-fit protocols in the analyzing software (Geomagic Control X, 3D Systems Inc., Rock Hill, SC, USA). For the alignment process, the intaglio surface was segmented from the remaining STL dataset, as shown in Figure 3. For the analysis of precision, the obtained STL files of each printed denture were matched to each other within each group.

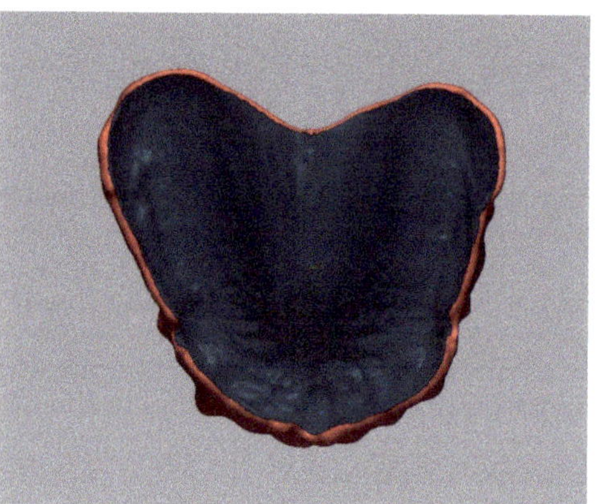

Figure 3. Segmentation of the reference dataset in Geomagic Control X (3D Systems) software for the further matching process. Only the intaglio surface (dark gray) was used for the best fit protocol.

For quantitative analysis of trueness and precision, the values were automatically calculated using the root mean square (RMS) error. RMS is recognized as a standard variable to measure differences between two 3D datasets [21]. The RMS deviation was calculated with the following formula:

$$RMS = \sqrt{\frac{\sum_{i=1}^{n}\left(x_{R,i} - x_{T,i}\right)^2}{n}} \qquad (1)$$

where n is the number of measured points, $x_{R,i}$ is the i-th measurement point of the reference model and $x_{T,i}$ is the measurement point of the dataset of the test model.

Furthermore, the mean and maximum deviations in mm were obtained for each dataset.

For qualitative analysis of trueness and precision, a heat map was generated for each dataset. The range of the maximum and minimum values was set to 1 mm. The tolerance level was set to ±0.025 mm as it represents the maximum z-axis resolution of the used AM methods of 0.05 mm.

For a better understanding of surface layering, the optical 3D metrology analysis with an optical scanner (Edge Master X, Alicona GmbH, Schönau am Königssee, Germany) of the palatal area of the intaglio surface was carried out. The scanning process was performed with $10\times$ magnification lens under standard light conditions and was analyzed in 3D Image Viewer software (Alicona GmbH, Schönau am Königssee, Germany).

All gathered data were statistically analyzed in statistic software (JMP 14, SAS Corp., Heidelberg, Germany). First, the data were tested for normality by goodness of fit with the Shapiro–Wilk test. For normally distributed data, the statistical difference was analyzed by using two-way analysis of variance (ANOVA) with printing techniques and orientations as two independent factors. Tukey's test was further performed for multiple comparison analysis. The threshold for significance was defined as a p-value less than 0.05.

3. Results

The mean differences between the RMS values, as well as mean and mean maximum deviation in mm for SLA and DLP denture bases, are shown in Table 1.

Table 1. The root mean square (RMS) values for trueness, precision and the average and maximum deviations in mm between various study groups.

		SLA (Stereolithography)			DLP (Direct Light Processing)		
		0°	45°	90°	0°	45°	90°
Trueness (RMS)	Mean	0.094	0.132	0.083	0.256	0.211	0.163
	SD	0.004	0.016	0.009	0.031	0.031	0.030
Precision (RMS)	Mean	0.087	0.094	0.098	0.134	0.048	0.044
	SD	0.042	0.034	0.037	0.028	0.023	0.023
Average + (mm)	Mean	0.082	0.099	0.055	0.166	0.101	0.066
	SD	0.011	0.015	0.009	0.027	0.010	0.010
Average − (mm)	Mean	−0.054	−0.089	−0.045	−0.187	−0.097	−0.065
	SD	0.006	0.018	0.010	0.024	0.008	0.006
Max + (mm)	Mean	0.613	0.630	0.533	0.547	0.573	0.500
	SD	0.136	0.119	0.030	0.169	0.163	0.071
Max − (mm)	Mean	−0.168	−0.294	−0.132	−0.366	−0.402	−0.416
	SD	0.005	0.143	0.020	0.058	0.027	0.048

The Shapiro–Wilk test revealed a normal distribution of the gathered data. As shown in Figures 4 and 5, a statistically significant interaction was found in the trueness ($F_{(2, 24)} = 10.78$, $p = 0.0005$). Additionally, each main effect showed significant differences: AM methods ($F_{(1, 24)} = 164.7$, $p < 0.0001$) and build angles ($F_{(2, 24)} = 16.39$, $p = 0.0744$, $p < 0.0001$). Furthermore, the overall tendency indicated that SLA denture bases had significantly higher trueness for most build angles compared to DLP ($p < 0.001$), confirmed by Tukey's multiple comparison tests.

Regarding the precision, there was a statistically significant interaction ($F_{(2, 18)} = 6.044$, $p = 0.0098$). Meanwhile, the main effect of build angles had significant differences ($F_{(2, 18)} = 4.061$, $p = 0.0350$) while AM methods showed no significant differences ($F_{(1, 18)} = 1.907$, $p = 0.1842$). Specifically, a post hoc multiple comparison test demonstrated significant greater precision in the DLP of 0° (0.134 ± 0.028) in comparison to the DLP of 45° (0.048 ± 0.023, $p = 0.0151$) and 90° (0.044 ± 0.023, $p = 0.0098$).

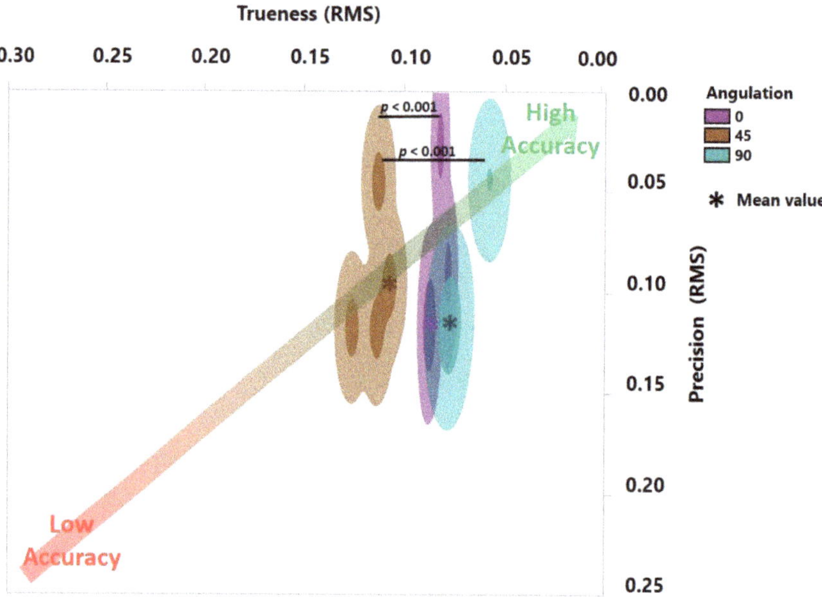

Figure 4. Accuracy of SLA-printed denture bases with three different build angles. Trueness is depicted horizontally and precision vertically. The *p*-value here with regard to trueness was calculated using Tukey's analysis. No statistically significant difference was detected for precision.

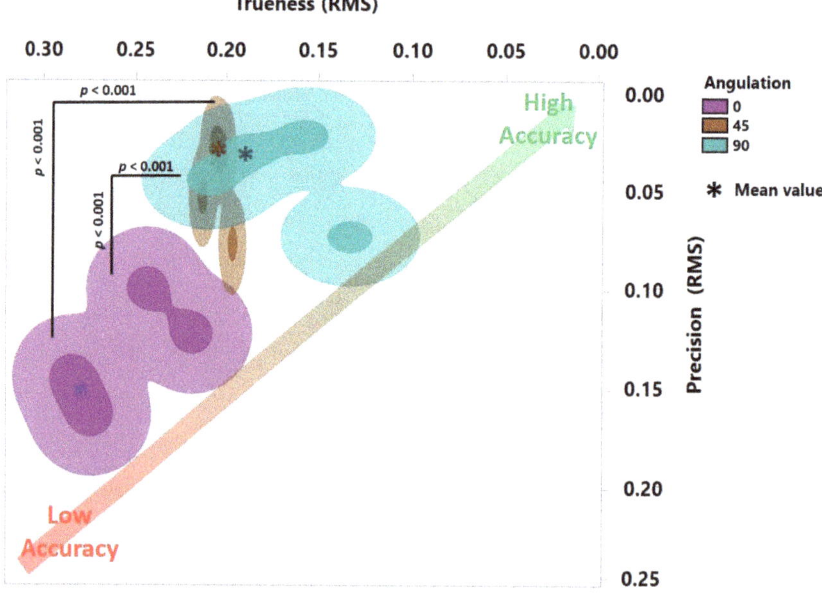

Figure 5. Accuracy of DLP-printed denture bases with three different build angles. Trueness is depicted horizontally and precision vertically. The *p*-value here with regard to trueness and precision was calculated using Tukey's analysis.

As shown in Figure 6, with regard to the mean deviation, no statistically significant differences were observed: interaction (F (2, 54) = 0.06212, p = 0.9398), AM methods (F (1, 54) = 0.1555, p = 0.6948) and build angles (F (2, 54) = 0.0022, p = 0.9978). However, in terms of the maximum deviation, significantly higher inaccuracies up to 0.5 mm were observed for the SLA group in 0° and 45° orientations compared to DLP (Figure 7). AM methods as the main factor showed statistically significant differences in mean maximum deviation (F (1, 54) = 7.053, p = 0.0104), confirmed by a two-way ANOVA.

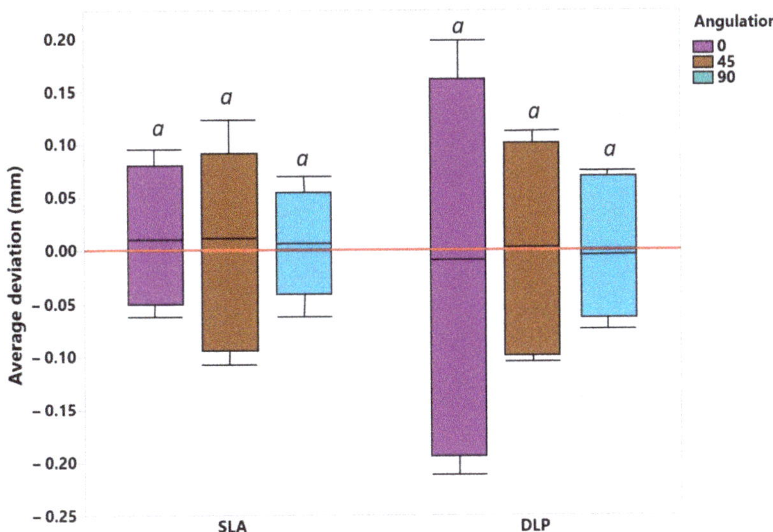

Figure 6. The mean average deviations between various study groups over the whole intaglio surface. *a* represents level of statistical significance.

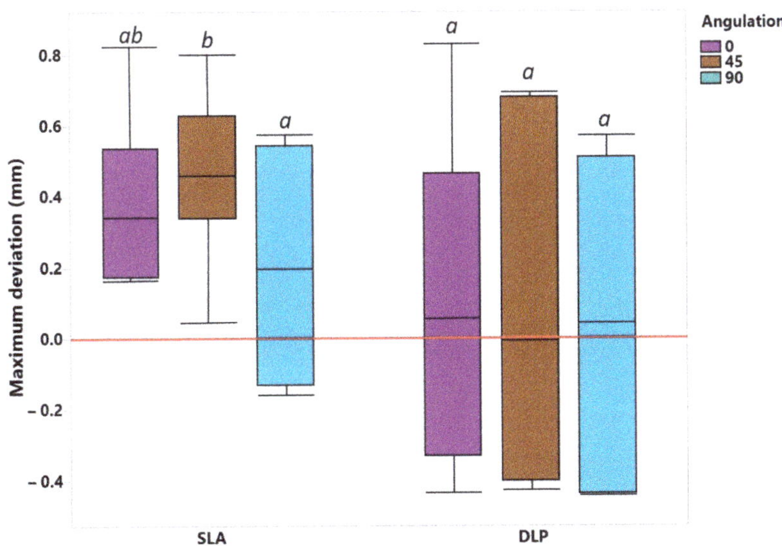

Figure 7. The mean maximum deviations between various study groups over the whole intaglio surface. *a*, *ab*, *b*—represent levels of statistical significance.

The trueness heat map demonstrated positive deviations (yellow to red) in the area of tuber maxillae and negative deviations (cyan to blue) in the palatal area for SLA bases printed with 0° and 45° orientations (Figures 8 and 9). Only shallow deviations could be observed in the 90° printed SLA bases. The intaglio surface of 0° printed DLP bases showed the poorest accuracy and was almost fully distorted in a positive way on the alveolar residual ridge and in a negative way on the palate and lingual slope. The 90° printed DLP bases showed the most uniform intaglio surface with fewer deviations.

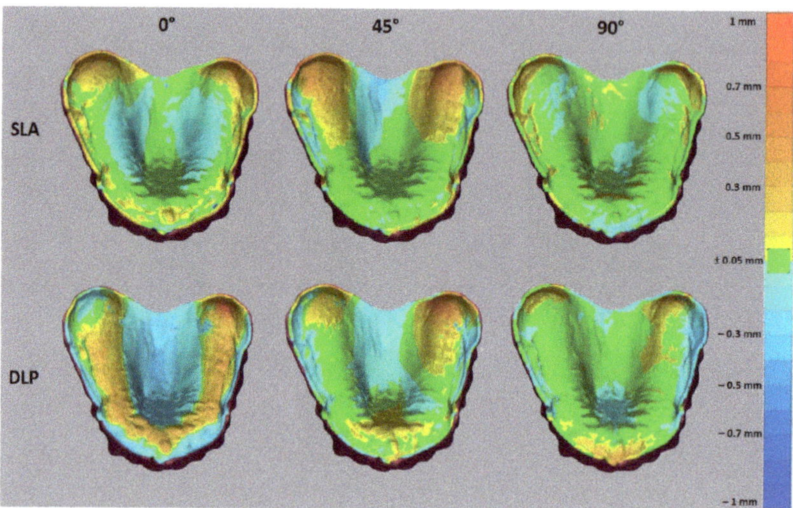

Figure 8. The heat map for the trueness of 3D printed denture bases. The tolerance level was set to ±0.05 mm. The yellow to red deviation shows the positive deviations and cyan to blue, negative.

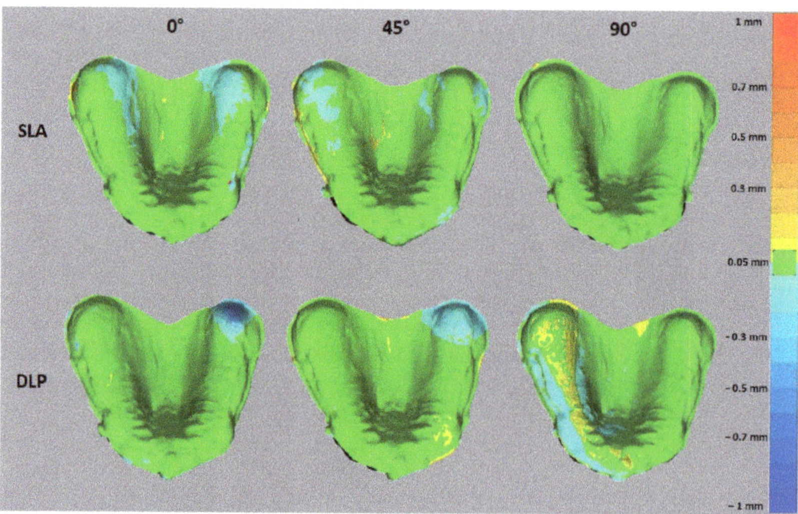

Figure 9. The heat map for the precision of 3D printed denture bases. The tolerance level was set to ±0.05 mm. The yellow to red deviation shows the positive deviations and cyan to blue, negative.

The optical 3D metrology test displayed a significant difference in surface structure (Figure 10). A strongly pronounced staircase effect was demonstrated for both SLA and DLP 0° printed specimens. No significant differences were observed for 45° specimens. Despite the same orientation, the 90° specimens showed isotropic surface structures, whereby the SLA base demonstrated the most uniform surface devoid of any staircase effect.

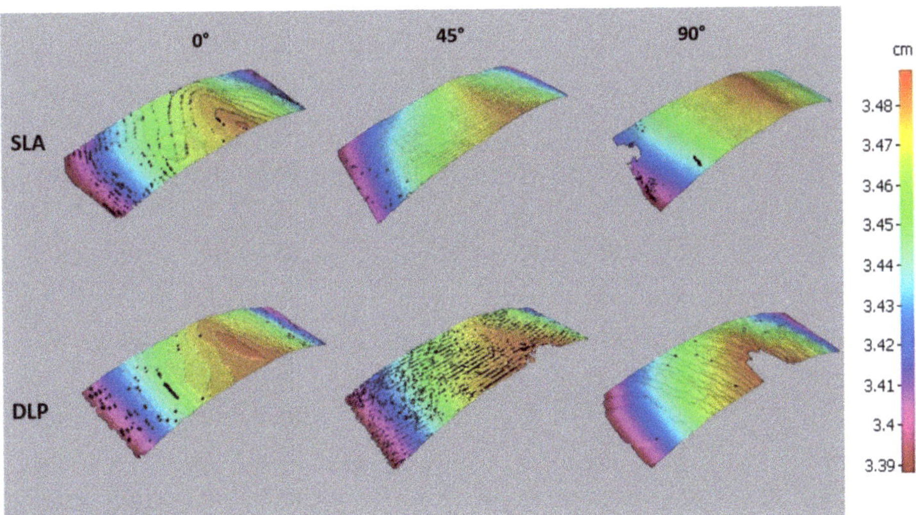

Figure 10. The optical 3D metrology showing differences in layer orientation for SLA and DLP specimens.

4. Discussion

4.1. Outcomes of the Accuracy Test

The accuracy analysis in the present study revealed greater trueness for SLA compared to DLP. Furthermore, the 90° orientation exhibited fewer deviations for both SLA and DLP methods. Therefore, both null hypotheses were rejected.

The greater deviations of the intaglio surface in the 0° and 45° groups may be attributed to the more pronounced staircase effect, which is related to the layer-wise building process. This exerted negative effects on the trueness of the palatal surface, including grooves and line angles [26]. In general, the large curved surfaces are more prone to the staircase effect than vertical surfaces, which leads to higher dimensional errors [27,28].

The optical 3D metrology analysis revealed that the SLA-printed bases demonstrated less of a staircase effect than DLP in all orientation groups. This fact may be attributed to the inherent process-related difference between these two vat polymerization methods. Thus, even if the layers are oriented perpendicular to the intaglio surface, the staircase effect may be caused by the light projection of the square-shaped 2D pixel patterns through the mirror device, which generates each voxel [29].

The majority of accuracy-related studies have been carried out using the DLP method. You et al. investigated the accuracy of an SLA-printed trial denture in beige material and reported the RMS values for the intaglio surface with 50 μm layer thickness in the order of 0.152 ± 0.01, which is in agreement with the results of the present study [26]. However, the heat map of You et al. revealed much higher centripetal shrinkage, as observed for SLA bases in the present case. An objective comparison between these two studies with regard to the localization of the deviations is restricted by the unclear build angle and utilization of the trial denture beige material in You et al.

Hada et al. investigated the accuracy of SLA-printed dentures using transparent material (Clear, Formlabs, Somerville, MA, USA) and a Form 2 printer (Formlabs, Somerville,

USA) with 100 µm layer thickness and reported the 45° orientation to provide the superior trueness [24]. It must be stressed that this material is not verified by the manufacturer for denture base fabrication nor for dental applications. For this reason, in the present study, the Denture Base OP Resin (Formlabs, Somerville, MA, USA) and Form 3B printer (Formlabs, Somerville, MA, USA) were utilized for comparative analysis with V-Print dentbase (VOCO GmbH, Cuxhaven, Germany) and the Solflex 350 Plus (W2P Engineering GmbH, Vienna, Austria) DLP method. Contrary to Hada et al., the outcomes of the present study propose the 90° angle to be an optimal build angle for denture base fabrication. Rubayo et al. investigated the accuracy of SLA-produced surgical guides using 100 µm layer thickness and revealed the 45° angle to perform best in terms of geometrical accuracy [22]. The contradiction between these studies and the present one may be attributed to an alternative layer thickness of 50 µm or the different materials, as used in this study. This assumption is further supported by the outcomes of Dalal et al., who postulated that 50 µm may be more accurate than 100 µm for the SLA printing process [23].

4.2. Clinical Interpretation

The observed discrepancies in RMS value between SLA and DLP methods need careful clinical interpretation. For this reason, the additional analysis of the mean deviation in mm was performed. This has shown that the mean deviation for SLA and DLP bases did not exceed a value of ±0.02 mm. This is contradictory with the study of Hwang et al., as they reported values of ±0.06 mm for DLP, and Yoon et al. reported values up to 0.5 mm. [11,13]. This might be due to the layer thickness of 0.1 mm, build angle of 100° and an alternative material used in both studies.

The mean maximum deviation reached a value of 0.5 mm for 0° and 45° SLA bases, which is significantly higher as compared to DLP (up to 0.1 mm). Furthermore, the SLA bases predominantly showed deviation of a positive manner in contrast to DLP.

The majority of negative deviations were found on the palatal area, which may compromise the posterior palatal seal. Further positive deviations were revealed in the stress-bearing areas such as residual alveolar ridge and tuber maxillae, which might cause compression and incongruence in these areas. According to the heat map, the 90° orientation may provide more accurate denture bases for both DLP and SLA, leading to a better tissue adaptation. This disagrees with the study of Jin et al., as they reported the 45° (135°) angle to be an optimal printing angle for DLP according to the heat map [12]. Here, it could be speculated that the different DLP devices used might generate different printing results. Furthermore, another factor might be the postprocessing method used. Studies have shown the influence on the mechanical properties and this could possibly also have an influence on the accuracy of the denture bases (development of negative residual stresses inside the printed parts) [30–32].

The present study concentrated on the intaglio surface, which does not necessarily reflect the general accuracy of the whole denture base. Thus, the studies of You et al. showed a certain discrepancy in the accuracy of intaglio, cameo and socketed surfaces for both SLA and DLP [26,33]. The manufacturing accuracy of the denture socketed area using SLA should be considered in future studies.

It must be emphasized that in the majority of the studies devoted to the accuracy of DLP-printed denture bases, a dental NextDent 3D printer was used with a layer thickness of 100 µm and a wavelength of 405 nm [7,8,12,26,34]. The printing hard- and software may vary depending on the manufacturer. So, there are more active factors than just the illumination source (SLA, DLP, liquid crystal display (LCD)). The DLP printer used in this study had a moving DLP projector with 385 nm, which increased the maximum usable area, but might lead to additional inaccuracy. In addition, different vat systems are now available on the market, which could also have an impact on precision, as pull-off forces, reduced light intensity due to "clouding" and unevenness in the vat bottom may also affect the result. A limitation of this study is also that multiple identical objects had to be printed in the same position, which can result in increased wear of the tray bottom, which can be reflected in the precision and accuracy. Further studies on the topic should include all available hardware options for DLP and SLA 3D printing of denture bases.

The part geometry also has a considerable influence on the accuracy, depending on the part alignment. The formation of suction cups during part generation can lead to considerable pull-off forces, which can result in subsequent distortion of the part. The prosthesis bases used in this study showed suction cup formation, especially in the 90° orientation. Therefore, accuracy may presumably depend on both component orientation and component geometry.

Additionally, the material itself is known to influence the geometrical accuracy [34]. The abovementioned facts cater for further research considering a comparison of various 3D printers within the one group of illumination technologies, utilization of diverse geometries of build parts and utilization of verified materials.

5. Conclusions

Within the limitations of the present in vitro study, the following conclusions were drawn:

- SLA may produce an intaglio denture surface with a better trueness than DLP;
- SLA and DLP demonstrated nearly the same precision for 3D printing of denture bases;
- The build orientation of 90° may provide the best trueness for both SLA and DLP;
- Besides the illumination source of a 3D printing hardware (SLA, DLP, LCD), the geometrical accuracy may be presumably influenced by part geometry and material type.

Author Contributions: Conceptualization, A.U., P.K.F. and S.S.; methodology, A.U.; software, A.U., P.K.F. and P.L.; validation, F.S., F.B. and S.S.; formal analysis, writing—original draft preparation, A.U.; writing—review and editing, P.L., P.K.F. and S.S.; visualization, A.U.; supervision, F.B., S.S.; project administration, S.S. All authors have read and agreed to the published version of the manuscript.

Funding: Pablo Kraemer Fernandez and Sebastian Spintzyk have a third-party funded project on the topic of digital dentures with VOCO. The stereolithography specimens were manufactured in the Department of Prosthodontics, Geriatric Dentistry and Craniomandibular Disorders (Charité-Universitätsmedizin Berlin).

Institutional Review Board Statement: Not applicable.

Informed Consent Statement: Not applicable.

Data Availability Statement: All study data is available on request.

Acknowledgments: The authors acknowledge the companies Formlabs and VOCO for providing their materials for this study.

Conflicts of Interest: The authors declare no conflict of interest.

References

1. Wemken, G.; Spies, B.C.; Pieralli, S.; Adali, U.; Beuer, F.; Wesemann, C. Do hydrothermal aging and microwave sterilization affect the trueness of milled, additive manufactured and injection molded denture bases? *J. Mech. Behav. Biomed. Mater.* **2020**, *111*, 103975. [CrossRef]
2. Schweiger, J.; Stumbaum, J.; Edelhoff, D.; Güth, J.-F. Systematics and concepts for the digital production of complete dentures: Risks and opportunities. *Int. J. Comput. Dent.* **2018**, *21*, 41–56.
3. Schweiger, J.; Güth, J.-F.; Edelhoff, D.; Stumbaum, J. Virtual evaluation for CAD-CAM-fabricated complete dentures. *J. Prosthet. Dent.* **2017**, *117*, 28–33. [CrossRef]
4. Lin, W.-S.; Harris, B.T.; Pellerito, J.; Morton, D. Fabrication of an interim complete removable dental prosthesis with an in-office digital light processing three-dimensional printer: A proof-of-concept technique. *J. Prosthet. Dent.* **2018**, *120*, 331–334. [CrossRef]
5. Unkovskiy, A.; Wahl, E.; Zander, A.T.; Huettig, F.; Spintzyk, S. Intraoral scanning to fabricate complete dentures with functional borders: A proof-of-concept case report. *BMC Oral Health* **2019**, *19*, 46. [CrossRef]
6. Bilgin, M.S.; Erdem, A.; Aglarci, O.S.; Dilber, E. Fabricating Complete Dentures with CAD/CAM and RP Technologies. *J. Prosthodont.* **2015**, *24*, 576–579. [CrossRef] [PubMed]
7. Cristache, C.M.; Totu, E.E.; Iorgulescu, G.; Pantazi, A.; Dorobantu, D.; Nechifor, A.C.; Isildak, I.; Burlibasa, M.; Nechifor, G.; Enachescu, M. Eighteen Months Follow-Up with Patient-Centered Outcomes Assessment of Complete Dentures Manufactured Using a Hybrid Nanocomposite and Additive CAD/CAM Protocol. *J. Clin. Med.* **2020**, *9*, 324. [CrossRef] [PubMed]

8. Inokoshi, M.; Kanazawa, M.; Minakuchi, S. Evaluation of a complete denture trial method applying rapid prototyping. *Dent. Mater. J.* **2012**, *31*, 40–46. [CrossRef] [PubMed]

9. Berman, B. 3-D printing: The new industial revolution. *Bus. Horiz.* **2012**, *55*, 155–162. [CrossRef]

10. Yoon, H.-I.; Hwang, H.-J.; Ohkubo, C.; Han, J.-S.; Park, E.-J. Evaluation of the trueness and tissue surface adaptation of CAD-CAM mandibular denture bases manufactured using digital light processing. *J. Prosthet. Dent.* **2018**, *120*, 919–926. [CrossRef]

11. Hwang, H.-J.; Lee, S.J.; Park, E.-J.; Yoon, H.-I. Assessment of the trueness and tissue surface adaptation of CAD-CAM maxillary denture bases manufactured using digital light processing. *J. Prosthet. Dent.* **2019**, *121*, 110–117. [CrossRef] [PubMed]

12. Jin, M.-C.; Yoon, H.-I.; Yeo, I.-S.; Kim, S.-H.; Han, J.-S. The effect of build angle on the tissue surface adaptation of maxillary and mandibular complete denture bases manufactured by digital light processing. *J. Prosthet. Dent.* **2020**, *123*, 473–482. [CrossRef]

13. Yoon, S.-N.; Oh, K.C.; Lee, S.J.; Han, J.-S.; Yoon, H.-I. Tissue surface adaptation of CAD-CAM maxillary and mandibular complete denture bases manufactured by digital light processing: A clinical study. *J. Prosthet. Dent.* **2020**, *124*, 682–689. [CrossRef]

14. Stansbury, J.W.; Idacavage, M.J. 3D printing with polymers: Challenges among expanding options and opportunities. *Dent. Mater.* **2016**, *32*, 54–64. [CrossRef] [PubMed]

15. Gibson, I.; Rosen, D.W.; Stucker, B. *Additive Manufacturing Technologies*, 2nd ed.; Springer: Boston, MA, USA, 2015; pp. 63–106.

16. Taormina, G.; Sciancalepore, C.; Messori, M.; Bondioli, F. 3D printing processes for photocurable polymeric materials: Technologies, materials, and future trends. *J. Appl. Biomater. Funct. Mater.* **2018**, *16*, 151–160. [CrossRef] [PubMed]

17. Unkovskiy, A.; Roehler, A.; Huettig, F.; Geis-Gerstorfer, J.; Brom, J.; Keutel, C.; Spintzyk, S. Simplifying the digital workflow of facial prostheses manufacturing using a three-dimensional (3D) database: Setup, development, and aspects of virtual data validation for reproduction. *J. Prosthodont. Res.* **2019**, *63*, 313–320. [CrossRef]

18. Unkovskiy, A.; Bui, P.H.-B.; Schille, C.; Geis-Gerstorfer, J.; Huettig, F.; Spintzyk, S. Objects build orientation, positioning, and curing influence dimensional accuracy and flexural properties of stereolithographically printed resin. *Dent. Mater.* **2018**, *34*, e324–e333. [CrossRef] [PubMed]

19. Ollison, T.; Berisso, K. Three-dimensional printing build variables that impact cylindricity. *J. Ind. Technol.* **2010**, *26*, 1–10.

20. Aretxabaleta, M.; Xepapadeas, A.B.; Poets, C.F.; Koos, B.; Spintzyk, S. Comparison of additive and subtractive CAD/CAM materials for their potential use as Tübingen Palatal Plate: An in-vitro study on flexural strength. *Addit. Manuf.* **2021**, *37*, 101693. [CrossRef]

21. Choi, J.-W.; Ahn, J.-J.; Son, K.; Huh, J.-B. Three-Dimensional Evaluation on Accuracy of Conventional and Milled Gypsum Models and 3D Printed Photopolymer Models. *Materials* **2019**, *12*, 3499. [CrossRef]

22. Rubayo, D.D.; Phasuk, K.; Vickery, J.M.; Morton, D.; Lin, W.-S. Influences of build angle on the accuracy, printing time, and material consumption of additively manufactured surgical templates. *J. Prosthet. Dent.* **2020**. [CrossRef] [PubMed]

23. Dalal, N.; Ammoun, R.; Abdulmajeed, A.A.; Deeb, G.R.; Bencharit, S. Intaglio Surface Dimension and Guide Tube Deviations of Implant Surgical Guides Influenced by Printing Layer Thickness and Angulation Setting. *J. Prosthodont.* **2020**, *29*, 161–165. [CrossRef] [PubMed]

24. Hada, T.; Kanazawa, M.; Iwaki, M.; Arakida, T.; Soeda, Y.; Katheng, A.; Otake, R.; Minakuchi, S. Effect of Printing Direction on the Accuracy of 3D-Printed Dentures Using Stereolithography Technology. *Materials* **2020**, *13*, 3405. [CrossRef] [PubMed]

25. You, S.-G.; Kang, S.-Y.; Bae, S.-Y.; Kim, J.-H. Evaluation of the adaptation of complete denture metal bases fabricated with dental CAD-CAM systems: An in vitro study. *J. Prosthet. Dent.* **2020**. [CrossRef]

26. You, S.-M.; Kang, S.-Y.; Bae, S.-Y.; Kim, J.-H. Evaluation of the accuracy (trueness and precision) of a maxillary trial denture according to the layer thickness: An in vitro study. *J. Prosthet. Dent.* **2021**, *125*, 139–145. [CrossRef] [PubMed]

27. Patzelt, S.B.; Bishti, S.; Stampf, S.; Att, W. Accuracy of computer-aided design/computer-aided manufacturing–generated dental casts based on intraoral scanner data. *J. Am. Dent. Assoc.* **2014**, *145*, 1133–1140. [CrossRef]

28. Choi, S.; Chan, A. A virtual prototyping system for rapid product development. *Comput. Des.* **2004**, *36*, 401–412. [CrossRef]

29. Formlabs. SLA vs. DLP: Guide to Resin 3D Printers. Available online: https://formlabs.com/blog/resin-3d-printer-comparison-sla-vs-dlp/ (accessed on 24 January 2021).

30. Xu, Y.; Xepapadeas, A.B.; Koos, B.; Geis-Gerstorfer, J.; Li, P.; Spintzyk, S. Effect of post-rinsing time on the mechanical strength and cytotoxicity of a 3D printed orthodontic splint material. *Dent. Mater.* **2021**. [CrossRef] [PubMed]

31. Reymus, M.; Stawarczyk, B. Influence of Different Postpolymerization Strategies and Artificial Aging on Hardness of 3D-Printed Resin Materials: An In Vitro Study. *Int. J. Prosthodont.* **2020**, *33*, 634–640. [CrossRef]

32. Reymus, M.; Lümkemann, N.; Stawarczyk, B. 3D-printed material for temporary restorations: Impact of print layer thickness and post-curing method on degree of conversion. *Int. J. Comput. Dent.* **2019**, *22*, 231–237. [PubMed]

33. You, S.-M.; Lee, B.-I.; Kim, J.-H. Evaluation of trueness in a denture base fabricated by using CAD-CAM systems and adaptation to the socketed surface of denture base: An in vitro study. *J. Prosthet. Dent.* **2020**. [CrossRef]

34. Kulkarni, P.; Marsan, A.; Dutta, D. A review of process planning techniques in layered manufacturing. *Rapid Prototyp. J.* **2000**, *6*, 18–35. [CrossRef]

Journal of
Clinical Medicine

Review

Accuracy of 3-Dimensionally Printed Full-Arch Dental Models: A Systematic Review

Yasaman Etemad-Shahidi, Omel Baneen Qallandar, Jessica Evenden, Frank Alifui-Segbaya and Khaled Elsayed Ahmed *

School of Dentistry and Oral Health, Griffith University, Griffith Health Centre (G40), Office: 7.59, Brisbane, QLD 4215, Australia; yasaman.etemadshahidi@griffithuni.edu.au (Y.E.-S.); omelbaneen.qallandar@griffithuni.edu.au (O.B.Q.); jessica.evenden@griffithuni.edu.au (J.E.); f.alifui-segbaya@griffith.edu.au (F.A.-S.)
* Correspondence: khaled.ahmed@griffith.edu.au; Tel.: +61-75-678-0596

Received: 8 September 2020; Accepted: 16 October 2020; Published: 20 October 2020

Abstract: The use of additive manufacturing in dentistry has exponentially increased with dental model construction being the most common use of the technology. Henceforth, identifying the accuracy of additively manufactured dental models is critical. The objective of this study was to systematically review the literature and evaluate the accuracy of full-arch dental models manufactured using different 3D printing technologies. Seven databases were searched, and 2209 articles initially identified of which twenty-eight studies fulfilling the inclusion criteria were analysed. A meta-analysis was not possible due to unclear reporting and heterogeneity of studies. Stereolithography (SLA) was the most investigated technology, followed by digital light processing (DLP). Accuracy of 3D printed models varied widely between <100 to >500 μm with the majority of models deemed of clinically acceptable accuracy. The smallest (3.3 μm) and largest (579 μm) mean errors were produced by SLA printers. For DLP, majority of investigated printers ($n = 6/8$) produced models with <100 μm accuracy. Manufacturing parameters, including layer thickness, base design, postprocessing and storage, significantly influenced the model's accuracy. Majority of studies supported the use of 3D printed dental models. Nonetheless, models deemed clinically acceptable for orthodontic purposes may not necessarily be acceptable for the prosthodontic workflow or applications requiring high accuracy.

Keywords: 3-dimensional printing; additive manufacturing; dental models; accuracy; systematic review; full-arch

1. Introduction

Three-dimensional (3D) printing is an additive manufacturing (AM) process that allows conversion of digital models into physical ones through a layer-by-layer deposition printing process. 3D printing has been adopted in dentistry at an increasing rate and construction of dental models is one of the main applications of this promising technology in prosthodontics, orthodontics, implantology and oral and maxillofacial surgery, amongst others [1]. An essential prerequisite of dental models is creating an accurate replication of teeth and the surrounding tissues to serve their intended purposes as diagnostic and restorative aids for assessment, treatment planning and fabrication of various dental appliances and prostheses. Currently, gypsum casts poured from conventional impressions (e.g., alginates silicones, poly-sulphurs, ethers) are considered the gold standard for constructing dental models [2]. However, these cast models suffer a number of limitations, including a need for expedited processing of impressions, depending on the impression material; storage space for resultant casts; the cost of human and laboratory resources involved in fabrication; poor structural durability; and a propensity to dimensional changes over time [3]. In contrast, 3D printed models could offer a more efficient workflow that can be manufactured on demand and are more resilient, less-labour intensive and potentially time-saving [4]. Nonetheless, 3D printed models also present a unique set of limitations. The accuracy of the resultant models depends on several factors that can introduce errors. This includes the data acquisition and image processing of the oral hard and soft tissues, and the myriad of parameters involved in the manufacturing and postprocessing processes [5]. Moreover, models acquired through vat polymerisation and material jetting are prone to shrinkage during the polymerisation stage as well as having stair-step surfaces due to the layering technique used in construction [6]. In addition, a recent study demonstrated that models exhibit dimensional changes postprocessing as they age with their dimensions reported to be significantly different after three-weeks of manufacturing [7].

At present, there is an array of printing technologies available utilising various techniques, with varying outputs and performances, and consequently confounding the issue of a standardised expectation of accuracy. The most commonly used techniques are stereolithography (SLA), digital light processing (DLP), material jetting (MJ) and fused filament fabrication (FFF). Other processes such as continuous liquid interface production (CLIP) and binder jetting (BJ) have also been utilised but are not as common [8]. The earliest and most widely adopted 3D printing technique is SLA, which utilises ultraviolet (UV) scanning laser to sequentially cure liquid photopolymer resin layers. Each layer is solidified in the x-y direction, and the build platform incrementally drops in the z-direction to be recoated by resin and cured [9]. The photopolymerisation of each new layer connects it to the prior layer resulting in models with good strength. DLP uses a conventional light source to polymerise photosensitive liquid resins. However, unlike SLA, each x-y layer is exposed to the light all at once using a selectively masked light source, resulting in shorter production time [10]. Both SLA and DLP are versatile techniques as they can be used with a wide variety of resin systems [11]. CLIP is an advanced form of DLP technology with the advantage of faster printing time. Additionally, this technique utilises a membrane, which allows oxygen permeation to inhibit radical polymerisation. MJ, similar to vat polymerisation techniques (SLA, DLP and CLIP) employs photopolymerisation. This technique allows for deposition of liquid photosensitive resin through multiple jet heads on a platform, which is then cured by UV light [12]. As opposed to SLA and DLP, this technique requires no post-curing. Unlike Vat polymerisation and MJ, which use photopolymer material, FFF relies on the melting of thermoplastic materials, extruded through a fine nozzle, to create objects through layering filaments [11]. BJ technology, on the other hand, utilises selectively deposited liquid bonding agents to fuse powdered material.

The International Organization for Standardization (ISO 5725-1:1994) identifies accuracy as a qualitative concept, with trueness and precision being its quantitative counterparts. Trueness is defined as the 'closeness of agreement between the arithmetic mean of a large number of test results and the true or accepted value'. Precision is defined as the 'closeness of agreement between test results' [13]. There is currently no systematic review of data published on accuracy of dental models manufactured using 3D printing technologies; henceforth, this review aims to investigate the existing literature and evaluate the accuracy of 3D printed dental models using different 3D printing technologies and identify the printing parameters influencing their accuracy.

2. Materials and Methods

2.1. Review Question

The review search question was formulated using the PICO principle (Population, Intervention, Control, Outcome) [14], with dental models as the population cohort, 3D printing as the intervention and accuracy as the outcome. No control was defined. Hence, the formulated question was, "What is the accuracy of dental models manufactured using 3D printing technologies?" The protocol was registered on PROSPERO (registration number: CRD42020164099). The PRISMA guidelines were followed, where applicable [15].

2.2. Eligibility and Search Strategy

An electronic databases search was performed for PubMed, Cochrane Database, Web of Science, Scopus, EMBASE, LILACS, Scientific Electronic Library Online (SciELO) and the first ten pages of Google Scholar, using keywords and MeSH terms (Table 1). The Peer Review of Electronic Search Strategies (PRESS) guidelines were followed with an independent peer-reviewing the suitability of the search strategy [16]. Additionally, hand searching and cross-referencing was performed to identify additional studies. All study designs were included, whether prospective, retrospective, experimental in-vivo or in-vitro. The studies were limited to those published in English in the past 15 years (from 1 January 2005 to 13 March 2020). Abstracts from conferences, letters to the editor and studies that did not assess the accuracy of human dentate dental arches were excluded.

Table 1. Search strategy.

1. Search (print * OR "rapid prototyping" OR "additive manufacturing" OR fabrication OR stereolithography OR "stereo-lithography" OR "stereo lithography" OR photopolymer * OR photopolymer * OR "fused deposition Ωmodelling" OR "fused filament fabrication" OR "material extrusion" OR "material jetting" OR photojet OR polyjet OR "photopolymer jetting" OR "multijet printing" OR "binder jetting" OR "digital light processing" OR "selective laser sintering" OR "continuous liquid interface production" OR photopolymer * OR RP OR AM OR SLA OR SL OR FDM OR FFF OR PPJ OR PJ OR MJP OR MJ OR DLP OR CLIP OR SLS)
2. Search ("dental cast *" OR "dental model *" OR edentulous * OR edentate * OR dentate OR "full arch" OR "replica cast *") AND (3 D OR 3D OR 3 dimensional OR three dimensional)
3. Search (accuracy OR accuracies OR applicability OR precision OR repeatability OR reproducibility OR trueness OR sensitivity OR specificity OR specificities OR validation OR validity OR value OR agreement OR "spatial error *" OR "geometric error *" OR "dimensional error *" OR correctness OR exactness)
4. Search ((#1 and #2 and #3)) Filters: Publication date from 01/01/2005 to 13/05/2020

Initial screening of the titles and abstracts was independently performed by two investigators (O.Q. and J.E.). A list of the selected papers was compiled and compared, and any disagreements were discussed with a third investigator (K.A.) until a consensus was reached. Thereafter, the full text of the selected articles was reviewed to confirm the fulfilment of the inclusion criteria.

2.3. Data Extraction

Inclusion criteria and trial quality of included articles were assessed individually by two investigators (O.Q. and J.E.). The selected data were independently extracted and then cross-checked between the investigators and discrepancies were resolved by referring to a third investigator (K.A). Data collection, extraction and synthesis of the included studies was performed according to the following criteria:

- Sample size;
- model type;
- the 3D printing technology used;
- resolution (x,y) and layer thickness (z) used;
- materials and postprocessing protocol;
- accuracy of intraoral/lab scanner;
- accuracy assessment methodology;
- measurement of dimensional accuracy over time;
- presence of a study control;
- findings (accuracy); and
- limitations.

The authors of the included studies were not contacted to provide missing data not reported in their published studies.

2.4. Statistical Analysis and Risk of Bias (Quality) Assessment

A quality assessment of the methodology of the included studies was performed using the quality assessment of diagnostic accuracy-2 (QUADAS-2) [17] to assess their risk of bias and applicability concerns. Each domain was assessed and ranked as high risk, low risk or unclear.

3. Results

A total of 2209 studies were initially identified after the databases search (Figure 1). Screening of the titles and abstracts, and removing duplicates, resulted in 39 studies being selected. Six additional studies were identified through cross-referencing. Excluded studies either did not assess full-arch dental model [18–27] or were not published in English [28–30]. Three additional studies were later removed as they assessed and compared the accuracy of different intraoral scanners [5,31,32]. In addition, one study [33] was excluded as it was a published abstract. Finally, twenty-eight studies fulfilled the inclusion criteria and were further synthesised.

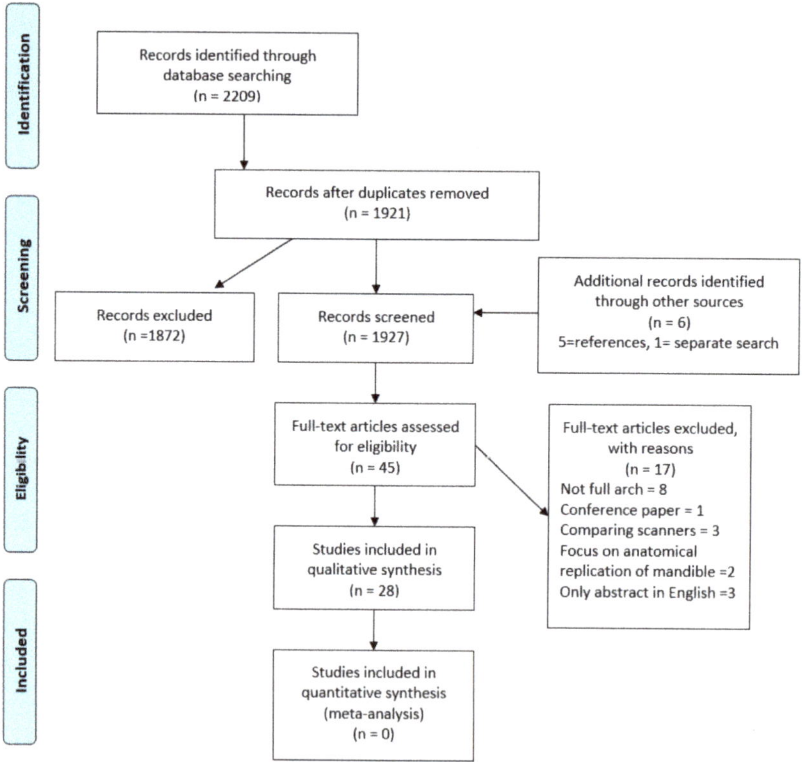

Figure 1. Flow chart for the selection of studies.

3.1. Study Characteristics

3.1.1. Sample Size and Reference Models

For this study, the sample size was determined based on the number of single dental arches manufactured by each printer. The majority of the studies (n = 19/28) assessed models of both maxillary and mandibular arches, and the remainder used either the maxillary (n = 8) or mandibular (n = 1) arches. The sample size ranged between one and sixty 3D printed single arch models per printer (Table 2).

Table 2. Details of studies included in the systematic review.

Authors/Date	3D printing Material	Model Data Source	Model Data Source	3D Printing System	3D Printer Details	Resolution x, y, z (μm)	Sample Size (Single Arch/Printer)	Assessment Method	Trueness (SD) (μm)	Precision (μm, ICC and IQR)
Aly and Mohsen, 2020 [34]	Photocurable polymer (liquid resin)	IOS scanned full dentate Typodont (Mx and Md)	IOS scanned Typodont	SLA	ProJet 6000, 3D Systems	Unclear	10	Digital callipers Tooth: MD, CH Arch: IC, IM	190 (100)	Unclear
Bohner et al. 2019 [35]	Unclear	Typodont (Mx, 7–7) containing implants at sites of 21, 24 and 26	Typodont (maxillary)	SLA	Unclear, Envisiontec	Unclear	10	Surveying software Arch: IP, IM	19.7 (13.3)	Unclear
Brown, Currier, Kadioglu and Kierl, (2018) [36]	Unclear	Patient IOS and alginate impressions (Mx and Md, min 6–6) 30 cases	Patient IOS and alginate impressions 30 cases	DLP MJ	Juell 3D Flash OC, Park Dental Research Objet Eden 260VS, Stratasys	z: 50, 100 z: 16	60	Digital callipers Arch: IC, IM, AD Tooth: MD, CH Occlusion: Unclear	70 80	Unclear
Burde et al. (2017) [37]	Poly-L-lactic acid wire Poly-L-lactic acid wire Grey light-curing resin	Patient stone model (Mx and Md) 10 cases. Unclear number of teeth present	Patient stone model 10 cases	FFF FFF SLA	Creatr HS, Leapfrog Custom RepRap, (based on a Prusal3 kit) Form 1+, Formlabs	z: 100 z: 100 z: 25	20	3D assessment Nominal ±11.51 Critical: ±230	156.2 (22.4) 128.3 (18.3) 207.9 (44.6)	Unclear
Camardella, de Vasconcellos Vilella and Breuning, (2017) [38]	Photopolymer resin Light-curing methacrylic resin (E-Denstone; Envisiontec)	Patient IOS 10 cases (Mx and Md, min 7–7)	Patient IOS 10 cases (mandibular)	MJ SLA	Objet Eden 260VS, Stratasys Ultra 3SP Ortho, Envisiontec	z: 16 z: 100	20	Surveying software Arch: IC, IP, IM Tooth: Unclear Occlusion: Unclear	Unclear	0.999ICC 0.998 ICC
Camardella, Vilella, van Hezel and Breuning, (2017) [39]	Light curing methacrylic resin (RC31, Envi-siontec)	Patients IOS and impressions (Mx and Md, min 6–6) 28 cases	Patients IOS and impressions 28 cases	SLA	Ultra 3SP, Envisiontec	Unclear	56	Digital callipers Arch: IC, IM Tooth: MD, CH Occlusion: OJ, OB AND 3D assessment Nominal: ±50 Critical: ±500	579 (1050)	Unclear
Cho, Schaefer, Thompson and Guentsch, (2015) [40]	Unclear	Lab scanned fully dentate Typodont (Mx) with 5 prepared teeth (16, 15, 21, 23, 26)	Lab scanned Typodont (maxillary)	SLA	Unclear	Unclear	5	3D assessment Nominal: ±50 Critical: ±500	27 (7)	91 (10)
Choi, Ahn, Son and Huh, (2019) [4]	Photopolymer Photopolymer	Typodont (Mx, 7–7) with prepared teeth (16, 11, 24 and 26)	Typodont (maxillary)	SLA DLP	ZENITH U, Dentis DIOPROBO, DIO	z: 50 z: 50	10	3D assessment Nominal: ±50 Critical: ±500	85.2 (13.1) 105.5 (22.5)	49.6 (12.1) 53.8 (17.5)
Cuperus et al. (2012) [41]	Epoxy Resin	IOS Dry human skull (min 6–6, with max 1 missing or deciduous tooth per skull) 10 cases Intra-oral scanner	IOS Dry human skull 10 cases	SLA	Unclear	Unclear	20	Digital callipers Arch: IC, IM Tooth: MD Occlusion: Unclear	100	Unclear
Dietrich, Ender, Baumgartnerand Mehl, (2017) [42]	Epoxy-based resin (Accura) photopolymer resins	Patient IOS 2 cases (Mx). Unclear number of teeth present	Patient IOS 2 cases (maxillary)	SLA MJ	Viper si2 SLA, 3D Systems Objet Eden 260, Stratasys	z: 100 at base and 50 at tooth level z: 16	10	3D assessment Nominal: ±20 Critical: ±100	92 (23) 62 (8)	20 (4) 38 (14)

Table 2. *Cont.*

Favero et al. (2017) [43]	Grey photopolymer resin (FLGPGR02; Formlabs). Unclear	Typodont (Mx, 7-7)	Typodont (maxillary)	SLA SLA DLP DLP MJ	Form 2, Formlabs Vector 3sp, Envisiontec Juell 3D, Park Dental Perfactory Desktop Vida, Envisiontec Objet Eden 260V, Stratasys	z: 25, 50, 100 z: 100 z: 100 z: 100 z: 28	12	3D assessment Nominal: ±20 Critical: ±250	64 79 44 56 85	Unclear
Hazeveld, Huddleston Slater and Ren, (2014) [44]	Unclear	Patient Stone model (Mx and Md, min 6–6) 6 cases	Patient Stone model 6 cases	DLP BJ MJ	Unclear, Envisiontec, Unclear, Z-Corp Unclear, Objet Geometries	Unclear	12	Digital callipers Arch: Unclear Tooth: MD, CH Occlusion: Unclear	Unclear	Unclear
Jin, Jeong, Kim and Kim, (2018) [45]	Unclear	Lab scanned Typodont (Mx, 7–7)	Lab scanned Typodont (maxillary)	MJ FFF	ProJet 3500 HDMax, 3D Systems Cube, 3D Systems	z: 31.97 z: 123.71 (thickness measured after printing)	10	3D assessment Nominal: ±50 Critical: ±500	129.1 (7.8) 149.0 (4.7)	44.6 (8.9) 52.1 (10.9)
Jin, Kim, Kim and Kim, (2019) [6]	Photocurable liquid resin Acrylic polymer	Lab scanned Typodont (Mx and Md, 7–7)	Lab scanned Typodont (maxillary)	SLA MJ	ProJet 6000, 3D Systems ProJet 3500 HD Max, 3D Systems	FMR FMR	10	3D assessment Nominal: ±50 Critical: ±500	114.3 (1.8) 124 (3.7)	59.6 (8.2) 41.0 (5.8)
Joda, Matthisson and Zitzmann, (2020) [7]	Light-curing polymer, (SHERAPrint-model plus "sand" UV, SHERA)	IOS Typodont (Mx, 7–7), with missing 25 and prepared 24 and 26)	IOS Typodont (maxillary)	SLA	P30, Straumann	Unclear	10	3D assessment Nominal: unclear Critical: unclear	3.3 (1.3)	Unclear
Kasparova et al. (2013) [46]	ABS plastic material, Clear resin	Patient stone model 10 cases. Unclear number of teeth present	Patient stone model 10 cases	FFF MJ	RepRap, Unclear ProJetHD3000, 3D Systems	x,y: 200, z: 0.35 Unclear	20 2	Digital callipers Tooth: CH Arch: IC	Unclear Unclear	Unclear Unclear
Keating, Knox, Bibb and Zhurov, (2008) [47]	Hybrid epoxy-based resin	Patient stone model 15 cases. Unclear number of teeth present	Patient stone model 15 cases	SLA	SLA-250/40, 3D Systems	z: 150	30	Digital callipers Tooth: CH Arch: IC, IP, IM	150 (160)	Unclear
Kim et al. (2018) [19]	Unclear	Lab scanned Typodont (Mx and Md, 7–7)	Lab scanned Typodont	SLA: DLP MJ FFF	ZENITH, Dentis M-One, MAKEX Technology Objet Eden 260VS, Stratasys Cubicon 3DP-110F, HyVISION System	x,y: 50 z: 50 x,y: 70 z: 75 z: 16 x,y: 100 z: 100	10	Surveying software Tooth: MD, BL, CH Arch: IC, IM	138 (79) 446 (46) 74 (39) 307 (61)	88 (14) 76 (14) 68 (9) 99 (14)
Kuo, Chen, Wong, Lu and Huang, (2015) [48]	Unclear	Patient IOS Patient impressions poured, and lab scanned (Md, 7–7) 1 case	Patient IOS Patient impressions poured, and lab scanned 1 case	MJ	Connex 350, Stratasys	Unclear	1	3D assessment Nominal: ±60 Critical: ±300	140	Unclear
Loflin et al. (2019) [49]	Grey photopolymer resin, (FLGPGR03; Formlabs)	Patient stone models (Mx and Md) 12 cases. Unclear number of teeth present	Patient stone model 12 cases	SLA	Form 2, Formlabs	z: 25, 50, 100	24	ABO tool Tooth: marginal ridge Occlusion: OJ, occlusal contacts	Unclear	Unclear

Table 2. *Cont.*

Nestler, Wesemann, Spies, Beuer and Bumann, (2020) [50]	Dental SG Optiprint Imprimo LC model ABS Polylactide	Cast in standard tessellation language (STL) format (Mx, 7–7) including 5 measuring cubes in areas 16, 26, 13, 23 and between 11 and 21)	Maxillary cast in standard tessellation language (STL) format	SLA SLA DLP FFF FFF	Forms 2, Formlabs Myrev140, Sisma Asiga Max UV, Asiga M2, Makergear Ultimaker 2+, Ultimaker	Unclear Unclear Xy: 62, Z: Unclear Unclear x,y: 12.5, z: Unclear	37 34 for Myrev140	Surveying software Arch: IC, IM, arch length	-80 (94 −175 (28) −16 (32) −55 (39) 12 (43)	134 28 47 55 56
Papaspyridakos et al., (2020) [51]	Photopolymer resin, dental model resin (Formlabs)	Lab scanned Patient stone model 1 case (Md) with 4 abutment-level implant analogs	Lab scanned Patient stone model 1 case (mandibular)	SLA	Form 2, Formlab	z: 25	25	3D assessment Nominal ±50 Critical: ±200	59 (16)	Unclear
Rebong, Stewart, Utreja and Ghoneima, (2018) [52]	Unclear	Patient stone models (Mx and Md, min 6–6) 12 cases	Patient stone model 12 cases	FFF SLA MJ	Makerbot Replicator, Makerbot Industries Projet 6000, 3DSystems Objet Eden 500V, Stratasys	z: 100 z: 50 z: 16	24	Digital calipers Arch: IC, IM Tooth: Unclear Occlusion: OJ, OB	110 (420) −20 (370) −190 (330)	Unclear
Rungrojwittayakul et al. (2020) [53]	Unclear	Lab scanned fully dentate Typodont (Mx,)	Lab scanned Typodont (maxillary	CLIP DLP	Carbon M2, Carbon MoonRay S100, SprintRay	Unclear	10	3D assessment Nominal: ±10 Critical: ±100	48 (44) 87 (57)	0.968 ICC 0.983 ICC
Saleh, Ariffin, Sherriff and Bister, (2015) [54]	Unclear	Lab scanned Typodont (Mx and Md, 7–7)	Lab scanned Typodont	MJ	Objet Eden 250, Stratasys	Unclear	8	Digital calipers Tooth: MD Arch: IC, IM Occlusion: OJ, OB	320 (156)	Unclear
Sherman, Kadioglu, Currier, Kierl and Li, (2020) [55]	Unclear	Patient IOS (Mx and Md, min 6–6) 15 cases	Patient IOS 15 cases	DLP	JUELL 3D Flash OC, Park Dental Research Corporation	z: 50, 100	30	Digital calipers Arch: IC, IM, AD Tooth: MD, CH Occlusion: Unclear	Unclear	Unclear
Wan Hassan, Yusoff and Mardi, 2017 [56]	High-performance composite (Zp151; 3D Systems).	Patient impression (Mx and Md, min 6–6) 10 cases	Patient impression 10 cases	BJ	Z Printer 450, 3D Systems	z: 89–102	30	Digital callipers Arch: IC, IP, IM Tooth: MD, CH, BL Occlusion: Unclear	−20	Unclear
Zhang, Li, Chu and Shen, (2019) [57]	Dental model resin (Formlabs) Model Ortho resin (Union Tec) Encashape, ENCA-Model resin Light curing methacrylate resin E-Denstone, EnvisionTEC	Patient IOS (Mx and Md, 7–7) 1 case	Patient IOS 1 case	SLA DLP DLP DLP	Form 2, Formlabs EvoDent, UnionTec EncaDent, Encashape Vida HD, EnvisionTec	x,y: 140 z:25, 30,10 z: 50,100 x,y: 58 z: 20, 30, 50,100 x,y: 50 z: 50, 100	2	3D assessment Nominal: ±50 Critical: ±250	34.4 23.3 26.5 31.7	Unclear

Mx = maxillary, Mn = mandinular, CH = crown height, BL = buccolingual width, MD = mesiodistal width, IC = intercanine width, IP = interpremolar width, IM = intermolar width, OB = overbite, OJ = overjet, SLA = stereolithography, MJ = material jetting, BJ = binder jetting, DLP = digital light processing, CLIP = continuous liquid interface production, FFF = fused filament fabrication, IOS = intraoral scanner, ABO = American Board of Orthodontics.

3.1.2. Sample Details and Controls

The inclusion criteria for the studies that collected patient samples (digital or physical impressions or models) varied slightly with the majority ($n = 14/25$) being full arch dentate post-orthodontic models, including up to permanent first molars [36–39,42,44,46–49,52,55–57]. One of the studies [42] also used a model with a shortened dental arch. Another used an edentulous mandibular cast with four multi-unit abutments for implant prosthodontic rehabilitation [51].

Twenty-four studies included reference models as controls in their methodology design [6,7,34,36–53,55–57]. The controls included were a dental stone cast ($n = 8$), a digital STL image of a dental stone cast ($n = 3$), typodont digital STL image ($n = 7$), typodont ($n = 1$), prefabricated resin model digital STL image ($n = 2$), patient intraoral scan image (2) or a dry human skull ($n = 1$). In addition, there were four studies [4,35,47,54] that did not include a reference model as a control, rather compared various printing technologies against each other.

3.2. Additive Manufacturing

3.2.1. D printing Technologies Assessed and Printing Parameters

An array of additive manufacturing systems were assessed in the included studies with several investigating more than one type of technology, printer brand or parameter settings (Table 2). The majority of studies investigated SLA ($n = 20$), MJ ($n = 11$), DLP ($n = 9$) and, to a lesser extent, FFF ($n = 6$), BJ ($n = 2$) and CLIP ($n = 1$). With regards to printing parameters, one study reported following the manufacturers' recommendations [6], while others explicitly detailed the printing parameters used [4,19,36–38,42,43,45–47,49–52,55–57]. In contrast, the remainder of the studies did not provide clear details regarding the printing parameters used.

3.2.2. Layer Thickness

The specified printing layer thickness (z-axis resolution) substantially varied amongst studies and ranged from 25–150 µm for SLA, 20–100 µm for DLP, 16–32 µm for MJ, 100–150 µm for FFF and 89–102 µm for BJ. The study using CLIP technology did not specify the layer thickness used [53]. Most studies did not specify the printing resolution in the x- and y-axes. However, in those that did, the x-y plane resolution ranged from 50–140 µm for SLA, 50–70 µm for DLP and 12.5–200 µm for FFF.

3.2.3. Materials Used

The materials used by 3D printers are broadly classified based on their printing technologies. Vat polymerisation technologies (SLA, DLP and CLIP) used liquid photopolymers, including acrylates and epoxides, 3D material extrusion technology (FFF) used polylactic acid (PLA), or acrylonitrile butadiene styrene (ABS). MJ technology used photopolymers resins (acrylates) in liquid form and BJ technology used polylactic acid powder. Eleven studies did not specify the material used for the corresponding technology [19,35,36,40,44,45,48,52–55]. Within the studies assessing stone models, four used Type IV dental stone [6,35,40,45], one used Type III dental stone [4] and one did not specify the stone type utilised [34].

3.2.4. Base Designs and Filling Patterns

The three types of base designs used in the studies were horseshoe-shaped bases [4,34,36–39,41,43,49–51,55], regular American Board of Orthodontics (ABO) [35,38,44,46,47,49,52,54,57] and horseshoe-shaped with a transverse supporting bar [7,38,48,53]. Six studies did not specify their base design [6,19,40,42,45,56]. Filling patterns employed in these studies were predominantly solid; however, hollow shelled [53,55] and honeycomb [37] were also utilised.

3.2.5. Postprocessing Protocol

The majority of studies did not specify the postprocessing protocol (n = 19/28) [6,19,34–36,40,41,44–48,50–52,54–57]. Nine studies reported their post-curing protocol for vat polymerisation techniques [4,7,37–39,42,43,49,53] which included cleaning the models with isopropyl alcohol [37,43,49,53], or ethanol [4] to remove uncured resin followed by curing with UV light. Three studies only used UV light to post-process SLA models [38,39,42]. One study placed the SLA models in an ultrasonic bath followed by using light with wavelengths of 280–580 nm for post-curing [7]. Two studies reported that MJ and FFF did not require post-curing [37,38], and one study rinsed the MJ printed models in a bath of caustic soda to clean them [42]. Additionally, three studies specified removing the support structures from the models [37,42,57].

3.3. Assessment Methodology

The assessment of the accuracy of 3D printed models was performed using either 3D deviation analyses or 2D linear measurements. For the 3D assessment, step-height measurements through iterative point-cloud surface-matching followed by 3D deviation assessment were performed. For 2D linear measurements, reference points were selected and measured either directly onto the physical model using digital callipers or indirectly on the model's digital image using surveying software. The majority of studies relied on 3D assessment [4,6,7,37,39,40,42,43,45,48,51,53,57] followed by digital callipers [34,36,39,41,44,46,47,52,54–56] and surveying software measurements [19,35,38,50]. One study [49] used the ABO cast-radiograph evaluation tool.

3.3.1. Surface Matching and 3D Deviation Analyses

The studies which performed 3D assessment used min/max nominal values ranging between ±10 to ±60 µm and min/max critical values of ± 100 to ± 500 µm. Before superimposition, the 3D-printed models were scanned and converted to standard tessellation language (STL) format. The scanners included desktop/laboratory scanners (n = 15) [4,6,7,19,35,37,38,40,42,43,45,48,51,53,57], intraoral scanners (n = 2) [36,41], and computerised tomography scanner (n = 1) [38]. Two studies did not specify the details of image acquisition [34,50]. While most studies did not specify the accuracy of the scanners nor mentioned calibrating the scanners before scan acquisition, the remaining studies reported a scanning accuracy <20 µm [4,6,19,40,45,46,57].

3.3.2. Linear Measurements of Physical and Digital Models

Studies that utilised digital callipers with physical models or measuring software with digital models relied on various reference points to perform 2D linear measurements. The reported accuracy of all callipers was 10 µm, and the ABO tool was 100µm. The selected reference points relied on varying tooth measurements (crown height, mesiodistal width, buccolingual width and marginal ridge width), arch measurements (intercanine width, interpremolar width and intermolar width) and occlusion measurements (overjet, overbite, occlusal contact and interarch sagittal relationships). Most studies used both tooth and arch measurements (n = 10) [19,34,36,39,41,43,46,47,55,56], while one study only used tooth measurements [44] and three studies only used arch measurements [35,38,40]. Moreover, five studies used occlusion measurements in addition to the arch measurements [19,39,49,52,54].

3.3.3. Time of Assessment

The time at which the 3D printed models were scanned or measured was reported by six studies [7,41,45–47,52]. Within those studies, five assessed the models after a week of printing [41,45–47,52] and one assessed the accuracy after one day, followed by weekly intervals for four consecutive weeks [7].

3.4. Outcomes Assessed

3.4.1. Clinical Acceptability

The clinically acceptable error defined in the studies varied widely from <100 μm [51,53], <200μm [6,45], <250 μm [43], <300 μm [44,48] and <500 μm [19,34–36,42,46,47,49,50,52,55–57]. One study [4] defined various acceptable ranges of error for different measurement points and seven studies did not define any clinically acceptable range [4,7,37,38,40,41,54]. From those, twelve assessed orthodontic models [19,36,39,42–44,47,49,52,55–57], five assessed fixed pros and implant models [6,34,35,51,53] and three did not specify [45,46,48].

3.4.2. Trueness

Overall, the mean deviations from the reference model across all studies ranged from 3.3 to 579 μm [7,39]. Studies which assessed the trueness of both 3D printed and stone models found that the mean error for the stone model was consistently lower than their 3D printed counterparts [4,6,34,35,40,45]. In contrast, one study [45] reported no statistical differences between stone and MJ models and another [6] found no statistical difference between SLA and stone. However, several studies did not fully report the details of the 3D printer/s used or their trueness results [38,40,41,44,46,49,55]. Nonetheless, six DLP printers, five SLA printers and one MJ printer had an error measurement of <100 μm for full-arch dental models, demonstrating high trueness (Figure 2). Similarly, the BJ printer (ZPrinter 450, 3D Systems, USA), CLIP printer (M2, Carbon, USA) and two FFF printers (Ultimaker 2+, Ultimaker B.V, Geldermalsen, The Netherlands; and M2, Makergear, USA) reported high trueness results (Table 2).

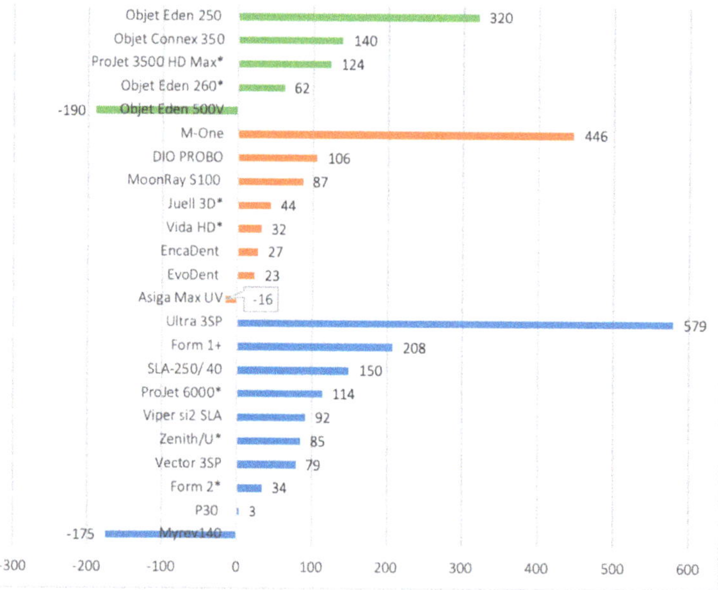

Figure 2. Reported trueness in microns for material jetting (MJ, green), digital light processing (DLP, orange) and stereolithography (SLA, blue) 3D printed full-arch dental models. * Asterisk denotes lowest mean error identified from different studies—other results in microns reported include: Form 2 = 59, 64 and −80; Zenith series = 138; Projet 6000 = 190; Juell 3D = 44, 70; Vida = 56; Objet Eden 260 series = 74, 80 and 85; Projet 3500 HD Max = 129. Data from studies that did not report details of 3D printer used or trueness data were not included in the figure.

All SLA printers consistently produced oversized 3D printed models compared to the control, excluding the Myrev 140 printer [50]. The P30 reported the lowest mean error of 3.3 µm [7] and the Form 2 printer followed with reported mean errors ranging between 34.4 to 64 µm [37,43,56]. The SLA Ultra 3SP demonstrated the highest mean error at 579 µm [39]. Similar results were found for DLP printers with the majority of printers producing oversized models, except the Asiga Max UV, which also reported the lowest mean error for DLP at—16 µm [50]. The Evodent was the second most accurate DLP printer with a 23.3 µm error, followed by the Encadent at 26.5 µm errors [50,57]. Furthermore, JUELL 3D FLASH OC, Vida HD and Vida had a reported mean error of 44 µm, 31.7 µm and 56µm, respectively [43,45,57]. The highest mean error for the DLP printing technology was the M-One printer with a mean error of 446 µm [19]. Within MJ printers, Objet Eden 260 series (V, VS) had the lowest mean errors ranging from 62 to 85 µm [19,36,42,43], whilst the highest mean error was 320 µm (Objet Eden 250) [54]. Ultimaker 2+ printer as FFF technology had the least deviation error of 12µm [50], while Cubicon 3DP 110F reported a mean error of 307 µm [19]. The two printers for BJ (Z printer 450 and unclear) and CLIP (Carbon M2) technologies had mean errors of—20 µm [56] and 48µm [53], respectively.

3.4.3. Precision

The precision of 3D printed models was assessed in 10 studies using either root mean square value (RMS) [4,6,19,40,42,45], the intraclass correlation coefficient (ICC) [38,53] and or interquartile range (IQR) [50]. The RMS value ranges for SLA, FFF, MJ and DLP were 23 to 91 µm, 52.1 to 99 µm, 38 to 68 µm and 53.8 to 76 µm, respectively. The range for ICC was 0.968 for CLIP and 0.999 for MJ. In addition, one study [50] reported IQR of 28 to 134 µm for SLA, 55 to 56 µm for FFF and 47 µm for DLP.

Two studies found 3D printed models to have equal or greater precision than conventional stone models [6,45]. By contrast, two studies [4,40] found conventional stone models to be more precise than the 3D printed models. Of note, studies that used ICC [38,53] to assess precision; demonstrated excellent reproducibility (>0.9 ICC value) of 3D printed models, according to the Koo and Li (2016) classification [58].

3.5. Statistical Analysis

The limited reporting, varying printing technologies, printing parameters, assessment methodology and statistical analysis employed in the included studies presented a heterogeneity that precluded from performing a meaningful meta-analysis.

3.6. Risk of Bias Assessment

The risk of bias and applicability concerns varied across the studies, which may have influenced the reliability of their results (Table 3). The reference standards used in almost all the studies had a low risk of bias and low concerns regarding applicability (27/28). The risk of the index test, however, was high for the majority of studies (21/28). This high risk was because the studies either did not use 3D superimposition, and therefore the mean error may not have been an accurate representation of the whole arch deviation, or the method of assessing the model's deviation introduced errors other than those arising from the CAM process. These errors include the use of full-arch intraoral scanning for data acquisition which may introduce scanner error in-addition to the 3D printing error. Similarly, lack of details of assessors and their calibration was a noted risk of bias in several studies. Finally, the majority of studies had a high risk of bias for sample selection. This high risk is attributed to the lack of details relating to sample size calculation, spectrum of selected samples and/or postprocessing protocol. However, most of the samples remained highly applicable with the measurement protocol employed in the studies appropriately described to allow the reviewer to answer the review question.

Table 3. Risk of bias and applicability concerns according to QUADAS-2 tool. Negative sign (–) denotes high risk of bias. Positive sign (+) denotes low risk of bias.

Study	Risk of Bias			Applicability Concerns		
	Patient (Sample) Selection	Index Test	Reference Standard	Patient (Sample) Selection	Index Test	Reference Standard
Aly and Mohsen, 2020 [34]	–	–	+	+	–	+
Bohner et al. 2019 [35]	–	–	+	–	–	+
Brown, Currier, Kadioglu and Kierl, 2018 [36]	–	–	+	+	–	+
Burde et al. 2017 [37]	–	+	+	+	+	+
Camardella, de Vasconcellos Vilella and Breuning, 2017 [38]	+	–	+	+	–	+
Camardella, Vilella, van Hezel and Breuning, (2017) [39]	–	–	+	+	–	+
Cho, Schaefer, Thompson and Guentsch, 2015 [40]	–	+	+	–	+	+
Choi, Ahn, Son and Huh, 2019 [4]	–	+	+	+	+	+
Cuperus et al. 2012 [41]	–	–	+	–	–	+
Dietrich, Ender, Baumgartner and Mehl, 2017 [42]	–	–	+	+	–	+
Favero et al. 2017 [43]	–	+	+	+	+	+
Hazeveld, Huddleston Slater and Ren, 2014 [44]	–	–	+	–	–	+
Jin, Jeong, Kim and Kim, 2018 [45]	–	+	+	+	+	+
Jin, Kim, Kim and Kim, 2019 [6]	–	+	+	+	+	+
Joda, Matthisson and Zitzmann, 2020 [7]	–	–	+	+	–	+
Kasparova et al. 2013 [46]	–	–	+	+	–	+
Keating, Knox, Bibb and Zhurov, 2008 [47]	–	–	+	+	–	+
Kim et al. 2018 [19]	–	+	+	+	+	+
Kuo, Chen, Wong, Lu and Huang, 2015 [48]	–	–	+	+	–	+
Loflin et al. 2019 [49]	–	–	+	+	–	+
Nestler, Wesemann, Spies, Beuer and Bumann, 2020 [50]	–	-	+	+	–	+
Papaspyridakos et al., 2020 [51]	+	–	+	+	–	+
Rebong, Stewart, Utreja and Ghoneima, 2018 [52]	–	–	+	+	–	+
Rungrojwittayakul et al. 2020 [53]	–	–	+	+	–	+
Saleh, Ariffin, Sherriff and Bister, 2015 [54]	–	–	+	+	–	+
Sherman, Kadioglu, Currier, Kierl and Li, 2020 [55]	–	–	+	+	–	+
Wan Hassan, Yusoff and Mardi, 2017 [56]	+	–	+	+	–	+
Zhang, Li, Chu and Shen, 2019 [57]	–	–	+	+	–	+

4. Discussion

Given 3D printing's promising potential and increased use in dentistry, it is essential to evaluate the accuracy of 3D printed dental models. This is the first systematic review, to the authors' knowledge, investigating the accuracy of dental models manufactured using 3D printing technology. The selection criteria for the included reference standards were high, subsequently the risk of bias and applicability concerns were low according to the QUADAS-2 tool. The findings of this review support the use of 3D printing for the fabrication of dental models and deem them as clinically acceptable with the majority of included studies (n = 20/28) establishing a clinically acceptable error range of <100 to 500 μm. 3D printed models were found to be a valid alternative to stone models when taking precision into account. Nonetheless, the study by Wan Hassan (2019) was an outlier which found BJ 3D printed models not clinically acceptable due to their discrepancy of >500 μm. It is, however, worth noting the included studies which used orthodontic models [19,34,36,42,46,47,49,50,52,55,57] had more relaxed thresholds for clinical acceptability (up to 500 μm), compared to those intended for prosthodontic applications (up to 200 μm) [6,51,53]. Indeed, in orthodontics, a measurement difference of <300 μm between orthodontic casts and 3D printed models has been reported to be clinically acceptable [59–61]. On the other hand, in prosthodontics, the accuracy needs of dental models for the fabrication of dental prostheses is generally considered higher. A recent study concluded that three-unit fixed partial dentures fabricated using 3D printed models, whilst demonstrating inferior fit when compared to those fabricated using stone casts [27], the detected marginal gaps remained within the clinically accepted threshold of 120 μm reported in the literature [62]. Such clinically relevant thresholds become more critical in complex prosthodontic treatment modalities. Implant-supported complete dental prostheses or hybrid bridges have a maximum acceptable threshold of fit between the prostheses platform and the dental implants ranging between 59–150 μm [63–65]. Accordingly, the choice of 3D printing technology must be determined by its intended application. Hence, it is reasonable to conclude that 3D printed models which are clinically acceptable for orthodontic purposes may not necessarily be acceptable for the prosthodontic workflow or other dental applications requiring high accuracy.

The most common 3D printing technology investigated by the included studies was SLA with the findings demonstrating that SLA and DLP achieved the best accuracy for full-arch models. Amongst the

SLA printers, Form 2 by Formlabs was investigated the most, and consistently produced clinically acceptable models. Although a wider range of mean errors was observed amongst SLA printed models, the Form 2 SLA desktop printer [43,49,51,57] also consistently produced models more accurate than MJ printers and was more cost-effective [43,44]. Moreover, the SLA printer P30 reported the most accurate models amongst all studies, followed by the DLP Asiga Max UV [7,50]. Additionally, SLA printers produced acceptable results regardless of their layer thickness, and therefore the layer thickness of 100 μm may be considered as an optimal thickness that balances accuracy and printing time when compared to 25 and 50 μm layers [49,57]. Moreover, it was suggested that a hollow or honeycomb infill could be indicated to reduce printing time and material-use with study models. Although no studies assessed the effect of using different resins with the same printer, using the manufacturer recommended resin was advised. In contrast, only one study assessed CLIP technology and used the Carbon M2 printer, which printed 3D models with deviations as small as 48 μm [53]. This study also concluded that the accuracy of 3D printed models was affected by the printing technique regardless of the base design. However, due to the limited studies that assessed the accuracy of BJ [56] and CLIP technologies [53], further investigation of these techniques is required to validate the viability of these printers. It is worth mentioning that some studies did not provide details of the sample size calculation, resin materials and/or post-curing protocols (Table 3), exposing them to high risk of bias and applicability concerns with regards to sample selection. As a result, no conclusions were drawn based on these parameters, other than those studies that reported using the manufacturer's recommendations.

The two studies which examined the Ultra printer by EnvisionTEC [38,39] reported that the SLA models with horseshoe bases were not accurate nor clinically acceptable due to contraction in the transversal dimension during the post-curing protocol. However, as the horseshoe base is favoured for appliance fabrication and reduces material use, the inclusion of a posterior connection bar was suggested to prevent this significant dimensional reduction in the posterior region of the SLA model [37,38]. Nevertheless, several studies assessing other SLA printers [4,34,37,41,43,50,51] contradicted these findings and concluded that models printed by SLA with a horseshoe base to be clinically acceptable.

When assessing DLP technology, apart from the M-One printer used by Kim et al. (2018), all other printers had accuracies comparable to SLA and MJ. The Asiga Max UV printer produced the lowest mean error (-16 μm) [50]. In addition, Sherman et al. (2020) and Zhang et al. (2019) assessed the accuracy of DLP printed models with various layer thicknesses ranging from 20–100 μm and suggested that all the printed models were clinically acceptable. Thus, similar to SLA printers, it can be inferred that a layer thickness of 100 μm can still produce models with clinically acceptable accuracies for DLP printers. In addition to layer thicknesses, two studies assessed different filling patterns for DLP printed models [53,55]. Altering the filling pattern from solid to hollow reduced material wastage, build time and cost with no statistically significant difference in mean error.

Most MJ printers could reproduce models with high levels of trueness and precision, regardless of their base design [38]. From those, Objet Eden 260 series [19,36,42,43], was the most commonly investigated printer and consistently produced models with the highest accuracies due to its smaller layer thickness of 16 μm followed by the Projet3500 HDMax [6,45]. These printers were used due to their relatively affordable price and ability to print in smaller layer thicknesses. It is worth mentioning that although the reduction in layer height resulted in smoother surface finish and greater detail, the printing time increased [43].

FFF desktop printers, albeit considered the most affordable printers [46,50], provided models with acceptable accuracy. The most accurate models were created by the Ultimaker 2+ printer (12 μm) [50]. Although the materials used by FFF printers, namely PLA or ABS were inexpensive; the resultant models had inferior surface properties compared to acrylates and epoxides which were used for vat polymerisation technologies (SLA, DLP and CLIP). Similar to SLA and DLP, studies assessing FFF suggested a layer thickness of 100 μm to be clinically acceptable. Moreover, Burde et al. (2017) printed

FFF models with a honeycomb pattern to reduce print time, material and cost with the resultant models deemed clinically acceptable.

There were very limited data to compare the results from 3D assessment to linear measurements for the same printers. However, it is worth noting that the highest risk of bias and applicability concerns for index test were recorded for studies that used linear measurements. This was reflective of the limited measuring points provided by those studies in comparison to a full arch deviation measurement by 3D superimposition. Additionally, some of the studies had a high risk of bias as human error may have been introduced by performing physical linear measurements with no information provided on the calibration of the examiners [19,49,50,53]. Furthermore, for 3D superimposition techniques, the risk of bias and applicability concerns were low for most studies as high accuracy desktop scanners were utilised and CAM was the only identified source of error. Nevertheless, studies that used intraoral scanners, made conventional impressions with or without pouring casts had a higher risk of bias due to the additional stages that may have introduced their own set of errors.

The Projet 6000 printed models were assessed using different methods [6,34]. The mean error calculated using full arch 3D superimposition (114.3 μm) was smaller than the intermolar width error measured by a surveying software (190 μm). Similarly, two studies assessed the Juell 3D printer [36,43], and the mean error calculated by full arch superimposition (44 μm) was smaller than the digital calliper measurements for the intermolar width (70 μm). On the other hand, two studies [19,36] assessed the Objet Eden 260VS model, using two different linear measurement methods. The mean errors calculated using surveying software and digital calliper were very similar (74 and 80 μm, respectively). These findings do highlight the need for a standardised measuring protocol to facilitate comparison of results across studies given the noted discrepancy between the different assessment techniques.

A potential limitation of this review is the assessment findings of the included studies in relevance to the measurement time of the 3D printed models. This limitation is due to the possible dimensional changes exhibited by printed models over time, with only six of the included studies identifying the time of model measurement. Joda et al. (2020) [7] assessed the effect of time on the accuracy of the printed models and was the solely identified study that reported assessing the models for more than one week. The results suggested that the accuracy of SLA printed models was time-dependent due to a statistically significant change in their dimensions after three weeks of storage, suggesting the use of SLA 3D printed models as single-use products with definitive prosthetic reconstructions. The lack of standardised reporting in included studies is also a limitation that may have resulted in a high risk of bias in terms of index test and sample selection.

Consequently, the evident heterogeneity of the included studies with varying techniques, manufacturing parameters, materials and assessment protocols, a meta-analysis was not feasible. It is also worth noting the limitations present in the literature which need to be addressed in future studies. Investigation of different layer thicknesses for FFF, MJ, BJ and CLIP printing technologies, the effect of time and storage conditions on the accuracy of different 3D printed models, as well as clinical patient outcomes, remain lacking. A standardised accuracy assessment protocol for 3D printing of dental models is also necessary to facilitate performance comparison. Future studies should also involve a standardised reporting protocol that details all printing parameters, materials used, postprocessing protocol and time of assessment.

5. Conclusions

The findings of this study support the use of 3D printed dental models, especially as orthodontic study models. Irrespective of the 3D printing technology, certain printers were able to demonstrate low errors and hence can be recommended for dental applications that require high accuracy models. Other factors such as layer thickness, base design, postprocessing and storage can equally influence the accuracy of the resultant 3D printed models. Nonetheless, the high risk of bias with regards to the lack of standardised testing of accuracy warrants careful interpretation of the findings.

Author Contributions: Y.E.-S.; methodology, writing—original draft, writing—review and editing, project administration. O.B.Q.; methodology, validation, formal analysis, investigation, writing—review and editing. J.E.; methodology, validation, formal analysis, investigation, writing—review and editing. F.A.-S.; methodology, writing—review and editing, supervision. K.E.A.; conceptualization, methodology, writing—original draft, writing—review and editing, project administration, supervision. All authors have read and agreed to the published version of the manuscript.

Funding: No internal or external funding was received.

Conflicts of Interest: The authors declare no conflict of interest.

References

1. Dawood, A.; Marti, B.M.; Sauret-Jackson, V. 3D printing in dentistry. *Br. Dent. J.* **2015**, *219*, 521–529. [CrossRef]
2. Ender, A.; Mehl, A. Accuracy of complete-arch dental impressions: A new method of measuring trueness and precision. *J. Prosthet. Dent.* **2013**, *109*, 121–128. [CrossRef]
3. Ahmed, K.E.; Whitters, J.; Ju, X.; Pierce, S.G.; MacLeod, C.N.; Murray, C.A. A Proposed Methodology to Assess the Accuracy of 3D Scanners and Casts and Monitor Tooth Wear Progression in Patients. *Int. J. Prosthodont.* **2016**, *29*, 514–521. [CrossRef]
4. Choi, J.-W.; Ahn, J.-J.; Son, K.; Huh, J.-B. Three-Dimensional Evaluation on Accuracy of Conventional and Milled Gypsum Models and 3D Printed Photopolymer Models. *Materials* **2019**, *12*, 3499. [CrossRef]
5. Patzelt, S.B.; Bishti, S.; Stampf, S.; Att, W. Accuracy of computer-aided design/computer-aided manufacturing–generated dental casts based on intraoral scanner data. *J. Am. Dent. Assoc.* **2014**, *145*, 1133–1140. [CrossRef]
6. Jin, S.-J.; Kim, D.-Y.; Kim, J.-H.; Kim, W.-C. Accuracy of Dental Replica Models Using Photopolymer Materials in Additive Manufacturing: In Vitro Three-Dimensional Evaluation. *J. Prosthodont.* **2018**, *28*, e557–e562. [CrossRef] [PubMed]
7. Joda, T.; Matthisson, L.; Zitzmann, N.U. Impact of Aging on the Accuracy of 3D-Printed Dental Models: An In Vitro Investigation. *J. Clin. Med.* **2020**, *9*, 1436. [CrossRef] [PubMed]
8. Oberoi, G.; Nitsch, S.; Edelmayer, M.; Janjić, K.; Müller, A.S.; Agis, H. 3D Printing—Encompassing the Facets of Dentistry. *Front. Bioeng. Biotechnol.* **2018**, *6*, 172. [CrossRef] [PubMed]
9. Alifui-Segbaya, F. Biomedical photopolymers in 3D printing. *Rapid Prototyp. J.* **2019**, *26*, 437–444. [CrossRef]
10. Ligon, S.C.; Liska, R.; Stampfl, J.; Gurr, M.; Mülhaupt, R. Polymers for 3D Printing and Customized Additive Manufacturing. *Chem. Rev.* **2017**, *117*, 10212–10290. [CrossRef] [PubMed]
11. Stansbury, J.W.; Idacavage, M.J. 3D printing with polymers: Challenges among expanding options and opportunities. *Dent. Mater.* **2016**, *32*, 54–64. [CrossRef] [PubMed]
12. Quan, H.; Zhang, T.; Xu, H.; Luo, S.; Nie, J.; Zhu, X. Photo-curing 3D printing technique and its challenges. *Bioact. Mater.* **2020**, *5*, 110–115. [CrossRef] [PubMed]
13. International Organization for Standardization. *Accuracy (Trueness and Precision) of Measurement Methods and Results—Part 1: General Principles and Definitions (ISO 5725-1)*; International Organization for Standardization: Geneva, Switzerland, 1994; Available online: https://www.iso.org/standard/11833.html (accessed on 25 July 2020).
14. Sayers, A. Tips and tricks in performing a systematic review. *Br. J. Gen. Pract.* **2008**, *58*, 136. [CrossRef] [PubMed]
15. Moher, D.; Liberati, A.; Tetzlaff, J.; Altman, D.G. Preferred reporting items for systematic reviews and meta-analyses: The PRISMA statement. *J. Clin. Epidemiol.* **2009**, *62*, 1006–1012. [CrossRef] [PubMed]
16. McGowan, J.; Sampson, M.; Salzwedel, D.M.; Cogo, E.; Foerster, V.; Lefebvre, C. PRESS Peer Review of Electronic Search Strategies: 2015 Guideline Statement. *J. Clin. Epidemiol.* **2016**, *75*, 40–46. [CrossRef]
17. Whiting, P.F.; Rutjes, A.W.; Westwood, M.E.; Mallett, S.; Deeks, J.J.; Reitsma, J.B.; Leeflang, M.M.; Sterne, J.A.; Bossuyt, P.M.M. QUADAS-2: A Revised Tool for the Quality Assessment of Diagnostic Accuracy Studies. *Ann. Intern. Med.* **2011**, *155*, 529–536. [CrossRef]
18. Jeong, Y.-G.; Lee, W.-S.; Lee, K.-B. Accuracy evaluation of dental models manufactured by CAD/CAM milling method and 3D printing method. *J. Adv. Prosthodont.* **2018**, *10*, 245–251. [CrossRef] [PubMed]

19. Kim, S.-Y.; Shin, Y.-S.; Jung, H.-D.; Hwang, C.-J.; Baik, H.-S.; Cha, J.-Y. Precision and trueness of dental models manufactured with different 3-dimensional printing techniques. *Am. J. Orthod. Dentofac. Orthop.* **2018**, *153*, 144–153. [CrossRef]

20. Al-Imam, H.; Gram, M.; Benetti, A.R.; Gotfredsen, K. Accuracy of stereolithography additive casts used in a digital workflow. *J. Prosthet. Dent.* **2018**, *119*, 580–585. [CrossRef]

21. Budzik, G.; Bazan, A.; Turek, P.; Burek, J. Analysis of the Accuracy of Reconstructed Two Teeth Models Manufactured Using the 3DP and FDM Technologies. *Strojniški Vestnik J. Mech. Eng.* **2016**, *62*. [CrossRef]

22. Ishida, Y.; Miyasaka, T. Dimensional accuracy of dental casting patterns created by 3D printers. *Dent. Mater. J.* **2016**, *35*, 250–256. [CrossRef] [PubMed]

23. Arnold, C.; Monsees, D.; Hey, J.; Schweyen, R. Surface Quality of 3D-Printed Models as a Function of Various Printing Parameters. *Materials* **2019**, *12*, 1970. [CrossRef] [PubMed]

24. Park, M.-E.; Shin, S.-Y. Three-dimensional comparative study on the accuracy and reproducibility of dental casts fabricated by 3D printers. *J. Prosthet. Dent.* **2018**, *119*, 861.e1–861.e7. [CrossRef] [PubMed]

25. Ayoub, A.; Rehab, M.; O'Neil, M.; Khambay, B.; Ju, X.; Barbenel, J.; Naudi, K. A novel approach for planning orthognathic surgery: The integration of dental casts into three-dimensional printed mandibular models. *Int. J. Oral Maxillofac. Surg.* **2014**, *43*, 454–459. [CrossRef]

26. Hatz, C.; Msallem, B.; Aghlmandi, S.; Brantner, P.; Thieringer, F.M. Can an entry-level 3D printer create high-quality anatomical models? Accuracy assessment of mandibular models printed by a desktop 3D printer and a professional device. *Int. J. Oral Maxillofac. Surg.* **2020**, *49*, 143–148. [CrossRef]

27. Jang, Y.; Sim, J.-Y.; Park, J.-K.; Kim, W.-C.; Kim, H.-Y.; Kim, J.-H. Accuracy of 3-unit fixed dental prostheses fabricated on 3D-printed casts. *J. Prosthet. Dent.* **2020**, *123*, 135–142. [CrossRef]

28. Zhang, H.-R.; Yin, L.-F.; Liu, Y.-L.; Yan, L.-Y.; Wang, N.; Liu, G.; An, X.-L.; Liu, B. Fabrication and accuracy research on 3D printing dental model based on cone beam computed tomography digital modeling. *West China J. Stomatol.* **2018**, *36*, 156–161.

29. Xiao, N.; Sun, Y.C.; Zhao, Y.J.; Wang, Y. A method to evaluate the trueness of reconstructed dental models made with photo-curing 3D printing technologies. *Beijing Da Xue Xue Bao Yi Xue Ban* **2019**, *51*, 120–130.

30. Zeng, F.-H.; Xu, Y.-Z.; Fang, L.; Tang, X.-S. Reliability of three dimensional resin model by rapid prototyping manufacturing and digital modeling. *Shanghai Kou Qiang Yi Xue Shanghai J. Stomatol.* **2012**, *21*, 53–56.

31. AlShawaf, B.; Weber, H.-P.; Finkelman, M.; El Rafie, K.; Kudara, Y.; Papaspyridakos, P. Accuracy of printed casts generated from digital implant impressions versus stone casts from conventional implant impressions: A comparative in vitro study. *Clin. Oral Implant. Res.* **2018**, *29*, 835–842. [CrossRef]

32. Dostálová, T.; Kasparova, M.; Kriz, P.; Halamova, S.; Jelinek, M.; Bradna, P.; Mendricky, J. Intraoral scanner and stereographic 3D print in dentistry—Quality and accuracy of model—New laser application in clinical practice. *Laser Phys.* **2018**, *28*, 125602. [CrossRef]

33. Burde, A.V.; VarvarÅ, M.; Dudea, D.; Câmpian, R.S. Quantitative evaluation of accuracy for two rapid prototyping systems in dental model manufacturing. *Clujul Med.* **2015**, *88*, S44.

34. Aly, P.; Mohsen, C. Comparison of the Accuracy of Three-Dimensional Printed Casts, Digital, and Conventional Casts: An In Vitro Study. *Eur. J. Dent.* **2020**, *14*, 189–193. [CrossRef] [PubMed]

35. Bohner, L.; Hanisch, M.; Canto, G.D.L.; Mukai, E.; Sesma, N.; Neto, P.T.; Tortamano, P. Accuracy of Casts Fabricated by Digital and Conventional Implant Impressions. *J. Oral Implant.* **2019**, *45*, 94–99. [CrossRef] [PubMed]

36. Brown, G.B.; Currier, G.F.; Kadioglu, O.; Kierl, J.P. Accuracy of 3-dimensional printed dental models reconstructed from digital intraoral impressions. *Am. J. Orthod. Dentofac. Orthop.* **2018**, *154*, 733–739. [CrossRef] [PubMed]

37. Burde, A.V.; Gasparik, C.; Baciu, S.; Manole, M.; Dudea, D.; Câmpian, R.S. Three-Dimensional Accuracy Evaluation of Two Additive Manufacturing Processes in the Production of Dental Models. *Key Eng. Mater.* **2017**, *752*, 119–125. [CrossRef]

38. Camardella, L.T.; Vilella, O.D.V.; Breuning, H. Accuracy of printed dental models made with 2 prototype technologies and different designs of model bases. *Am. J. Orthod. Dentofac. Orthop.* **2017**, *151*, 1178–1187. [CrossRef] [PubMed]

39. Camardella, L.T.; Vilella, O.V.; Van Hezel, M.M.; Breuning, K.H. Accuracy of stereolithographically printed digital models compared to plaster models Genauigkeit von stereolitographisch gedruckten digitalen Modellen im Vergleich zu Gipsmodellen. *J. Orofac. Orthop. Fortschritte Kieferorthopädie* **2017**, *40*, 162–402. [CrossRef]
40. Cho, S.-H.; Schaefer, O.; Thompson, G.A.; Guentsch, A. Comparison of accuracy and reproducibility of casts made by digital and conventional methods. *J. Prosthet. Dent.* **2015**, *113*, 310–315. [CrossRef]
41. Cuperus, A.M.R.; Harms, M.C.; Rangel, F.A.; Bronkhorst, E.M.; Schols, J.G.; Breuning, K.H. Dental models made with an intraoral scanner: A validation study. *Am. J. Orthod. Dentofac. Orthop.* **2012**, *142*, 308–313. [CrossRef]
42. Dietrich, C.A.; Ender, A.; Baumgartner, S.; Mehl, A. A validation study of reconstructed rapid prototyping models produced by two technologies. *Angle Orthod.* **2017**, *87*, 782–787. [CrossRef]
43. Favero, C.S.; English, J.D.; Cozad, B.E.; Wirthlin, J.O.; Short, M.M.; Kasper, F.K. Effect of print layer height and printer type on the accuracy of 3-dimensional printed orthodontic models. *Am. J. Orthod. Dentofac. Orthop.* **2017**, *152*, 557–565. [CrossRef] [PubMed]
44. Hazeveld, A.; Slater, J.J.H.; Ren, Y. Accuracy and reproducibility of dental replica models reconstructed by different rapid prototyping techniques. *Am. J. Orthod. Dentofac. Orthop.* **2014**, *145*, 108–115. [CrossRef]
45. Jin, S.-J.; Jeong, I.-D.; Kim, J.-H.; Kim, W.-C. Accuracy (trueness and precision) of dental models fabricated using additive manufacturing methods. *Int. J. Comput. Dent.* **2018**, *21*, 107–113. [PubMed]
46. Kasparova, M.; Grafova, L.; Dvorak, P.; Dostalova, T.; Prochazka, A.; Eliasova, H.; Prusa, J.; Kakawand, S. Possibility of reconstruction of dental plaster cast from 3D digital study models. *Biomed. Eng. Online* **2013**, *12*, 49. [CrossRef]
47. Keating, A.P.; Knox, J.; Bibb, R.; Zhurov, A.I. A comparison of plaster, digital and reconstructed study model accuracy. *J. Orthod.* **2008**, *35*, 191–201. [CrossRef]
48. Kuo, R.-F.; Chen, S.-J.; Wong, T.-Y.; Lu, B.-C.; Huang, Z.-H. Digital Morphology Comparisons between Models of Conventional Intraoral Casting and Digital Rapid Prototyping. In Proceedings of the 5th International Conference on Biomedical Engineering in Vietnam, Ho Chi Minh, Viet Nam, 16–18 June 2014; Springer International Publishing: Cham, Switzerland, 2015.
49. Loflin, W.A.; English, J.D.; Borders, C.; Harris, L.M.; Moon, A.; Holland, J.N.; Kasper, F.K. Effect of print layer height on the assessment of 3D-printed models. *Am. J. Orthod. Dentofac. Orthop.* **2019**, *156*, 283–289. [CrossRef] [PubMed]
50. Nestler, N.; Wesemann, C.; Spies, B.C.; Beuer, F.; Bumann, A. Dimensional accuracy of extrusion- and photopolymerization-based 3D printers: In vitro study comparing printed casts. *J. Prosthet. Dent.* **2020**. [CrossRef]
51. Papaspyridakos, P.; Chen, Y.-W.; AlShawaf, B.; Kang, K.; Finkelman, M.; Chronopoulos, V.; Weber, H.-P. Digital workflow: In vitro accuracy of 3D printed casts generated from complete-arch digital implant scans. *J. Prosthet. Dent.* **2020**. [CrossRef]
52. Rebong, R.E.; Stewart, K.T.; Utreja, A.; Ghoneima, A.A. Accuracy of three-dimensional dental resin models created by fused deposition modeling, stereolithography, and Polyjet prototype technologies: A comparative study. *Angle Orthod* **2018**, *88*, 363–369. [CrossRef]
53. Rungrojwittayakul, O.; Kan, J.Y.; Shiozaki, K.; Swamidass, R.S.; Goodacre, B.J.; Goodacre, C.J.; Lozada, J.L. Accuracy of 3D Printed Models Created by Two Technologies of Printers with Different Designs of Model Base. *J. Prosthodont.* **2019**, *29*, 124–128. [CrossRef] [PubMed]
54. Saleh, W.K.; Ariffin, E.; Sherriff, M.; Bister, D. Accuracy and reproducibility of linear measurements of resin, plaster, digital and printed study-models. *J. Orthod.* **2015**, *42*, 301–306. [CrossRef] [PubMed]
55. Sherman, S.L.; Kadioglu, O.; Currier, G.F.; Kierl, J.P.; Li, J. Accuracy of digital light processing printing of 3-dimensional dental models. *Am. J. Orthod. Dentofac. Orthop.* **2020**, *157*, 422–428. [CrossRef] [PubMed]
56. Hassan, W.W.; Yusoff, Y.; Mardi, N.A. Comparison of reconstructed rapid prototyping models produced by 3-dimensional printing and conventional stone models with different degrees of crowding. *Am. J. Orthod. Dentofac. Orthop.* **2017**, *151*, 209–218. [CrossRef]
57. Zhang, Z.-C.; Li, P.-L.; Chu, F.-T.; Shen, G. Influence of the three-dimensional printing technique and printing layer thickness on model accuracy. *J. Orofac. Orthop. Fortschritte Kieferorthopädie* **2019**, *80*, 194–204. [CrossRef]
58. Koo, T.K.; Li, M.Y. A Guideline of Selecting and Reporting Intraclass Correlation Coefficients for Reliability Research. *J. Chiropr. Med.* **2016**, *15*, 155–163. [CrossRef]

59. Hirogaki, Y.; Sohmura, T.; Satoh, H.; Takahashi, J.; Takada, K. Complete 3-D reconstruction of dental cast shape using perceptual grouping. *IEEE Trans. Med. Imaging* **2001**, *20*, 1093–1101. [CrossRef]
60. Bell, A.; Ayoub, A.F.; Siebert, P. Assessment of the accuracy of a three-dimensional imaging system for archiving dental study models. *J. Orthod.* **2003**, *30*, 219–223. [CrossRef]
61. Naidu, D.; Freer, T.J. Validity, reliability, and reproducibility of the iOC intraoral scanner: A comparison of tooth widths and Bolton ratios. *Am. J. Orthod. Dentofac. Orthop.* **2013**, *144*, 304–310. [CrossRef]
62. McLean, J.W.; Von Fraunhofer, J.A. The estimation of cement film thickness by an in vivo technique. *Br. Dent. J.* **1971**, *131*, 107–111. [CrossRef]
63. Jemt, T.; Book, K. Prosthesis misfit and marginal bone loss in edentulous implant patients. *Int. J. Oral Maxillofac. Implant.* **1996**, *11*, 620–625.
64. Jemt, T. In vivo measurements of precision of fit involving implant-supported prostheses in the edentulous jaw. *Int. J. Oral Maxillofac. Implant.* **1996**, *11*, 151–158.
65. Papaspyridakos, P.; Chen, C.-J.; Chuang, S.-K.; Weber, H.-P.; Gallucci, G.O. A systematic review of biologic and technical complications with fixed implant rehabilitations for edentulous patients. *Int. J. Oral Maxillofac. Implant.* **2012**, *27*, 102–110.

Publisher's Note: MDPI stays neutral with regard to jurisdictional claims in published maps and institutional affiliations.

Journal of
Clinical Medicine

Article

Impact of Aging on the Accuracy of 3D-Printed Dental Models: An In Vitro Investigation

Tim Joda *, Lea Matthisson and Nicola U. Zitzmann

Department of Reconstructive Dentistry, University Center for Dental Medicine Basel, University of Basel, 4058 Basel, Switzerland; lea.matthisson@unibas.ch (L.M.); n.zitzmann@unibas.ch (N.U.Z.)
* Correspondence: tim.joda@unibas.ch; Tel.: +41-61-267-2636

Received: 23 April 2020; Accepted: 11 May 2020; Published: 12 May 2020

Abstract: The aim of this in vitro study was to analyze the impact of model aging on the accuracy of 3D-printed dental models. A maxillary full-arch reference model with prepared teeth for a three-unit fixed dental prosthesis was scanned ten times with an intraoral scanner (3Shape TRIOS Pod) and ten models were 3D printed (Straumann P-Series). All models were stored under constant conditions and digitized with a desktop scanner after 1 day; 1 week; and 2, 3, and 4 weeks. For accuracy, a best-fit algorithm was used to analyze the deviations of the abutment teeth (GFaI e.V Final Surface®). *Wilcoxon Rank Sum Tests* were used for comparisons with the level of significance set at $\alpha = 0.05$. Deviation analysis of the tested models showed homogenous intragroup distance calculations at each timepoint. The most accurate result was for 1 day of aging (3.3 ± 1.3 µm). A continuous decrease in accuracy was observed with each aging stage from day 1 to week 4. A time-dependent difference was statistically significant after 3 weeks ($p = 0.0008$) and 4 weeks ($p < 0.0001$). Based on these findings, dental models should not be used longer than 3 to 4 weeks after 3D printing for the fabrication of definitive prosthetic reconstructions.

Keywords: rapid prototyping; 3D printing; accuracy; dental materials science; digital workflow

---

1. Introduction

Digitalization is en vogue: *#WhatCanBeDigitalWillBe*. Therefore, it is not surprising that the demands of clinicians and patients are also changing in the field of dentistry. Scan technology has opened the possibility to digitize the patient's dental situation: either lab-side scanning of conventional gypsum casts or directly chairside with an intraoral scanning device (IOS). With both methods, the patient-specific situation can be captured optically and stored as a three-dimensional (3D) surface file, namely a standard tessellation language (STL) file [1]. Scan technology is currently of great interest in all dental disciplines, in particular in prosthodontics for the manufacturing of fixed dental prostheses (FDP) [2].

IOS enables fully digital chairside workflows, incorporating computer-aided-design and computer-aided-manufacturing (CAD/CAM) without any physical models [3]. IOS meets the ubiquitous trend of digitalization in the society, supports more convenient treatments [4], and will successively displace conventional impression taking in dentistry [5]. Whenever possible, complete digital workflows will be used in dental medicine in the future [6].

The typical IOS domain has been single-unit restorations. Complete digital workflows have been proven for tooth- and implant-supported monolithic single-unit restorations, especially in posterior sites [7,8]. From an economic point of view, clinical and technical protocols for single crowns can be streamlined to achieve time-efficient therapy outcomes with a reasonable cost–benefit ratio and a high quality of CAD/CAM-processed restorations [9,10].

At present, not all prosthetic indications can be addressed with model-free workflows, i.e., manually veneered multi-unit FDPs. Even though technical progress is rapid and the combination of IOS and laser-melting seems to be promising for CAD/CAM processing of frameworks of removable partial dentures (RPD), a dental model is still required for the finalization of the RPD [11,12]. With the increase in performance of IOS devices, however, the desire to expand the range of indications from single crowns to more complex prosthetic reconstructions including removable restorations with edentulous mucosal tissues has grown [13]. Clinical and technical protocols (combining IOS and dental model fabrication) are required to cover such complex indications in a digital workflow. Rapid prototyping is a technique to construct and build any geometry using 3D printing [14]. The 3D printing process is a promising solution to generate dental models out of polymers based on lab-side or intraorally acquired STL files [15]. Dental models reconstructed by 3D printing were considered clinically acceptable in terms of accuracy and reproducibility compared to classical stone casts [16]; while compared to CAD/CAM milling, 3D-printed models demonstrated even higher accuracy [17].

Three-dimensional printing is a relatively new technique in dentistry, and consequently, detailed information on the dimensional accuracy and stability of 3D printed models with regard to time and storage is not available [18]. Initial laboratory studies evaluated the accuracy of 3D-printed full-arch dental models [19–22] but did not consider the impact of aging of the models.

Therefore, this in vitro study aimed to investigate the impact of model aging on the dimensional stability of 3D-printed dental models. The hypothesis was that the period of storage time has no significant influence on the accuracy of printed models.

2. Experimental Section

2.1. Study Setup

A maxillary full-arch reference model (Model ANA-4, Frasaco GmbH, Tettnang, Germany) with abutment tooth preparation for a three-unit FDP in positions 24–26 was scanned ten times using an IOS device (TRIOS Pod, version 19.2.4, 3Shape, Copenhagen, Denmark). All IOSs were performed by an experienced single operator. The IOS system was calibrated prior to each scan, and the scan strategy followed the manufacturer's instructions. Scan data were directly exported as STL data sets ($n = 10$).

Afterwards, TRIOS STL files were converted into 3D printable data sets using the built-in model builder tool of the desktop scanner (Netfabb, version 2020.2, Institut Straumann AG, Basel, Switzerland). Standardized parameters were defined for 3D-printed models with a base height of 4 mm and a thickness of 3 mm with two stabilizing bars. A base plate with hexagonal cell design with a height of 2 mm, wall thickness of 0.8 mm, and cell size of 1.5 mm was selected for all models (Figure 1).

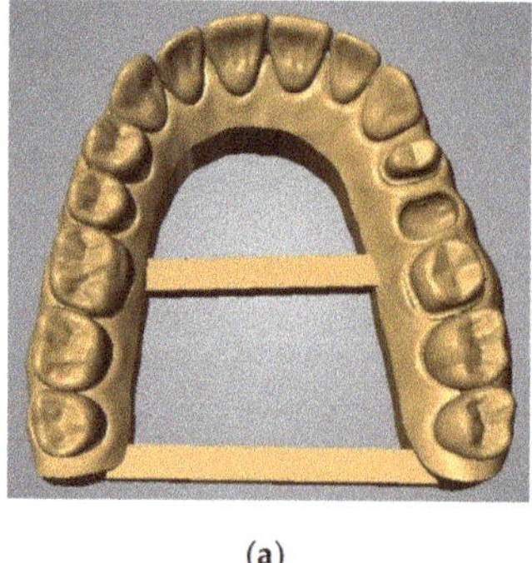

(a)

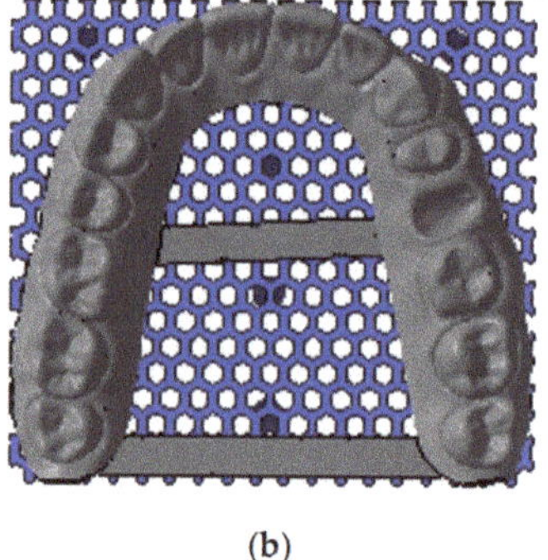

(b)

Figure 1. Conversion of TRIOS standard tessellation language (STL) files into 3D printable data sets with (**a**) model builder software including two stabilizing bars (**b**) virtual preparation for 3D printing using a base plate with hexagonal cell design.

Based on each of these IOS data sets from the TRIOS STL files, ten dental models were 3D printed (P 30, version 2019.2.11, Institut Straumann AG, Basel, Switzerland) using a light-curing 3D printing material with 385-nm wavelength technology (SHERAPrint-model plus "sand" UV, SHERA, Lemförde, Germany). This polymer is specifically formulated for the production of high-precision dental models. The printing model platform was cleaned with isopropanol and placed in the 3D printer. Afterwards, the models were removed from the platform and placed in an ultrasonic bath for cleaning. The models were then blown dry with compressed air, checked for excess material, cleaned again if required, and left to rest for 30 min before further processing. Finally, the models were exposed to a burst of light with wavelengths of 280–580 nm to cure the polymer.

The 3D-printed models were stored under constant conditions at 20 °C and 50% humidity without direct light exposure and successively digitized with laboratory desktop scanner (Series 7, version 13.1.3.33179, Institut Straumann AG, Basel, Switzerland) after storage periods of 1 day, 1 week, 2 weeks, 3 weeks, and 4 weeks.

2.2. Accuracy Analysis

A total of 50 STL files (ten 3D-printed dental models scanned at five timepoints) were evaluated for accuracy by means of trueness (means) and precision (standard deviations). For accuracy measurements, the original maxillary full-arch reference model was digitized using the same laboratory desktop scanner that was used to digitize the 3D-printed models after aging. Based on the manufacturer's information, the power of the laser diode is 5 mW with a laser wavelength of 660 nm and the accuracy is specified with 7 µm.

The STL file of the reference model was imported into a 3D analysis software and matched pairwise with the 50 STL files of the 3D-printed models (Final Surface® version 2019.0, GFaI e.V., Berlin, Germany). A best-fit algorithm of the 3D analysis software was applied for accuracy testing of the superimposed model pairings using a 2D distance analysis of the abutment teeth in areas 24 and 26 at indexed landmarks at the finishing lines. For visualization, a color mapping function of the 3D analysis software was used with a graduate scale in µm (Figure 2).

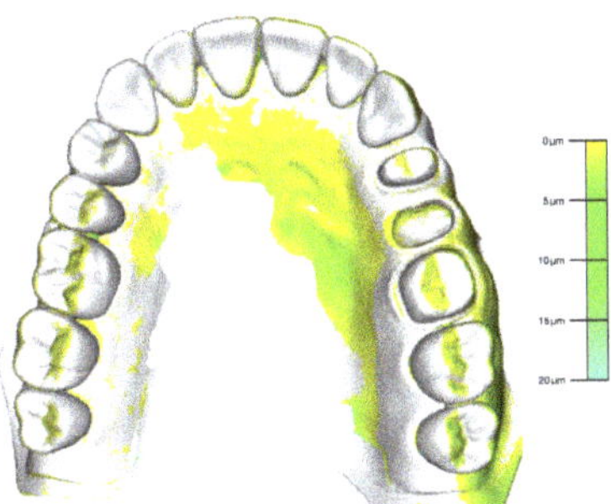

Figure 2. Visualization of the superimposed 3D-printed model with the reference by means of color mapping (Final Surface® version 2019.0, GFaI e.V., Berlin, Germany).

2.3. Statistical Analysis

Statistical analysis was carried out to evaluate the impact of aging on the accuracy of the 3D-printed dental models. Descriptive statistics were calculated for means with standard deviations (SD) including minimum and maximum values. *Wilcoxon Rank Sum Tests* were used for all comparisons. The level of significance was set at $\alpha = 0.05$. Calculations were made with the open-source software "GraphPad Software" (http://www.graphpad.com).

3. Results

The descriptive statistics are shown in Table 1. The deviation analysis of the 3D-printed models #01–#10 compared to the reference model demonstrated overall homogenous intragroup results for distance calculations at each isolated timepoint. Taking into account the entire investigation period, the range of mean deviations was very close for all tested 3D-printed models, revealing minimum to maximum distance values of 1 to 12 µm.

Table 1. Deviation (in µm) of the 3D-printed models #01–#10 from the reference model after aging of 1 day; 1 week; and 2, 3, and 4 weeks (SD = standard deviation, Min = minimum, and Max = maximum).

	1 Day	1 Week	2 Weeks	3 Weeks	4 Weeks
#01	2	2	3	5	6
#02	1	2	2	3	5
#03	2	3	2	7	8
#04	2	2	3	4	9
#05	4	3	6	6	9
#06	4	6	5	7	10
#07	5	7	6	8	12
#08	5	5	7	7	9
#09	3	4	7	8	9
#10	5	6	8	9	12
Mean	3.3	4.0	4.9	6.4	8.9
SD	1.3	1.9	2.2	1.9	2.2
Min	1	2	2	3	5
Max	5	7	8	9	12

Considering the factor of aging of the 3D-printed models, the most accurate result was for 1 day with a mean deviation of 3.3 ± 1.3 µm. A continuous decrease in accuracy was observed with each further aging stage of the tested 3D-printed models from 1 day up to 4 weeks (Figure 3). The time-dependent difference was statistically significant after 3 weeks ($p = 0.0008$) and 4 weeks ($p < 0.0001$), respectively, when comparing with 1 day.

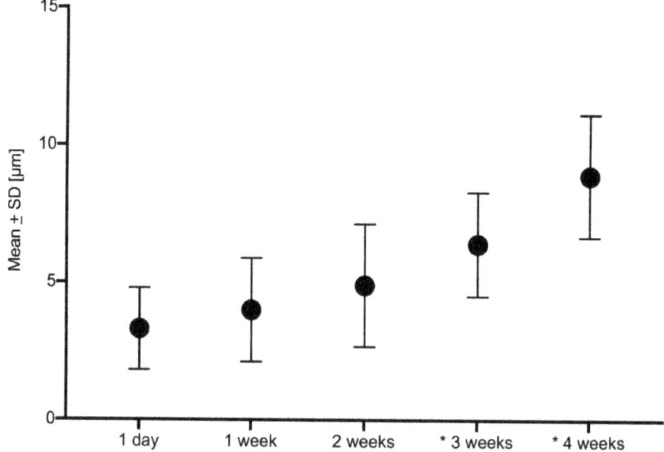

Figure 3. Deviation analysis of the tested 3D-printed models after defined aging represented by the five timepoints for mean values including standard deviations (SD) in micrometer (µm) with significant differences after * 3 weeks ($p = 0.0008$) and * 4 weeks ($p < 0.0001$).

4. Discussion

This in vitro investigation aimed to analyze the impact of aging on the dimensional stability of 3D-printed dental models. The present findings revealed a time-dependent significant change in dimensions after 3 and 4 weeks of aging. Therefore, the hypothesis that the period of storage time has no significant influence on the accuracy of 3D printed models had to be rejected.

The present study setup was carried out with an exemplary tooth-supported three-unit FDP in regions 24 to 26 for deviation analysis of the prepared abutment teeth. For fabrication of a manually veneered FDP, the overall time required-comprising the sum of clinical and technical work steps, including impression taking, try-in, potential adjustments, and seating of the reconstruction, is approximately 2–3 weeks [23]. During this period of time, the dental technician must rely on the dimensional stability of the dental cast. Based on this supposed working schedule, the present investigation considered five timepoints up to 4 weeks of aging for accuracy analysis of 3D-printed models.

The focus of the present study was to investigate the dimensional changes of the 3D-printed models related to the prepared abutment teeth representing short-span analysis rather than full-arch comparisons. The obtained results for accuracy during aging up to 4 weeks were very consistent for intragroup comparisons at each isolated timepoint. Analyzing longer spans, i.e., cross-arch reconstructions, the accuracy testing might reveal different results for intra- as well as intergroup comparisons. Even though significant differences of the 3D-printed models were observed after 3 and 4 weeks of aging, examination of the distance analysis revealed negligible changes of 1 to 12 μm for minimum to maximum values. Therefore, the question of clinical relevance remains, provided that the prosthetic reconstruction can be completed on the model within 3 or 4 weeks and that the models have been properly stored during this time (constant conditions at 20 °C and 50% humidity without direct light exposure). It must be critically emphasized that the dimensional changes of the 3D-printed models were small and comparable to investigations analyzing the accuracy of dental stone casts obtained from classical impression taking [24,25].

Ideally, dental master casts for the manufacturing of retrievable prosthetic reconstructions, such as screw-retained implant-supported FDPs or removable dental prostheses, can be reused in case of a potential emergency in the future. Tested 3D-printed models demonstrated a continuous decrease of dimensional stability. We cannot extrapolate beyond the 4-week duration of this study; therefore, it is not possible to predict with certainty whether these changes will continue or stabilize somehow over time. Thus, 3D-printed dental master models should be considered as single-use product, at least for the manufacturing of definitive prosthetic reconstructions. Nevertheless, the existing digital data sets can be stored and reused for the production of "fresh" dental models if necessary.

The bottleneck of accuracy is the process chain of STL files. The interface management guaranteeing a loss-free data transfer from the IOS device to the software of the virtual model builder and to the 3D printer is the key for success. The findings are therefore only representative for the combination of the equipment and materials used in this investigation and cannot be transferred to workflows of other manufacturers. Further investigations are necessary to analyze different setups of IOS > software > 3D printer including materials used and additional prosthetic indications. The following has to be considered: what are the impacts of the polymer and the 3D-printing technology, or is it a combination of both?

In general, the translation of laboratory findings into clinical (routine) protocols must proceed with caution. In the present study, the in vitro study setting can only be transferred with digital impressions in the upper jaw. It must be taken into account that in vivo digital impressions are different considering the localization. In contrast to the maxilla, the mandible exhibits characteristic inherent mobility during dynamic movements, in particular during mouth opening. The transfer of laboratory results from full-arch scans to an in vivo patient situation is difficult, specifically in the mandible.

While in the past, conventional workflows with classical impression techniques and plaster model production have been continuously optimized, there are still no explicit recommendations for digital workflows with IOS, further STL processing, and consecutive fabrication of 3D-printed dental models. Due to the various commercially available 3D printers with different quality levels and the diverse light-curing polymers in the material segment, the field of digitally produced dental models is very complex and subject to constant change without generally defined standards [15,26]. Kim et al. (2018) reported on significant differences for the analysis of full-arch dental models manufactured with different 3D printing techniques [27]. In addition, Nestler et al. (2020) have shown that inexpensive 3D printers were no less accurate than more expensive ones [19]. However, the authors did not clarify what is the meaning of "inexpensive" compared to "expensive", especially when a global market is considered. Three-dimensional printing generates not only enthusiasm but also great uncertainty. Therefore, future research must focus on the definition and establishment of evidenced-based standards in the field of 3D printing techniques in dentistry. Otherwise it is not possible for both dental technicians and dentists to distinguish which workflows with which equipment and material combination can deliver reliable results [28].

5. Conclusions

Three-dimensionally printed dental models for the production of three-unit FDPs demonstrated very accurate results. However, a significant decrease in dimensional stability of the models was observed after 3 weeks of aging under constant conditions. Based on these findings, it can be concluded that 3D-printed dental master models should not be used for the fabrication of definitive prosthetic reconstructions more than 3 to 4 weeks after 3D printing. Nevertheless, the changes observed due to aging in this study were small and comparable to the variation seen in conventional plaster cast models that are used in routine practice today.

Author Contributions: Conceptualization and methodology, T.J.; software; validation, formal analysis, and investigation, T.J. and L.M.; resources, T.J. and N.U.Z.; data curation, T.J. and L.M.; writing—original draft preparation, T.J.; writing—review and editing, T.J. and N.U.Z.; visualization, supervision, and project administration, T.J.; funding acquisition, n/a. All authors have read and agreed to the published version of the manuscript.

Acknowledgments: The authors express their gratitude to Isabell Wiestler for lab-side digitalization of the 3D-printed models at the different timepoints and to James Ashman for proofreading the final manuscript.

Conflicts of Interest: The authors declare no conflict of interest.

References

1. Chiu, A.; Chen, Y.W.; Hayashi, J.; Sadr, A. Accuracy of CAD/CAM Digital Impressions with Different Intraoral Scanner Parameters. *Sensors* **2020**, *20*, 1157. [CrossRef] [PubMed]
2. Joda, T.; Bornstein, M.M.; Jung, R.E.; Ferrari, M.; Waltimo, T.; Zitzmann, N.U. Recent Trends and Future Direction of Dental Research in the Digital Era. *Int. J. Environ. Res. Public Health* **2020**, *17*, 1987. [CrossRef] [PubMed]
3. Blatz, M.B.; Conejo, J. The Current State of Chairside Digital Dentistry and Materials. *Dent. Clin. N. Am.* **2019**, *63*, 175–197. [CrossRef] [PubMed]
4. Joda, T.; Bragger, U. Patient-centered outcomes comparing digital and conventional implant impression procedures: A randomized crossover trial. *Clin. Oral Implant. Res.* **2016**, *27*, e185–e189. [CrossRef]
5. Guo, D.N.; Liu, Y.S.; Pan, S.X.; Wang, P.F.; Wang, B.; Liu, J.Z.; Gao, W.H.; Zhou, Y.S. Clinical Efficiency and Patient Preference of Immediate Digital Impression after Implant Placement for Single Implant-Supported Crown. *Chin. J. Dent. Res.* **2019**, *22*, 21–28.
6. Joda, T.; Zarone, F.; Ferrari, M. The complete digital workflow in fixed prosthodontics: A systematic review. *BMC Oral Health* **2017**, *17*, 124. [CrossRef]
7. Joda, T.; Bragger, U. Time-efficiency analysis of the treatment with monolithic implant crowns in a digital workflow: A randomized controlled trial. *Clin. Oral Implant. Res.* **2016**, *27*, 1401–1406. [CrossRef]
8. Joda, T.; Ferrari, M.; Bragger, U. Monolithic implant-supported lithium disilicate (LS2) crowns in a complete digital workflow: A prospective clinical trial with a 2-year follow-up. *Clin. Implant Dent. Relat. Res.* **2017**, *19*, 505–511. [CrossRef]

9. Joda, T.; Bragger, U. Time-Efficiency Analysis Comparing Digital and Conventional Workflows for Implant Crowns: A Prospective Clinical Crossover Trial. *Int. J. Oral Maxillofac. Implant.* **2015**, *30*, 1047–1053. [CrossRef]

10. Joda, T.; Bragger, U. Digital vs. conventional implant prosthetic workflows: A cost/time analysis. *Clin. Oral Implant. Res.* **2015**, *26*, 1430–1435. [CrossRef]

11. Almufleh, B.; Emami, E.; Alageel, O.; de Melo, F.; Seng, F.; Caron, E.; Nader, S.A.; Al-Hashedi, A.; Albuquerque, R.; Feine, J.; et al. Patient satisfaction with laser-sintered removable partial dentures: A crossover pilot clinical trial. *J. Prosthet. Dent.* **2018**, *119*, 560–567. [CrossRef] [PubMed]

12. Gintaute, A.; Straface, A.; Zitzmann, N.U.; Joda, T. Removable Dental Prosthesis 2.0: Digital from A to Z? *Swiss Dent. J.* **2020**, *130*, 229–235. [PubMed]

13. Joda, T.; Ferrari, M.; Gallucci, G.O.; Wittneben, J.G.; Bragger, U. Digital technology in fixed implant prosthodontics. *Periodontology 2000* **2017**, *73*, 178–192. [CrossRef] [PubMed]

14. Alharbi, N.; Wismeijer, D.; Osman, R.B. Additive Manufacturing Techniques in Prosthodontics: Where Do We Currently Stand? A Critical Review. *Int. J. Prosthodont.* **2017**, *30*, 474–484. [CrossRef]

15. Quan, H.; Zhang, T.; Xu, H.; Luo, S.; Nie, J.; Zhu, X. Photo-curing 3D printing technique and its challenges. *Bioact. Mater.* **2020**, *5*, 110–115. [CrossRef]

16. Hazeveld, A.; Huddleston Slater, J.J.; Ren, Y. Accuracy and reproducibility of dental replica models reconstructed by different rapid prototyping techniques. *Am. J. Orthod. Dentofac. Orthop.* **2014**, *145*, 108–115. [CrossRef]

17. Jeong, Y.G.; Lee, W.S.; Lee, K.B. Accuracy evaluation of dental models manufactured by CAD/CAM milling method and 3D printing method. *J. Adv. Prosthodont.* **2018**, *10*, 245–251. [CrossRef]

18. Zhang, Z.C.; Li, P.L.; Chu, F.T.; Shen, G. Influence of the three-dimensional printing technique and printing layer thickness on model accuracy. *J. Orofac. Orthop.* **2019**, *80*, 194–204. [CrossRef]

19. Nestler, N.; Wesemann, C.; Spies, B.C.; Beuer, F.; Bumann, A. Dimensional accuracy of extrusion- and photopolymerization-based 3D printers: In vitro study comparing printed casts. *J. Prosthet. Dent.* **2020**. [CrossRef]

20. Rungrojwittayakul, O.; Kan, J.Y.; Shiozaki, K.; Swamidass, R.S.; Goodacre, B.J.; Goodacre, C.J.; Lozada, J.L. Accuracy of 3D Printed Models Created by Two Technologies of Printers with Different Designs of Model Base. *J. Prosthodont.* **2020**, *29*, 124–128. [CrossRef]

21. Sim, J.Y.; Jang, Y.; Kim, W.C.; Kim, H.Y.; Lee, D.H.; Kim, J.H. Comparing the accuracy (trueness and precision) of models of fixed dental prostheses fabricated by digital and conventional workflows. *J. Prosthodont. Res.* **2019**, *63*, 25–30. [CrossRef] [PubMed]

22. Dietrich, C.A.; Ender, A.; Baumgartner, S.; Mehl, A. A validation study of reconstructed rapid prototyping models produced by two technologies. *Angle Orthod.* **2017**, *87*, 782–787. [CrossRef] [PubMed]

23. Muhlemann, S.; Benic, G.I.; Fehmer, V.; Hammerle, C.H.F.; Sailer, I. Randomized controlled clinical trial of digital and conventional workflows for the fabrication of zirconia-ceramic posterior fixed partial dentures. Part II: Time efficiency of CAD-CAM versus conventional laboratory procedures. *J. Prosthet. Dent.* **2019**, *121*, 252–257. [CrossRef] [PubMed]

24. Valente Vda, S.; Zanetti, A.L.; Feltrin, P.P.; Inoue, R.T.; de Moura, C.D.; Padua, L.E. Dimensional accuracy of stone casts obtained with multiple pours into the same mold. *ISRN Dent.* **2012**, *2012*, 730674. [CrossRef]

25. Vitti, R.P.; da Silva, M.A.; Consani, R.L.; Sinhoreti, M.A. Dimensional accuracy of stone casts made from silicone-based impression materials and three impression techniques. *Braz. Dent. J.* **2013**, *24*, 498–502. [CrossRef]

26. Barazanchi, A.; Li, K.C.; Al-Amleh, B.; Lyons, K.; Waddell, J.N. Additive Technology: Update on Current Materials and Applications in Dentistry. *J. Prosthodont.* **2017**, *26*, 156–163. [CrossRef]

27. Kim, S.Y.; Shin, Y.S.; Jung, H.D.; Hwang, C.J.; Baik, H.S.; Cha, J.Y. Precision and trueness of dental models manufactured with different 3-dimensional printing techniques. *Am. J. Orthod. Dentofac. Orthop.* **2018**, *153*, 144–153. [CrossRef]

28. Jockusch, J.; Ozcan, M. Additive manufacturing of dental polymers: An overview on processes, materials and applications. *Dent. Mater. J.* **2020**. [CrossRef]

Journal of
Clinical Medicine

Article

Accuracy of Three-Dimensional (3D) Printed Dental Digital Models Generated with Three Types of Resin Polymers by Extra-Oral Optical Scanning

Eugen S. Bud [1], Vlad I. Bocanet [2], Mircea H. Muntean [2], Alexandru Vlasa [1,*], Sorana M. Bucur [3], Mariana Păcurar [1], Bogdan R. Dragomir [4,*], Cristian D. Olteanu [5] and Anamaria Bud [1]

[1] Faculty of Dental Medicine, University of Medicine and Pharmacy, Science and Technology George Emil Palade, 540139 Târgu-Mureș, Romania; Eugen.bud@umfst.ro (E.S.B.); marianapac@yahoo.com (M.P.); Anamaria.bud@umfst.ro (A.B.)
[2] Technical University of Cluj-Napoca, 400114 Cluj-Napoca, Romania; vlad.bocanet@tcm.utcluj.ro (V.I.B.); mircea.muntean@spectromas.ro (M.H.M.)
[3] Faculty of Medicine, University Dimitrie Cantemir, 540545 Targu-Mures, Romania; bucursoranamaria@gmail.com
[4] Faculty of Dental Medicine, University of Medicine and Pharmacy Grigore T. Popa, 700115 Iași, Romania
[5] Faculty of Dental Medicine, University of Medicine and Pharmacy, 400010 Cluj-Napoca, Romania; olteanu.cristian@umfcluj.ro
* Correspondence: alexandru.vlasa@umfst.com (A.V.); radu.dragomir@umfiasi.ro (B.R.D.); Tel./Fax: +40-742825920 (A.V.); +40-744437881 (B.R.D.)

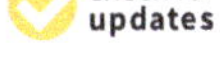

Citation: Bud, E.S.; Bocanet, V.I.; Muntean, M.H.; Vlasa, A.; Bucur, S.M.; Păcurar, M.; Dragomir, B.R.; Olteanu, C.D.; Bud, A. Accuracy of Three-Dimensional (3D) Printed Dental Digital Models Generated with Three Types of Resin Polymers by Extra-Oral Optical Scanning. *J. Clin. Med.* **2021**, *10*, 1908. https://doi.org/10.3390/jcm10091908

Academic Editor: Gianrico Spagnuolo

Received: 27 February 2021
Accepted: 21 April 2021
Published: 28 April 2021

Publisher's Note: MDPI stays neutral with regard to jurisdictional claims in published maps and institutional affiliations.

Abstract: Digital impression devices are used alternatively to conventional impression techniques and materials. The aim of this study was to evaluate the precision of extraoral digitalization of three types of photosensitive resin polymers used for 3D printing with the aid of a digital extraoral optical scanner. The alignment of the scans was performed by a standard best-fit alignment. Trueness and precision were used to evaluate the models. The trueness was evaluated by using bias as a measure and the standard deviation was used to evaluate the precision. After assessing the normality of the distributions, an independent Kruskal–Wallis test was used to compare the trueness and precision across the material groups. The Mann–Whitney test was used as a post-hoc test for significant differences. The result of the analysis showed significant differences ($U = 66$, $z = -2.337$, $p = 0.019$) in trueness of mesiodistal distances. Upon visual inspection of the models, defects were noticed on two out of nine of the models printed with a photosensitive polymer. The defects were presented as cavities caused by air bubbles and were also reflected in the scans. Mean precision did not vary too much between these three photosensitive polymer resins, therefore, the selection of 3D printing materials should be based on the trueness and the required precision of the clinical purpose of the model.

Keywords: 3D printed dental models; polymer resin; extra-oral scanning

1. Introduction

In-office dental 3D printing helps improve the efficiency of forward-thinking practices all over the world. By leveraging existing technologies that exist in digital dentistry, 3D printing enables better responsiveness to patient needs, significantly reduces manufacturing times, and opens up new treatment options. With low operating costs, minimal maintenance, and user-friendly design, these products make it easy to bring digital dentistry and 3D printing together in dental practice. 3D printing dental models and digital wax-ups reveal anatomical details, high precision for exceptional measurements, patient education, and dental laboratory collaboration. A large selection of dental 3D printing materials are designed, developed, and tested for high performance in digital dentistry. 3D printing resins are built to achieve equivalent or superior results than conventional dental materials while providing better value for money [1]. Digital impression making using intraoral and extraoral scanners may be an approach to improve the accuracy of dental restorations, as, by their nature, these processes tend to eliminate the error caused by conventional impression making and gypsum model casting [2]. There are only a few studies published on the accuracy of printed models compared with plaster models [3–6]. These studies concluded that the printed models can be used as a replacement for plaster models, but it is unclear whether the samples used in these studies were sufficient to draw definitive conclusions. Given 3D printing's promising potential and increased use in dentistry, it is essential to evaluate the accuracy of 3D printed dental models [7]. An accurate printed model is fundamental for dental diagnostic purposes.

With the introduction of computer-aided design (CAD) and computer-aided manufacturing (CAM) technologies in dentistry, virtual models of teeth are required. Digital processes are applied for prosthetic-driven backward planning of implant surgery, orthodontic measurements, and treatment planning combined with surgical planning. Data acquired by intraoral scanning, computed tomography, cone-beam computed tomography, and extraoral surface scanning can be fused [8–10]. Digital three-dimensional (3D) models are created by scanning impression and plaster models using desktop scanners or otherwise by cone-beam computed tomography. These methods have been widely accepted in clinical orthodontics and are advantageous due to the compact storage space, their potential to expand applications for treatment planning, and their easy customization [11,12].

The majority of the literature has focused on the reproducibility error of obtaining the 3D datasets either indirectly via sequential dental models [13–16] or directly using digital scanners when analyzing the accuracy of the 3D modes. However, the superimposition or alignment of the two datasets is not trivial and is also prone to error [17]. The alignment of the scans is performed by minimizing the mesh distance error between each corresponding data point. In our study, a standard best-fit alignment [18] was used. This method uses an iterative closest point (ICP) algorithm to align scans, with each software using a slightly different algorithm and does not involve operator-based decisions. The alignment is performed by minimizing the mesh distance error between each corresponding data point [18–20].

The precision of a 3D Printer NextDent™ using three different photosensitive polymer resins for three-dimensional (3D) printings with the help of GOM ATOS Capsule™ structured light optical scanner was examined in this study.

2. Materials and Method

A power study assuming 80% power and an alpha of 0.05 showed that 3 pairs of printed models for each material group were needed to show statistical differences of 0.5 mm in measurements with a 0.2 mm standard deviation [21].

A typodont model (Frasaco™ Gmbh, Tettnang, Germany) containing 16 mandibular permanent teeth (Figure 1) was chosen as a reference model. With the aid of GOM ATOS Capsule (Zeiss™ Gmbh, Braunschweig, Germany), which is an optical precision measuring machine, the reference model was scanned. This device used 2 12Mp CCD cameras and a fringe blue light projector to scan the surface. Spatial referencing was done via uncoded markers, while the stereo camera technology provided an overdetermined system of equations for each measurement. It was able to measure the reference markers with a deviation of 3 μm to 5 μm. Its result was a 3D mesh created by polygonizing the large number of triangulated points captured by the cameras. The scan of the reference model was used as a benchmark for comparison later in the study.

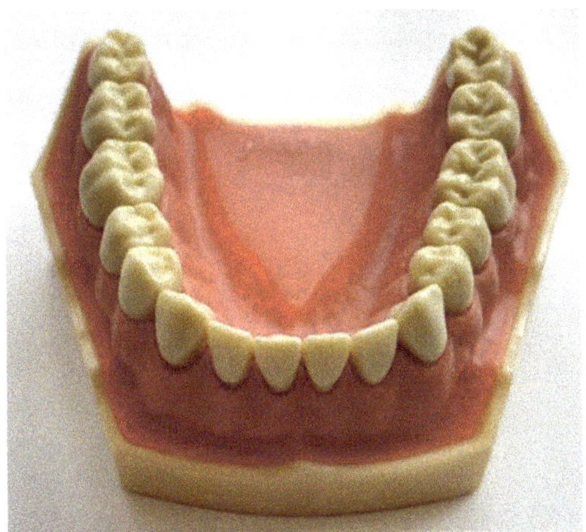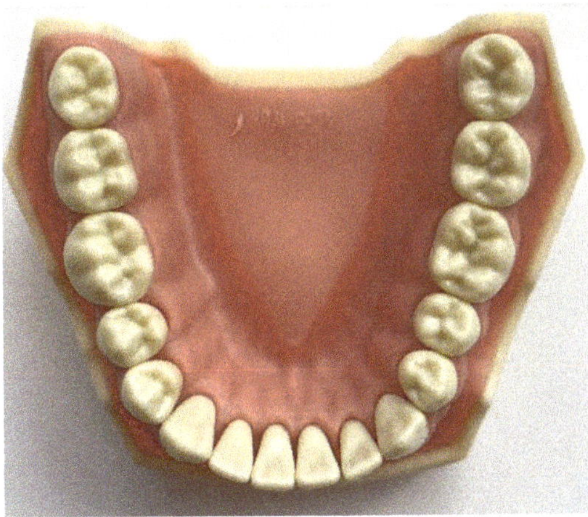

Figure 1. Model used for the reference comparison.

Three photosensitive resins were used in the study from 3 different producers. The name and brand of producers were intentionally omitted because of commercial purposes and were generally named Producer 1, Producer 2, Producer 3. The material from Producer 1 had a price range of 300$ to 400$ per 1 kg container, Producer 2 from 200$ to 300$ per 1 kg container, and Producer 3 under 100$ per 1 kg container. Three models were printed from each material using the reference model scan.

A NextDent 5100 (Soesterberg, The Netherlands) 3D printer was used, having the following settings: Build volume 124.8 × 70.2 × 196 mm (4.9 × 2.8 × 7.7 in), resolution 1920 × 1080 pixels, pixel pitch 65 microns (0.0025 in) (390.8 effective PPI), wavelength 405 nm. This printer used light (wavelength of 405 nm) to cure the resin. The resulting 9 models were scanned with the same 3D scanner as the reference model in similar light and temperature conditions. An observer repeated the measurements at a 1-week interval. As the model made by Producer 3 was shinier, the decision was made to cover the models with an antireflective powder. To preserve the measurement conditions, all models were covered irrespective of the material. The meshes resulting from scanning were exported in STL format. The STL file resulting from the scan of the reference model was used for printing the models from the 3 materials.

Later in the study, the meshes of models were compared with the reference scan, and distances were measured using the GOM Inspect 2020™ (Braunschweig, Germany) software package. The printed model and the reference model surfaces were pre-aligned through a standard best fit method. This method globally minimizes the deviations between the 2 surfaces. As this method only superimposes the 2 surfaces to get a globally acceptable deviation, a local best fit method was used to align the 2 surfaces in the teeth area (marked with red in Figure 2). This method minimized the deviations between the 2 entities in the region of interest for this study.

In the subsequent stage of the study, a comparison between the 3D printed model surface and the reference model surface was performed, outlining the deviations between them. Figure 3 shows these comparisons for each model grouped by material. Positive deviations were noticeable in the molar regions (yellow and red), while negative deviations can be visible in the incisor and canine regions (blue). On average, these deviations ranged from −0.06 to 0.06 mm.

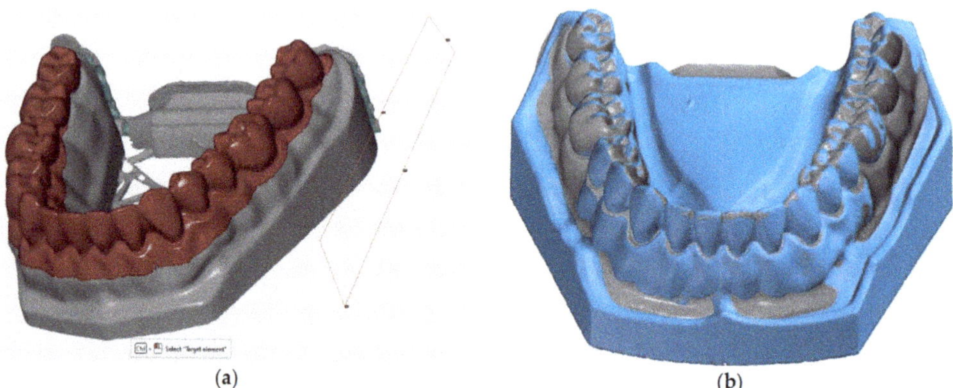

Figure 2. (**a**) Local best-fit alignment done in the region of interest between the 3D printed model and the reference surface; (**b**) the scanned surface (grey) and the reference surface (blue) overlapped.

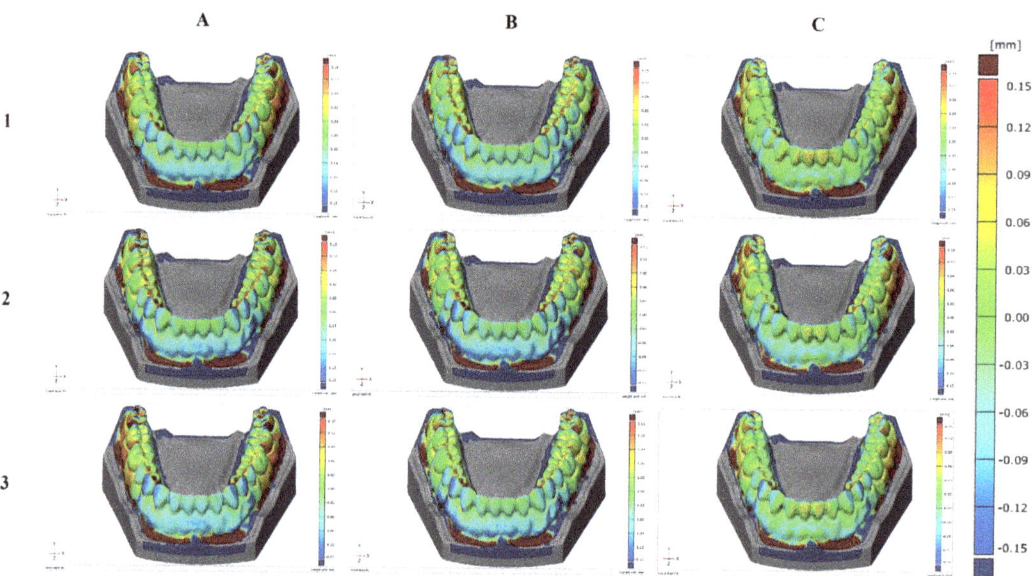

Figure 3. Surface comparison of the three printed models with the reference model for each material: (**A**) Producer 1; (**B**) Producer 2; (**C**) Producer 3.

For each model, the buccolingual width and mesiodistal width of each tooth were measured and the length of the arch curve. The dental arch width was also measured (the inter-canine, inter-premolar, and inter-molar distances) from the interior and exterior surfaces of each tooth and the height of each tooth (Figure 4).

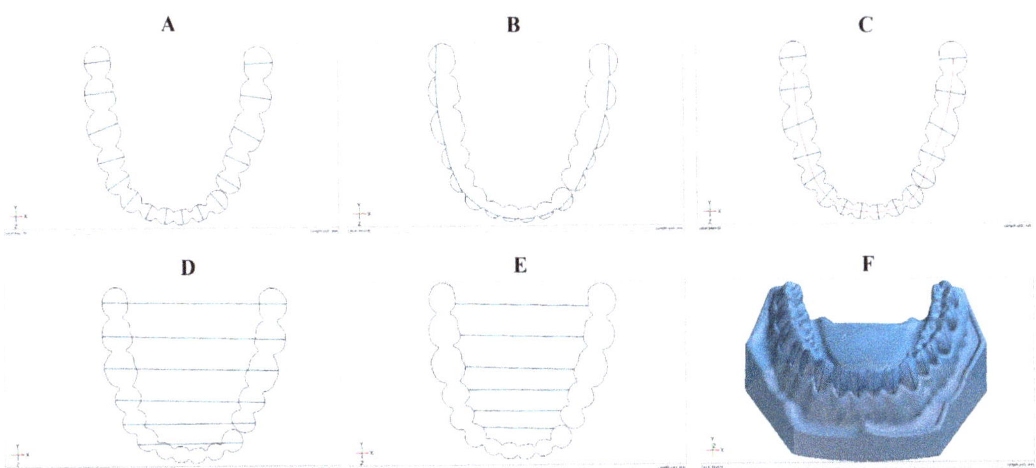

Figure 4. Measurements done on the models: (**A**) The buccolingual width, (**B**) the mesiodistal width, (**C**) the arch curve length, (**D**) arch width (exterior), (**E**) arch width (interior).(**F**) 3D printed model surface

The buccolingual tooth width was measured using a section plane to better quantify the deviations between the models (Figure 4A). To obtain the section plane, the crown heights of molars 38, 48, and incisor 41 were measured. The midpoints of the distances from the mucogingival junction to the occlusal surface of each of these 3 teeth were used to create the plane (Figure 5).

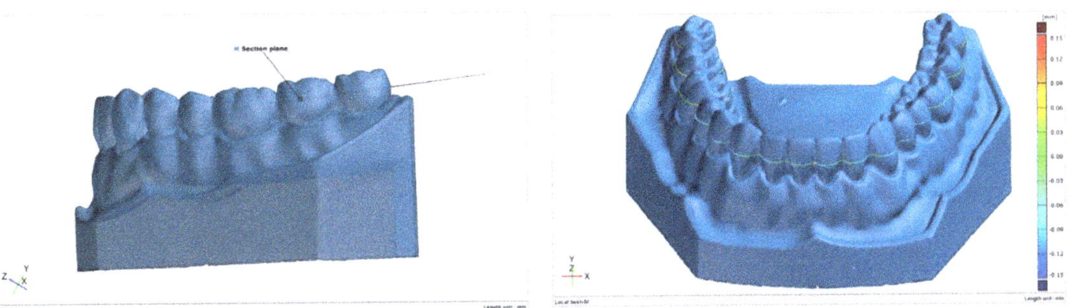

Figure 5. Section plane used for the buccolingual width measurements.

Surface points were placed on opposite sides of each tooth in the buccolingual direction, and the distance between the points was measured (Figure 6). Repeatability was ensured by using the same measurement program for all models.

The mesiodistal distance (Figure 4B) was more difficult to determine because of the way the model was scanned and printed. The stereoscopic scanner that was used was not able to scan until the point where the teeth touch on the real reference model. As a result, the gaps in the mesh were filled, creating a continuous surface. An exception was the distance between incisor 33 and premolar 34 that was wide enough to permit the creation of 2 distinct surfaces on the reference model. The mesiodistal distance was determined such that it is on either side of the measured tooth but not in the filling area from between the teeth. Measurements were made only in the midplane created for the buccolingual measurements.

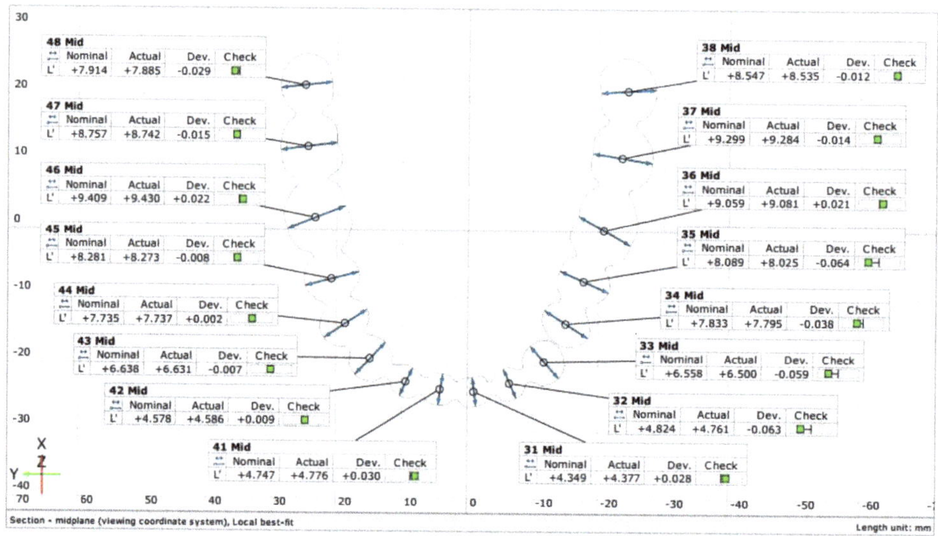

Figure 6. Section measurements in the buccolingual direction for each tooth.

The arch curve was determined through the midpoints of the buccolingual distance of each tooth on the midplane (Figure 4C).

The dental arch width was determined by measuring the inter-canine, inter-premolar, and inter-molar distances both from the exterior and interior surfaces of each tooth, as presented in Figure 4D,E. The height of each tooth was measured from the marginal gingiva to the occlusal surface of each tooth (Figure 4F).

Trueness and precision were used to evaluate each variable of the arch and tooth measurements. According to ISO 5725, trueness refers to "the closeness of agreement between the arithmetic mean of a large number of test results and the true or accepted reference value" while "precision refers to the closeness of agreement between test results" [21]. Trueness is a measure of systematic error while the precision of random error [22–24]. The standard deviation was used to evaluate the precision, while bias was used to evaluate trueness:

$$s = \sqrt{\frac{\sum_{i=1}^{n}(x_i - \bar{x})^2}{n}}$$

$$ias = abs(\bar{x} - x_{nom})$$

where: x_i is the measured value on a model for a specific characteristic. $\bar{x}$ the mean of measurements for the 3 models printed from the same material. $n = 3$ is the number of measurements. x_{nom} is the nominal value from the reference model.

3. Statistical Analysis

The Shapiro–Wilk test was used to assess the normality of the distributions. An independent-samples Kruskal–Wallis test was used to compare the precision and trueness values across the material groups. In the cases of significant results, a Mann–Whitney test was performed between each group pair. The level of significance chosen was $\alpha = 0.05$. The analysis was done in IBM SPSS v26, and the data were preprocessed in Microsoft Excel 365.

4. Results

Upon visual inspection of the models, defects were noticed on two out of three of the models printed with material from Producer 3. The defects presented as cavities caused by air bubbles. The defects were also reflected in the scans, as presented in Figure 7. As a result of the placement of these defects, some values of the mesiodistal distances were unable to be computed as the defects were in the exact region of the measurements.

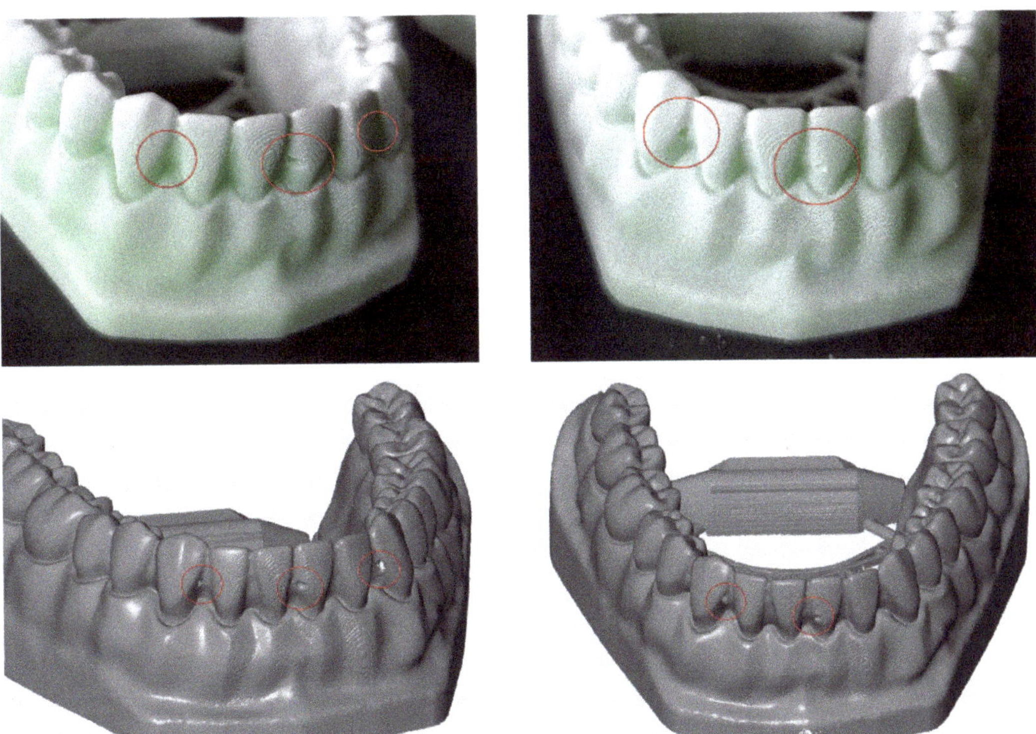

Figure 7. Defects, shown in red circles, in two models printed with the material from Producer 3 visible in the resulting scan.

A Kruskal–Wallis test was conducted to determine if there were differences in precision and trueness values for arch distances between groups with materials from different manufacturers: Producer 1, Producer 2, and Producer 3. Distributions of precision values were not similar for all groups, as assessed by visual inspection of a boxplot (Figure 8). Although there were variations, median precision values were not statistically significantly different between the different material groups, for either precision ($\chi^2(2) = 1.428$, $p = 0.490$) or trueness ($\chi^2(2) = 0.202$, $p = 0.904$).

The differences in precision and trueness values for buccolingual distances between groups with materials from different manufacturers: Producer 1, Producer 2, and Producer 3 were assessed. Through visual inspection of the boxplot, it was concluded that the distributions of precision values were similar for all groups. Although there were variations, median precision values were not statistically significantly different between the different material groups, for either precision ($\chi^2(2) = 2.327$, $p = 0.312$) or trueness ($\chi^2(2) = 4.349$, $p = 0.114$).

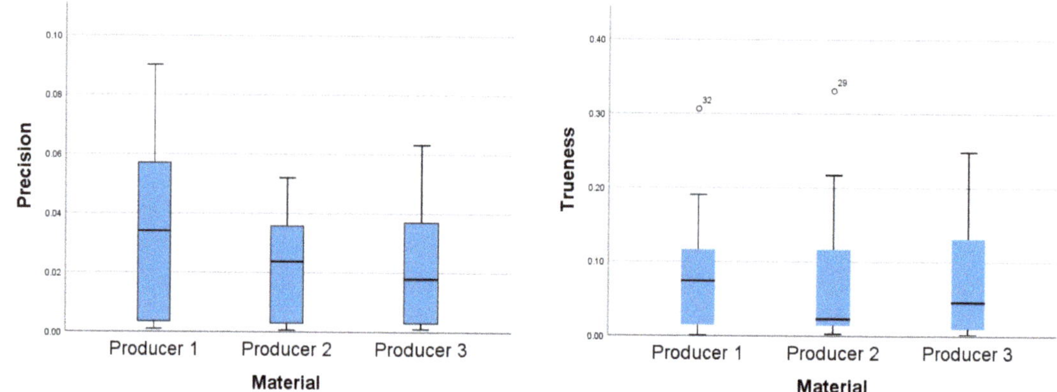

Figure 8. Distributions of precision values (**left**) and trueness values (**right**) for arch distances between the three material groups.

The same test was used to determine if there were differences in precision and trueness values for tooth height between groups with materials from different manufacturers: Producer 1, Producer 2, and Producer 3. Distributions of trueness values were similar for all groups but different for precision values, as assessed by visual inspection of the boxplots (Figure 9). Median precision values were not statistically significantly different between the different material groups, for either precision ($\chi^2(2) = 0.391$, $p = 0.822$) or trueness ($\chi^2(2) = 0.145$, $p = 0.930$). The precision and trueness mean (with the 95% CI) and standard deviations for the measurements are summarized in Table 1.

Table 1. Descriptive statistics for precision and trueness for the three materials for each type of measurement.

Measurement	Indicator	Variable	Producer 1	Producer 2	Producer 3
Arch distances	Trueness	Mean	0.083	0.073	0.080
		(95% CI)	[0.042, 0.124]	[0.027, 0.119]	[0.042, 0.118]
		Std. Dev.	0.083	0.092	0.077
	Precision	Mean	0.033	0.022	0.024
		(95% CI)	[0.018, 0.048]	[0.013, 0.031]	[0.13, 0.034]
		Std. Dev.	0.030	0.018	0.021
Buccolingual measurements	Trueness	Mean	0.024	0.040	0.023
		(95% CI)	[0.015, 0.033]	[0.25, 0.55]	[0.015, 0.030]
		Std. Dev.	0.017	0.028	0.014
	Precision	Mean	0.015 [0.011, 0.019]	0.017	0.020
		(95% CI)		[0.012, 0.021]	[0.015, 0.026]
		Std. Dev.	0.007	0.008	0.01
Mesiodistal measurements	Trueness	Mean	0.087	0.103	0.061
		(95% CI)	[0.060, 0.113]	[0.067, 0.138]	[0.014, 0.108]
		Std. Dev.	0.050	0.066	0.088
	Precision	Mean	0.022	0.019	0.056
		(95% CI)	[0.009, 0.034]	[0.009, 0.029]	[−0.008, 0.120]
		Std. Dev.	0.023	0.019	0.112
Tooth height	Trueness	Mean	0.046	0.043	0.045
		(95% CI)	[0.029, 0.063]	[0.027, 0.058]	[0.027, 0.062]
		Std. Dev.	0.032	0.029	0.032
	Precision	Mean	0.014	0.011	0.010
		(95% CI)	[0.009, 0.019]	[0.007, 0.015]	[0.007, 0.013]
		Std. Dev.	0.010	0.007	0.006
Arch curve length	Trueness	Length	0.143	0.121	0.029
	Precision	Length	0.036	0.012	0.01

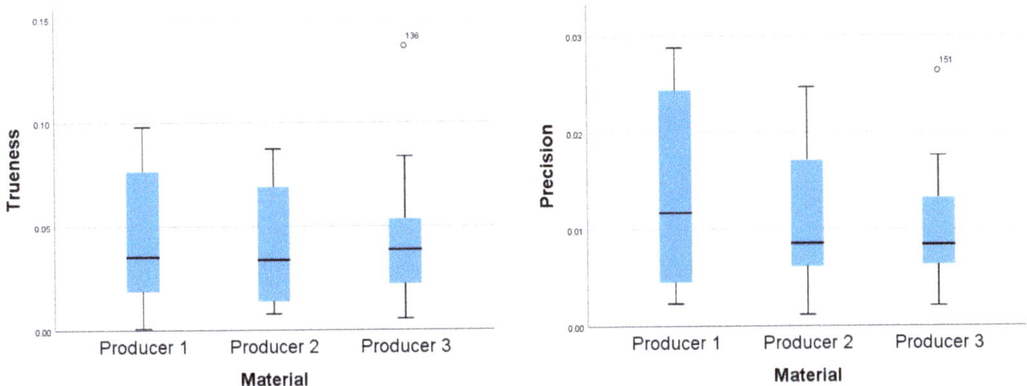

Figure 9. Distributions of precision values (**left**) and trueness values (**right**) for tooth height between the three material groups.

The results for the buccolingual and mesiodistal measurements had a lower spread for precision than for trueness, except for the material from the third producer for the mesiodistal measurements (as seen in Figure 10).

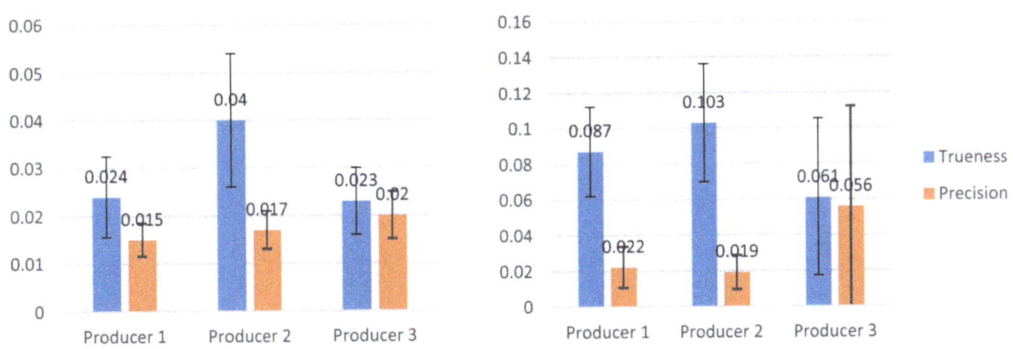

Figure 10. Buccolingual (**left**) and mesiodistal (**right**) measurements for the three material types.

The distributions of values for precision and trueness were tested for normality using both the Kolmogorov–Smirnov and the Shapiro–Wilk tests. The results of the tests were mixed, some values being normally distributed while others not. In addition, some outliers were identified. As a result, the decision was made to use non-parametric methods for further analysis as they do not make any assumptions regarding the distribution of data, and they are more robust to the presence of outliers than parametric methods.

Differences in precision values for the mesiodistal length between groups with materials from different manufacturers: Producer 1, Producer 2, and Producer 3 were assessed. Precision values were not distributed similarly for all groups, as assessed by visual inspection of a boxplot. Median precision values were statistically significantly different between the different material groups ($\chi^2(2) = 6.370$, $p = 0.041$. Subsequently, pairwise comparisons were performed using Dunn's procedure with a Bonferroni correction for multiple comparisons. This post-hoc analysis revealed statistically significant differences in precision values between the material from Producer 1 and Producer 3 and between Producer 2 and Producer 3, but when applying the Bonferroni correction, the statistical significance disappeared. The decision was made to follow up with a Mann–Whitney test for the two pairs that resulted in a significant difference.

A Mann–Whitney U test was run to determine if there were differences in precision values for the mesiodistal distance between materials from Producer 1 and Producer 3.

Precision values were statistically significantly higher for the material from Producer 3 (Mean rank = 20.38) than for the material from Producer 1 (Mean rank = 12.63), U = 66, $z = -2.337$, $p = 0.019$, using an exact sampling distribution for U [23].

The same test was run to determine if there were differences in precision values for the mesiodistal distance between materials from Producer 2 and Producer 3. Precision values were not statistically significantly different, U = 76, $z = -1.960$, $p = 0.050$. Although the precision values for the material from Producer 3 were higher (Mean rank = 19.75) than for the material from Producer 1 (Mean rank = 13.25), the p-value was marginally significant.

Distributions of the precision values were not similar, as assessed by visual inspection for either of the two tests.

Differences in trueness values for the mesiodistal length between groups with materials from different manufacturers: Producer 1, Producer 2, and Producer 3 were tested. After a visual inspection of the distributions of trueness values, it was determined that they were not similar. Median precision values were statistically significantly different between the different material groups, ($\chi^2(2) = 10.050$, $p = 0.007$. Adjusted p-values are presented. Subsequently, pairwise comparisons were performed using Dunn's procedure with a Bonferroni correction for multiple comparisons. This post-hoc analysis revealed statistically significant differences in trueness values between the material from Producer 1 (Mean rank = 27.69) and Producer 3 (Mean rank = 15.56) ($p = 0.043$) and between Producer 2 (Mean rank = 30.25) and Producer 3 (0.009), but not between materials from Producer 1 and Producer 2 ($p = 1.000$). This shows that the material from Producer 3 has a significantly lower bias for the mesiodistal distances than materials from Producer 1 and Producer 2.

5. Discussion

In this study, printed models obtained from digital scans made with an extraoral scanner were used because extraoral scanning is increasingly used to make digital dental models, and some of the errors that can occur in the traditional impression-taking procedure can be avoided. Digital models have several advantages compared with plaster models, such as ease of data storage and data transmission, provide both visual and tactile information, and can be used for diagnostic, therapeutic, and education purposes. The goal of this study was to assess the trueness and precision of dental models obtained by the extraoral scanning technique, fabricated using three different types of polymer resin with a 3D printer. The accuracy of various 3D printed models has been validated only by a few studies. In one such study, Hazeveld et al. [6] decided to fabricate dental models using three types of rapid prototyping in order to analyze the accuracy of these models. They used digital calipers to measure the size of the teeth, focusing on measuring the mesiodistal height and width. It failed to measure the buccolingual width, which is also influenced by the method of printing and polymerization. This might have affected the fit of orthodontic appliances and individualized trays. In another study, Murugesan et al. [25] also made dental models using three types of rapid prototyping and also used digital calipers to measure the teeth in order to assess the accuracy of the models. The use of digital calipers in measuring the teeth might have led to errors in measurement due to the difficulty in finding the tooth of a reference point. Their models were printed by different printers, which might have also led to inconsistencies and errors in measurements. We tried to address both shortcomings by applying 3D software to validate the trueness and precision of the dental models made by the 3D printers and by establishing more clear reference points on the gingival areas and the teeth. We consider that the careful establishment of clear reference points played an important and decisive role in getting good results. Highly repeatable measurements have been reported by Salmi et al. [26], but in this study, they used a 10.0 mm reference point.

Another important factor when scanning 3D printed models is represented by the scan spray (antireflective powder). This powder helps in lowering the reflection of light of the printed models. In the 3D color map of the experimental group, the labial and buccal surfaces of the 3D printed models in all experimental groups displayed a homogenous pattern of the blue area, which represented shrinkage. This might be explained by the characteristic of the surface in these areas, which was usually smooth and allowed the polymers to contact evenly. The opposite situation was present on the occlusal and interdental surface with its pit and groove regions, where an uneven pattern of shrinkage was observed.

As far as we know, the differences between 3D printed models and a reference model have never been carefully examined in order to prove they are clinically acceptable. One possibility might be that the dimension differences have no impact on clinical applications [3,27]. As specified in Hirogaki, Y [28] from a clinical perspective, a 0.3 mm dimension difference in dental models might be accurate enough. Depending on the treatment method, different clinical standards should be used for determining the accuracy and adequacy of dental models. The choice of 3D printing technology must be determined by its intended application. Hence, it is reasonable to conclude that 3D printed models, which are clinically acceptable for orthodontic purposes, may not necessarily be acceptable for the prosthodontic workflow or other dental applications requiring high accuracy [26,28].

When using 3D superimposition techniques, the risk of bias and applicability concerns are low as high accuracy desktop scanners are utilized, and CAM is the only identified source of error. However, it is worth noting that increased risk of bias and applicability concerns for index tests are recorded for studies that use linear measurements because of human error when performing physical linear measurements with no information provided on the calibration of the examiners. We believe that the dental models produced via 3D printing may be good enough for clinical purposes. It can be expected that the costs of printing dental models will decrease, and the costs will possibly become comparable with the conventional fabrication of plaster models. Increased use of CAD/CAM techniques for making customized orthodontic appliances with appliance printing techniques can be expected. However, in order to fully analyze the clinical efficacy and the accuracy of the 3D models, more studies are needed.

6. Conclusions

There were significant discrepancies in the trueness of mesiodistal distance measurements between the 3D printing polymer resins. Producer 1 and Producer 2 were more precise than Producer 3 material, with the Producer 1 photosensitive polymer displaying the highest accuracy. Our results show that the material from Producer 3 has a significantly lower precision (a higher spread) for the mesiodistal measurements than the material from Producer 1. For the comparison between the materials from Producer 2 and Producer 3, the results are inconclusive as the p-value is marginally significant ($p = 0.05$). Mean precision does not vary too much between these three photosensitive polymer resins, therefore, the trueness and the required precision of the clinical purpose of the model should be deciding factors in choosing the proper 3D printing materials.

Author Contributions: Conceptualization, A.V., S.M.B. and A.B.; data curation, E.S.B., S.M.B. and M.P.; investigation, E.S.B., V.I.B. and M.H.M.; methodology, A.V. and C.D.O.; project administration, M.H.M., M.P. and B.R.D.; resources, V.I.B., M.H.M. and C.D.O.; software, V.I.B., B.R.D. and A.B.; supervision, M.P.; visualization, C.D.O.; writing—original draft, A.V.; writing—review and editing, E.S.B. and B.R.D. All authors contributed equally to this research. All authors have read and agreed to the published version of the manuscript.

Funding: This research received no external funding.

Institutional Review Board Statement: Not applicable.

Informed Consent Statement: Not applicable.

Conflicts of Interest: The authors declare that they have no conflict of interest regarding this manuscript.

References

1. Sason, G.K.; Mistry, G.; Tabassum, R.; Shetty, O. A comparative evaluation of intraoral and extraoral digital impressions: An in vivo study. *J. Indian Prosthodont. Soc.* **2018**, *18*, 108–116. [CrossRef]
2. Patzelt, S.B.; Vonau, S.; Stampf, S.; Att, W. Assessing the feasibility and accuracy of digitizing edentulous jaws. *J. Am. Dent. Assoc.* **2013**, *144*, 914–920. [CrossRef] [PubMed]
3. Kasparova, M.; Grafova, L.; Dvorak, P.; Dostalova, T.; Prochazka, A.; Eliasova, H. Possibility of reconstruction of dental plaster cast from 3D digital study models. *Biomed. Eng. Online* **2013**, *12*, 49. [CrossRef] [PubMed]
4. Saleh, W.K.; Ariffin, E.; Sherriff, M.; Bister, D. Accuracy and reproducibility of linear measurements of resin, plaster, digital and printed study-models. *J. Orthod.* **2015**, *42*, 301–306. [CrossRef] [PubMed]
5. Keating, A.P.; Knox, J.; Bibb, R.; Zhurov, A.I. A comparison of plaster, digital and reconstructed study model accuracy. *J. Orthod.* **2008**, *35*, 191–201; discussion 175 . [CrossRef]
6. Hazeveld, A.; Huddleston Slater, J.J.; Ren, Y. Accuracy and reproducibility of dental replica models reconstructed by different rapid prototyping techniques. *Am. J. Orthod. Dentofac. Orthop.* **2014**, *145*, 108–115. [CrossRef]
7. Etemad-Shahidi, Y.; Qallandar, O.B.; Evenden, J.; Alifui-Segbaya, F.; Ahmed, K.E. Accuracy of 3-Dimensionally Printed Full-Arch Dental Models: A Systematic Review. *J. Clin. Med.* **2020**, *9*, 3357. [CrossRef] [PubMed]
8. Russell Mullen, S.; Martin, C.A.; Ngan, P.; Gladwin, M. Accuracy of space analysis with emodels and plaster models. *ALODO* **2007**, *132*, 346–352.
9. Asquith, J.A.; McIntyre, G.T. Dental arch relationships on three-dimensional digital study models and conventional plaster study models for patients with unilateral cleft lip and palate. *Cleft Palate Craniofacial J.* **2012**, *49*, 530–534. [CrossRef]
10. Metzger, M.C.; Hohlweg-Majert, B.; Schwarz, U.; Teschner, M.; Hammer, B.; Schmelzeisen, R. Manufacturing splints for orthognathic surgery using a three-dimensional printer. *Oral Surg. Oral Med. Oral Pathol. Oral Radiol. Endodontology* **2008**, *105*, 1–7. [CrossRef]
11. Vogel, A.B.; Kilic, F.; Schmidt, F.; Rübel, S.; Lapatki, B.G. Dimensional accuracy of jaw scans performed on alginate impressions or stone models: A practice-oriented study. *J. Orofac. Orthop.* **2015**, *76*, 351–365. [CrossRef]
12. Vogel, A.B.; Kilic, F.; Schmidt, F.; Rübel, S.; Lapatki, B.G. Optical 3D scans for orthodontic diagnostics performed on fullarch impressions. Completeness of surface structure representation. *J. Orofac. Orthop.* **2015**, *76*, 493–507. [CrossRef] [PubMed]
13. Rodriguez, J.M.; Austin, R.S.; Bartlett, D.W. A method to evaluate profilometric tooth wear measurements. *Dent. Mater.* **2012**, *28*, 245–251. [CrossRef] [PubMed]
14. Tantbirojn, D.; Pintado, M.R.; Versluis, A.; Dunn, C.; Delong, R. Quantitative analysis of tooth surface loss associated with gastroesophageal reflux disease. *J. Am. Dent. Assoc.* **2012**, *143*, 278–285. [CrossRef] [PubMed]
15. Pintado, M.R.; Anderson, G.C.; DeLong, R.; Douglas, W.H. Variation in tooth wear in youngadults over a two-year period. *J. Prosthet. Dent.* **1997**, *77*, 313–320. [CrossRef]
16. Chadwick, R.G.; Mitchell, H.L.; Manton, S.L.; Ward, S.; Ogston, S.; Brown, R. Maxillary incisor palatal erosion: No correlation with dietary variables? *J. Clin. Pediatr. Dent.* **2005**, *29*, 157–164. [CrossRef] [PubMed]
17. Besl, P.; McKay, N. A Method for Registration of 3-D Shapes. *IEEE Trans. Pattern Anal. Mach. Intell.* **1992**, *14*, 239–256. [CrossRef]
18. Becker, K.; Wilmes, B.; Grandjean, C.; Drescher, D. Impact of manual control point selection accuracy on automated surface matching of digital dental models. *Clin. Oral Investig.* **2018**, *22*, 801–810. [CrossRef] [PubMed]
19. Wulfman, C.; Koenig, V.; Mainjot, A.K. Wear measurement of dental tissues and materials in clinical studies: A systematic review. *Dent. Mater.* **2018**, *34*, 825–850. [CrossRef] [PubMed]
20. O'Toole, S.; Osnes, C.; Bartlett, D.; Keeling, A. Investigation into the accuracy and measurement methods of sequential 3D dental scan alignment. *Dent. Mater.* **2019**, *35*, 495–500. [CrossRef]
21. Kane, S.P. Sample Size Calculator. ClinCalc. Updated 24 July 2019. Available online: https://clincalc.com/stats/samplesize.aspx (accessed on 13 April 2021).
22. ISO 5725-1:1994 (en), Accuracy (Trueness and Precision) of Measurement Methods and Results—Part 1: General Principles and Definitions. Available online: https://www.iso.org/obp/ui/#iso:std:iso:5725:-1:ed-1:v1:en (accessed on 20 January 2021).
23. Dinneen, L.C.; Blakesley, B.C. Algorithm AS 62: A Generator for the Sampling Distribution of the Mann—Whitney U Statistic. *J. Roy. Stat. Soc. Ser. C* **1973**, *22*, 269–273. [CrossRef]
24. Menditto, A.; Patriarca, M.; Magnusson, B. Understanding the meaning of accuracy, trueness and precision. *Accredit. Qual. Assur.* **2007**, *12*, 45–47. [CrossRef]
25. Murugesan, K.; Anandapandian, P.A.; Sharma, S.K. Comparative Evaluation of Dimension and Surface Detail Accuracy of Models Produced by Three Different Rapid Prototype Techniques. *J. Indian Prosthodont. Soc.* **2011**, *12*, 16–20. [CrossRef] [PubMed]
26. Salmi, M.; Paloheimo, K.S.; Tuomi, J.; Wolff, J.; Mäkitie, A. Accuracy of medical models made by additive manufacturing (rapid manufacturing). *J. Cranio Maxillofac. Surg.* **2013**, *41*, 603–609. [CrossRef]
27. Choi, J.Y.; Choi, J.H.; Kim, N.K.; Kim, Y.; Lee, J.K.; Kim, M.K. Analysis of errors in medical rapid prototyping models. *Int. J. Oral Maxillofac. Surg.* **2002**, *31*, 23–32. [CrossRef] [PubMed]
28. Hirogaki, Y.; Sohmura, T.; Satoh, H.; Takahashi, J.; Takada, K. Complete 3-D reconstruction of dental cast shape using perceptual grouping. *IEEE Trans. Med. Imaging* **2001**, *20*, 1093–1101. [CrossRef] [PubMed]

Journal of
Clinical Medicine

Efficacy of Plasma-Polymerized Allylamine Coating of Zirconia after Five Years

Nadja Rohr [1,2,*], Katja Fricke [3], Claudia Bergemann [2], J Barbara Nebe [2] and Jens Fischer [1]

1   Biomaterials and Technology, Department of Reconstructive Dentistry, University Center for Dental Medicine, University of Basel, 4058 Basel, Switzerland; jens.fischer@unibas.ch

2   Department of Cell Biology, Rostock University Medical Center, 18057 Rostock, Germany; claudia.bergemann@med.uni-rostock.de (C.B.); barbara.nebe@med.uni-rostock.de (J.B.N.)

3   Leibniz Institute for Plasma Science and Technology e.V. (INP), 17489 Greifswald, Germany; k.fricke@inp-greifswald.de

*   Correspondence: nadja.rohr@unibas.ch; Tel.: +41-612-672-799

Received: 16 July 2020; Accepted: 24 August 2020; Published: 27 August 2020

Abstract: Plasma-polymerized allylamine (PPAAm) coatings of titanium enhance the cell behavior of osteoblasts. The purpose of the present study was to evaluate a PPAAm nanolayer on zirconia after a storage period of 5 years. Zirconia specimens were directly coated with PPAAm (ZA0) or stored in aseptic packages at room temperature for 5 years (ZA5). Uncoated zirconia specimens (Zmt) and the micro-structured endosseous surface of a zirconia implant (Z14) served as controls. The elemental compositions of the PPAAm coatings were characterized and the viability, spreading and gene expression of human osteoblastic cells (MG-63) were assessed. The presence of amino groups in the PPAAm layer was significantly decreased after 5 years due to oxidation processes. Cell viability after 24 h was significantly higher on uncoated specimens (Zmt) than on all other surfaces. Cell spreading after 20 min was significantly higher for Zmt = ZA0 > ZA5 > Z14, while, after 24 h, spreading also varied significantly between Zmt > ZA0 > ZA5 > Z14. The expression of the mRNA differentiation markers collagen I and osteocalcin was upregulated on untreated surfaces Z14 and Zmt when compared to the PPAAm specimens. Due to the high biocompatibility of zirconia itself, a PPAAm coating may not additionally improve cell behavior.

Keywords: zirconia implant; human osteoblasts; cell viability; cell spreading; gene expression; plasma-polymerized allylamine; X-ray photoelectron spectroscopy

1. Introduction

To replace missing teeth, dental implants made of titanium are a valuable treatment option. However, in recent years, titanium implants have been critically discussed regarding the release of titanium particles and biologic complications [1]. There are some indications that Ti ions released from the implant surface upregulate the expression of chemokines and cytokines in human osteoclasts and osteoblasts. Consequently, osteoclastogenesis is induced, which may contribute to the pathomechanism of aseptic loosening [2,3]. Dental implants made of zirconia can be considered promising alternatives to titanium implants [4–6]. Clinical data are available, reporting survival rates of 95.4% after 3 years [4] and 98.4% after 5 years in situ [5].

Permanent osseointegration, indicated by the formation of a direct bone–implant contact, is the most important requirement for the clinical success of an implant [7]. The endosseous part of the implant is shaped as screw to achieve a certain primary stability after insertion. Additionally, most implant surfaces are micro-structured, which is reported to enhance osseointegration [8]. For zirconia implants, different approaches are undertaken to structure the endosseous surface such as sandblasting, acid-etching, laser structuring, additive sintering or injection molding [9–11]. The currently available surfaces providing long-term clinical data for zirconia implants are sandblasted followed by acid etching [4,12] and, optionally, heat treated [5].

Another approach is the creation of a biologically active implant surface by applying an additional functional layer, which has been done for titanium surfaces [13,14]. Nitrogen-rich surface chemistry is known to promote cellular attachment because it contains polar groups [15]. The most common plasma precursors used to generate amine functionalities on biomaterials are allylamine [16–18], ethylendiamine [19], cyclopropylamine [20,21] as well as mixtures of hydrocarbon-containing gases and molecular nitrogen [22]. Due to the presence of positively charged carriers such as NH_2 groups on the surface coating [23,24], the net negative charged eukaryotic cells are attracted. For instance, plasma-polymerized allylamine (PPAAm) coatings have been applied on titanium [25–30], titanium alloy (Ti6Al4V) [22], porous calcium phosphate [31] and yttria-stabilized zirconia (Y-TZP) [32] to improve their hydrophilic properties by generating positively charged amine groups. The resulting zeta potential changed from negative into positive values, e.g., untreated titanium: −82.3 mV and PPAAm-coated Ti: +8.6 mV (pH 7.4) [33]. On all tested materials, the cell spreading of human osteoblastic cells MG-63 was accelerated by the PPAAm coating. PPAAm-coated titanium plates have also been inserted in the muscular neck tissue of rats, revealing lower macrophage-related reactions in the mid (14 d) and late (56 d) phases of the study than uncoated titanium specimens [29]. However, the PPAAm coating is susceptible to aging. Within 7 days after coating, 70% of the primary amino groups of the PPAAm layer were already converted into amides. Zeta potential remained positive and even increased with prolonged storage of 200 d from 13.9 ± 1.2 mV to 26.3 ± 0.5 mV (pH 6.0), probably due to the increased density of imines, nitriles and acid amides [30]. Nevertheless, the cell spreading of human osteoblasts on PPAAm-coated titanium alloys that were stored over 360 d was accelerated compared to uncoated specimens [30]. To evaluate the differentiation behavior of osteoblastic cells, gene expression of differentiation markers such as alkaline phosphatase (ALP), collagen type 1 (COL) or osteocalcin (OCN) are measured. COL and ALP are considered early differentiation markers in the osteoblast lineage, while the transcription of OCN is enhanced in a later differentiation stage. The purpose of the present study is to test whether a PPAAm coating on zirconia is stable up to 5 years, which is the common shelf life of ready-for-sale implants. The reaction of human osteoblasts to the PPAAm coating on zirconia has therefore been assessed by evaluating cell viability, spreading, cell morphology and gene expression.

2. Materials and Methods

Zirconia discs with a diameter of 13 mm and a height of 2 mm were produced. The discs were machine overdimensioned in the green state, sintered and isostatically hot pressed in order to get disc-shaped specimens. The zirconia was composed of 93.0 wt% ZrO_2, 5.0 wt% Y_2O_3, 0.1 wt% Al_2O_3, 1.9 wt% HfO_2; its grain size was 0.3 µm (MZ111, CeramTec, Plochingen, Germany). Four different surfaces were produced according to Table 1: ZA0: as-sintered zirconia, heat treated for 1 h at 1250 °C, PPAAm coating, ZA5: as-sintered zirconia, heat treated for 1 h at 1250 °C, PPAAm coating, aged 5 years in sealed package, Zmt: as-sintered zirconia, heat treated for 1 h at 1250 °C, Z14: sandblasted Al_2O_3 105 µm, etched for 1 h in hydrofluoric acid 38–40%, heat treated for 1 h at 1250 °C. Z14 is the endosseous surface of a clinically tested implant [5] (cer.face 14, Vita, Bad Säckingen, Germany) and served as the clinically relevant control. Z14 displayed the following roughness parameters: arithmetical mean (Ra) = 1.47 ± 0.0.6 µm, maximum height of profile (Rz) = 10.85 ± 0.67 µm [34]. Specimens were heat treated at 1250 °C for 1 h to achieve a higher tetragonal phase of zirconia and consequently increase its resistance to aging. The surface of Zmt (Ra = 0.33 ± 0.0.2 µm, Rz = 2.71 ± 0.24 µm [34]) served as a substrate for the specimens treated with PPAAm (ZA0 and ZA5). Those specimens were coated with a thin (approximately 40 nm) PPAAm layer using a low-pressure plasma reactor (V55G, Plasma Finish, Germany) according to the following two-step procedure: (1) activation of the substrates by a continuous wave oxygen/argon plasma (500 W, 50 Pa, 1000-sccm O2/5 sccm Ar) for 60 s and (2) deposition of PPAAm by microwave-excited (2.45 GHz) pulsed plasma (500 W, 50 Pa, 50 sccm Ar) for 480 s (effective treatment time). Prior to flushing the reactor with allylamine, the precursor was carefully purified of air by evacuating and purging with N_2. Substrates were treated in a downstream position 9 cm from the microwave coupling window. ZA0 and ZA5 were then immediately stored in aseptic packaging until use. Specimens of ZA5 were stored in aseptic packaging at room temperature for a period of 5 years. Prior to all experiments, Z14 and Zmt specimens were cleaned in an ultrasonic bath, 70% ethanol for 5 min, distilled water for 5 min, sterilized in a heating chamber at 200 °C for 2 h (FED-240, Binder, Tuttlingen, Germany) and stored in sterile petri dishes that were wrapped with aluminum foil for at least 2 weeks. The specimen surfaces were then characterized in terms of their elemental composition and visualized using scanning electron microscopy (SEM).

Table 1. Pretreatment of zirconia surfaces of the respective groups.

Group	Surface Pretreatment
ZA0	heat treated for 1 h at 1250 °C, plasma-polymerized allylamine coating September 2018
ZA5	heat treated for 1 h at 1250 °C, plasma-polymerized allylamine coating August 2013
Zmt	heat treated for 1 h at 1250 °C
Z14	sandblasted Al_2O_3 105 µm, etched 1 h hydrofluoric acid 38–40%, heat treated for 1 h at 1250 °C

2.1. Specimen Characterization

2.1.1. Elemental Composition (XPS)

The elemental surface composition was analyzed by high-resolution scanning XPS. The spectra were acquired using an Axis Supra delay-line detector (DLD) electron spectrometer (Kratos Analytical, Manchester, UK) equipped with a monochromatic Al K_α source (1486.6 eV). The analysis area was approximately 250 µm in diameter during the acquisition, obtained by using the medium magnification lens mode (field of view 2) and by selecting the slot mode. The core level spectra of each element, which were identified in the survey spectra, were collected at a pass energy of 80 eV by applying an emission current of 10 mA and a high voltage of 15 kV. Charge neutralization was implemented by a low-energy electron injected into the magnetic field of the lens from a filament located directly atop the sample. For each sample, spectra were recorded on three different spots and randomly distributed. Data processing was carried out using CasaXPS software, version 2.3.22PR1.0 (Casa Software Ltd., Teighnmouth, UK). Due to sample charging, the binding energy scale was corrected for all samples by setting the carbon C1s binding energy to 285.0 eV. Concentrations are provided in atomic percent (at%). The labeling of primary amino groups was performed with 4-trifluoromethyl-benzaldehyde (TFBA, Alfa Aesar, Haverhill, MA, USA) at 40 °C in a saturated gas phase for 2 h. The density of the amino groups, the ratio of NH_2 to carbon atoms (NH_2/C), was determined from the fluorine elemental fraction.

2.1.2. SEM Imaging

The specimens' surfaces were gold-sputtered and visualized with a scanning electron microscope (SEM) using mixed secondary electrons (SE) and backscattered electrons (BSE) modes at 15 kV (ESEM XL30, Philips, Eindhoven, the Netherlands).

2.2. Cell Behavior

2.2.1. Cell Cultivation

The human osteoblastic cell line MG-63 (American Type Culture Collection ATCC, CRL1427) was cultivated in Dulbecco's modified Eagle medium (DMEM + GlutaMAX-1 + 4.5 g/L DGlucose + Pyruvate; gibco, Thermo Fisher Scientific, Waltham, MA, USA) with the addition of 10% fetal calf serum (FCS superior standardized S0615 0879F, Biochrom, Berlin, Germany) and 1% antibiotic (gentamicin, ratiopharm, Ulm, Germany) to 70–80% confluency up to passages 8–21 [35] at 37 °C in a humidified atmosphere with 5% CO_2. Cells were detached with 0.05% trypsin/0.02% ethylenediaminetetraacetate (EDTA, PAA Laboratories GmbH) for 5 min at 37 °C. After stopping trypsinization by the addition of a complete cell culture medium, an aliquot of 100 µL was put into 10 mL of CASY ton buffer solution (Roche Innovatis, Reutlingen, Germany) and the cell number was measured in the counter CASY Model DT (Schärfe System, Reutlingen, Germany). Specimens were seeded with the appropriate cell number and incubated in 24-well plates (Greiner Bio-One, Frickenhausen, Germany) for the respective time intervals. All cell experiments were performed independently three times using different cell passages.

2.2.2. Cell Viability

The mitochondrial dehydrogenase activity of MG-63 cells on the respective specimens was measured by 3-(4,5-dimethylthiazol-2-yl)-2,5-diphenyltetrazolium bromide (MTS) assay to determine cell viability. A drop of 120 µL cell culture medium containing 5×10^4 MG-63 cells was carefully placed on each specimen (n = 2 per group) and incubated for 20 min to ensure cell attachment on the specimens. Afterwards, 1 mL of cell culture medium was added per well and the specimens were incubated for 24 h at 37 °C. Specimens were transferred to a new 24-well plate with MTS solution (CellTiter 96 ONE-Solution Cell Proliferation Assay, Promega, Madison, WI, USA) and culture medium (1:5) was added to each specimen. Blanks containing a specimen of each group with culture medium but without cells and a control group with cells growing on polystyrene were additionally tested. After 80 min, supernatants were transferred to a 96-well plate (for each specimen 3×80 µL were analyzed). The optical density (OD) was recorded at 490 nm with a micro-plate reader (Anthos, Mikrosysteme, Krefeld, Germany). Relative cell viability was calculated using the following equation:

$$\text{Relative cell viability} = (OD_{\text{specimen}} - OD_{\text{blank specimen}})/(OD_{\text{control}} - OD_{\text{blank control}})$$

2.2.3. Cell Spreading

Cell spreading was assessed on all surfaces after 20 min and 24 h, respectively. In total, 10^6 cells were suspended in 250 µL diluent C and their cell membranes were stained with PKH-26, a lipophilic membrane dye (PKH-26 general cell linker kit, Sigma-Aldrich, Steinheim, Germany) for 5 min at 37 °C using a dilution of 2 µL PKH-26 + 248 µL diluent C. After stopping the staining reaction using FCS, cells were washed with Dulbecco's phosphate buffered saline (PBS), resuspended in cell culture medium and 3×10^4 cells were seeded per specimen. After 20 min or 24 h, cells were rinsed twice with Dulbecco's phosphate buffered saline (PBS) (Sigma-Aldrich), fixed with 4% paraformaldehyde for 10 min at room temperature (RT), rinsed with PBS and embedded with mounting medium (Fluoroshield with DAPI, Sigma-Aldrich) and a cover slip. Cells were examined with a water immersion objective (C Apochromat 40×, 1.2 W, Carl Zeiss, Oberkochen, Germany) at a wavelength of 546 nm using a confocal laser scanning microscope (LSM780, Carl Zeiss, Oberkochen, Germany; ZEN 2011 software black version, Carl Zeiss, Oberkochen, Germany). The mean spreading area in μm^2 of 40 cells per specimen was then calculated using image processing software (ImageJ, v2.0.0, National Institutes of Health, Bethesda, Maryland, USA).

2.2.4. Cell Morphology

The morphology of 4×10^4 cells on the respective specimens after 20 min and 24 h was visualized using SEM. Cells on the specimens were rinsed with PBS after the respective time intervals, fixed with 2.5% glutaraldehyde (Merck KGaA, Darmstadt, Germany) for 30 min at 4 °C, rinsed with PBS, dehydrated with ethanol (30%, 50%, 70%, 90%, abs.), dried in a desiccator with silica gel and gold-sputtered.

2.2.5. Gene Expression

On each specimen, 3×10^4 cells were seeded (n = 2 per group) and cultivated for 24 h or 3 d, respectively. Total RNA was purified using the NucleoSpin RNA kit (Machery-Nagel, Düren, Germany) after the cell lysate of the 2 specimens per group was pooled. The RNA concentration for each group was measured using NanoDrop 1000 (Peqlab/VWR, Erlangen, Germany).

After isolating total RNA, first-strand cDNA was synthesized from at least 400 ng total RNA by reverse transcription with SuperScript II (Life Technologies, Darmstadt, Germany) using 2.5-μM random hexamers (Life Technologies) (MiniCycler, MJ Research/Biozym Diagnostik, Hess, Germany). cDNA of each group, resulting from the reverse transcription, was diluted with RNase free H_2O 1:2.5. Twelve-μL Mastermix. composed of 10-μL TaqMan Universal PCR Master Mix (Life Technologies), 1-μL RNase free H_2O and 1-μL Assays-on-Demand gene expression assay mix (Life Technologies) for the detection of either alkaline phosphatase (ALP, #Hs00758162_m1ALPL), collagen type 1 (COL I, #Hs00164004_m1COLA1) or osteocalcin (OCN, #Hs01587813_g1BGLAP) and for glyceraldehyde 3-phosphate dehydrogenase as an endogenous control (GAPDH, #Hs99999905_m1GAPDH, housekeeping gene) was analyzed with 8 μL of cDNA.

Quantitative real-time PCR assays were performed with a 3×20-μL reaction mix per group, marked and monitored with the ABI PRISM 7500 sequence detection system (Applied Biosystems, Darmstadt, Germany). Relative mRNA expression for each marker protein was calculated based on the comparative $\Delta\Delta CT$-method, normalized to GAPDH as an endogenous control and calibrated to the control cells grown on polystyrene after 24 h.

2.3. Statistical Analysis

Data are presented as the mean and its standard deviation. Values were analyzed for normal distribution using the Shapiro–Wilk test. For normally distributed data, one-way ANOVA was applied followed by a post-hoc Fisher least significant difference (LSD) test to determine differences between groups. Values of gene expression were analyzed with Student's *t*-test. The level of significance was set to $\alpha = 0.05$.

3. Results

3.1. Specimen Characterization

The elemental surface compositions of the coated samples, ZA0 and ZA5, determined with XPS, are listed in Table 2. The PPAAm coating on ZA0 is mainly composed of carbon and nitrogen, which were the constituents of the precursor used for the plasma polymerization (except for hydrogen, which cannot be analyzed by XPS), as well as a marginal fraction of oxygen that originates from post-oxidation processes. For the aged PPAAm coating (ZA5), a remarkably higher portion of oxygen was determined compared to ZA0 and, furthermore, traces of zirconium, silicon, fluorine and chloride at a total amount of <1 at% were detected. The amino group density of NH_2/C was found to be 3.4% for the as-deposited PPAAm coating (ZA0) and 0.3% for the 5-year aged layer (ZA5).

Table 2. Elemental composition of plasma-polymerized allylamine (PPAAm)-coated zirconia surfaces (ZA0) and aged surfaces (ZA5) determined with XPS.

	C (at.%)	N (at.%)	O (at.%)	N/C (%)	O/C (%)	NH_2/C (%)
ZA0	74.5 ± 2.1	22.9 ± 2.3	2.6 ± 0.3	30.7 ± 3.8	3.5 ± 0.3	3.4 ± 0.1
ZA5	73.9 ± 0.8	13.9 ± 0.8	11.4 ± 0.3	18.8 ± 1.3	15.4 ± 0.1	0.3 ± 0.1

The high-resolution XPS C1s spectra of ZA0 and ZA5 are shown in Figure 1. The PPAAm C1s peak of ZA0 can be fitted with three components: one at 285.0 eV, characteristic for C–H or/and C–C aliphatic bonds, another at 285.9 eV assigned to C–NH, and a third component at 286.8 eV, which corresponds to C–O, C–O–C, C = N or nitriles. In contrast, the highly resolved C1s spectrum of ZA5 shows drastic changes in the shape of the C1s peak with two further components at 287.9 eV and 289.0 eV attributed to C = O and O–C = O, respectively.

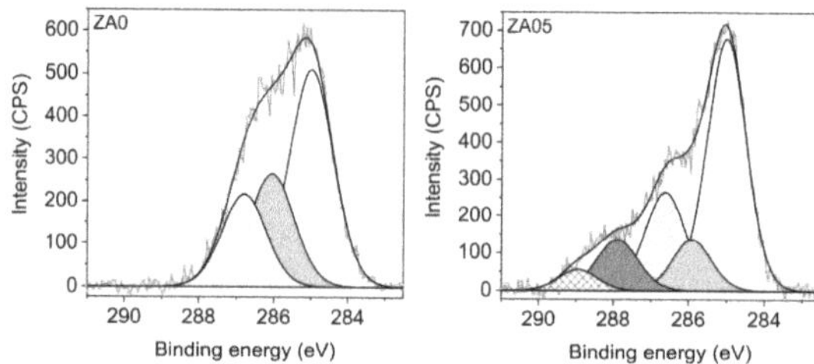

Figure 1. XPS C1s high resolution spectra of PPAAm after preparation (ZA0) and aging in a sealed aseptic packing at ambient conditions for 5 years (ZA5).

SEM images of specimens are displayed in Figure 2. No differences between the surfaces of Zmt, ZA0 and ZA5 could be observed in SEM images. Granules can be observed on all surfaces. Z14 displayed a rougher surface with micro-rough lacunae due to sandblasting with Al_2O_3 particles (105 µm) and hydrofluoric acid etching.

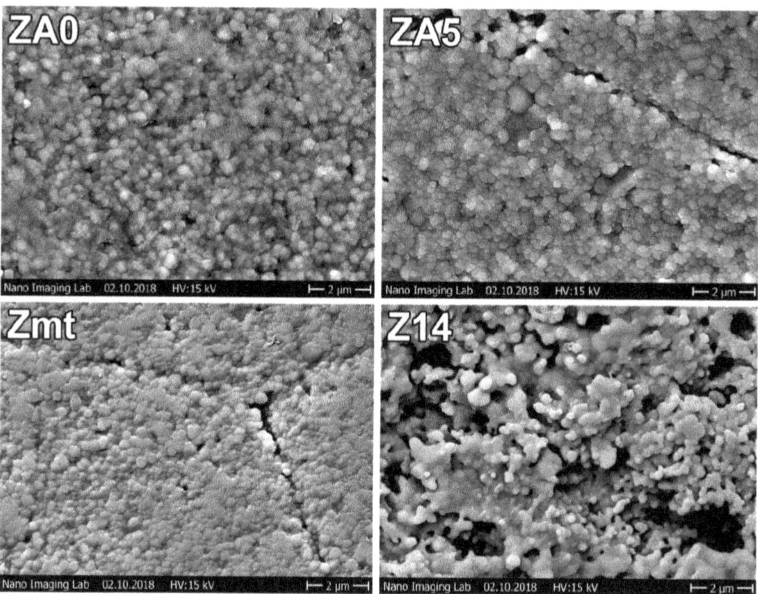

Figure 2. Specimen surface morphologies (SEM 10,000×).

3.2. Cell Behavior

Cell viability after 24 h was significantly higher for Zmt than for all other specimens ($p < 0.001$) (Figure 3a). Cell spreading after 20 min was significantly highest for Zmt = ZA0 > ZA5 > Z14, while, after 24 h, spreading was also significantly different between Zmt > ZA0 > ZA5 > Z14 ($p < 0.001$) (Figure 3b), possibly influenced by the increased roughness of Z14.

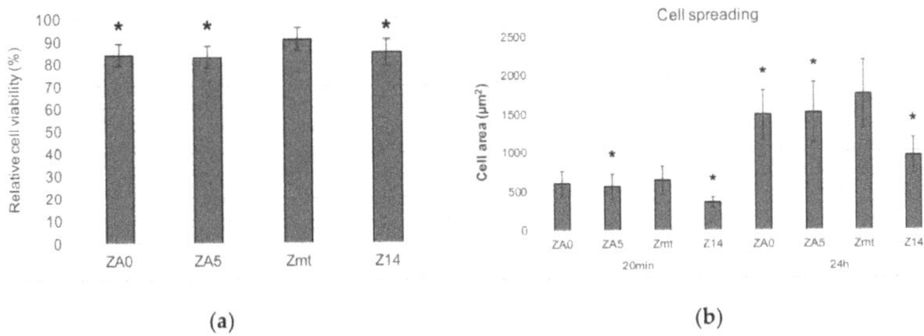

(a)   (b)

Figure 3. (**a**) Mean relative cell viability and standard deviation after 24 h in% normalized to the control cells grown on well bottoms. Statistically significant differences in specimens compared to uncoated zirconia specimens (Zmt), determined with a post-hoc Fisher LSD test, are indicated with * ($p < 0.001$). (**b**) Human osteoblastic cell (MG-63) area after 20 min and 24 h on differently treated zirconia specimens. Statistically significant differences in specimens compared to the control Zmt of the respective group of either 20 min or 24 h, determined with a post-hoc Fisher LSD test, are indicated with * ($p < 0.001$), n = 40 cells per group × 3 independent experiments, mean ± standard deviations.

The cell morphology visible in SEM images in Figure 4a was in accordance with the spreading determined with LSM. After 20 min, cells start to change from spherical into planar shapes; this process proceeds even further on the smooth surfaces of ZA0, ZA5 and Zmt compared to Z14. After 24 h, cells appear to be spread further on Zmt than on all other surfaces. Exposed nucleoli can be observed in the center of the cells on surfaces ZA0, ZA5 and Zmt. Due to the higher roughness on Z14, cells are less spread, but they are extended into the microstructures, where they anchor their filopodia (Figure 4b).

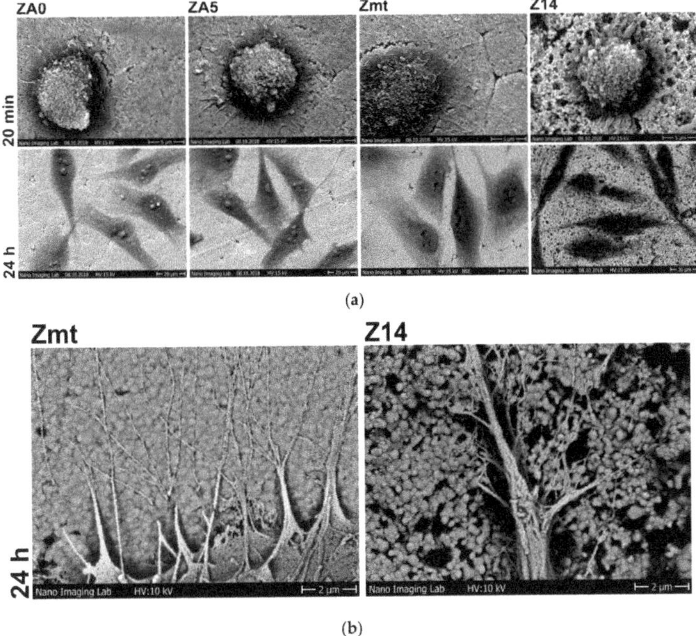

(a)

(b)

Figure 4. (**a**) MG-63 cells on differently treated zirconia surfaces. First row: spreading after 20 min (SEM, 5000×, bar 5 µm); second row: spreading after 24 h (SEM, 1000×, bar 20 µm). (**b**) MG-63 cell filopodia formation and interaction with the substrates Zmt and the micro-structured endosseous surface of a zirconia implant (Z14) (SEM, 10,000×, bar 2 µm).

The gene expression of early osteogenic marker ALP and late markers COL and OCN is displayed in Figure 5. The relative mRNA of ALP was significantly reduced on all specimens after 3 d when compared to the control cells grown on well bottoms for 24 h; COL remained stable and OCN was significantly increased for all specimens except ZA5. Significant differences when compared to Zmt at the respective time intervals are displayed in Figure 5 ($p < 0.05$).

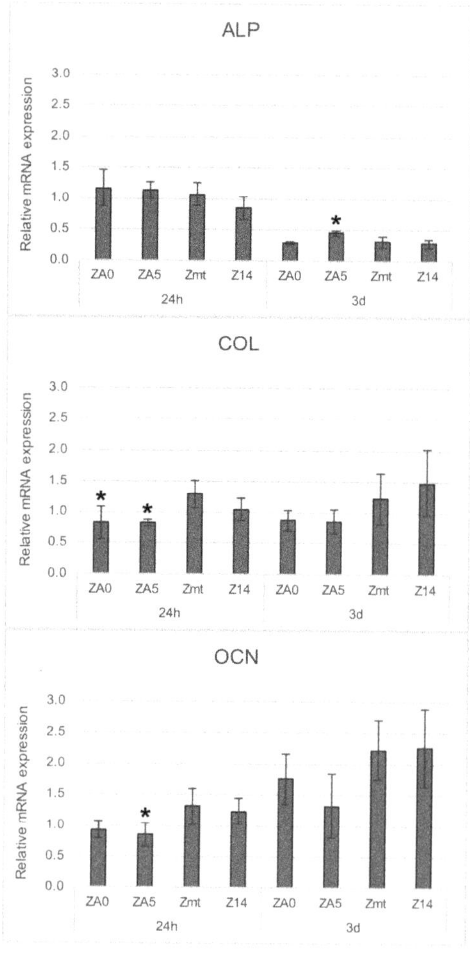

Figure 5. Relative mRNA expression of alkaline phosphatase (ALP), collagen type 1 (COL) and osteocalcin (OCN) in MG-63 cells on zirconia surfaces ZA0, ZA5, Zmt and Z14. The relative mRNA expression is normalized to the control cells grown on well bottoms after 24 h (=1.0), statistically significant differences in Zmt after 24 h or 3 d, respectively, determined with Student's t-test, are indicated with * ($p < 0.05$).

4. Discussion

The purpose of the present study was to determine whether a PPAAm coating on zirconia is stable for up to 5 years and still able to improve the osteoblast reactions compared to previously reported results on titanium [25–29], porous calcium phosphate [31] and Y-TZP [32]. Surprisingly, the control surface of the as-sintered and heat-treated zirconia (Zmt) was the substrate that accelerated initial cell behavior. In contrast to previous findings for Y-TZP [32], in this study, a coating with PPAAm on the

Zmt surface did not have an additional positive effect on the cells and even reduced their viability and spreading capability. The five-year aging of the PPAAm surfaces resulted in the oxidation of the coating, which, however, did not affect cell behavior differently than for freshly coated specimens, as also previously reported for titanium [30].

PPAAm films on ZA0 exhibited a N/C value of almost 32%—close to the theoretical N/C value of 33% for the precursor allylamine—which indicates the presence of nitrogen-containing functional groups at the surface. Additionally, XPS analysis revealed no further elements originating from the zirconia substrate (i.e., Al, Hf, Y, Zr), which confirms the homogeneous coverage of the substrate with a nanometer-thin PPAAm film. Amine-bearing plasma polymer coatings are susceptible to oxidation when stored under ambient conditions [30]. Hence, for PPAAm-coated specimen ZA5, stored for 5 years, the uptake of the oxygen content and a considerable depletion of the amino group density were determined by XPS.

The SEM images of the specimens revealed a surface structure with rounded granules sized around 100 nm. Due to the sandblasting and etching with hydrofluoric acid of Z14, niches were formed and surface roughness consequently increased. Since the thickness of the PPAAm layer is around 40 nm, the surface textures of ZA0 and ZA5 are comparable to the control Zmt.

Cell viability was significantly highest for Zmt than for all other surfaces. It has been previously seen that viability on smooth surfaces is increased compared to micro-structured surfaces [34]. The presence of the PPAAm layer also reduced cell viability when compared to the control without a coating. However, the viability of cells on the PPAAm-coated surfaces was comparable to Z14 and was above 80%; consequently, no toxic effect was initiated by the PPAAm coating when considering ISO standard 10993-5, which indicates no toxic effects for cell viability > 75%. The biocompatibility of a PPAAm coating on titanium has previously been tested in a rat model and no increased local inflammation compared to uncoated specimens was observed [29].

The initial cell spreading of MG-63 osteoblasts has been identified as the main factor that is highly accelerated by the PPAAm coating on titanium surfaces [27,30,31]. However, spreading on zirconia could not be improved when the coating was applied in the present study and was even lower than on uncoated specimens after 20 min as well as after 24 h. In contrast to the present study using 10% FCS, previously, no serum was added to the cell culture medium when cells were seeded on the PPAAm-coated specimens. Serum proteins improve initial cell adhesion and the spreading of osteoblasts on zirconia [36] and the addition of fetal calf serum to the culture medium may have masked the potential effects of PPAAm. That spreading was generally higher on smooth than on micro-roughened surfaces has been previously observed for osteoblasts on zirconia [37,38], as well as on titanium [39]. Adequate spreading is a crucial factor for the proliferation of adherent cells because maximum extracellular matrix contact with the whole cell body is aspired to maintain osteoblastic function [40,41].

The cell morphology visible in SEM images was in accordance with the measured cell areas in LSM images. On Z14 surfaces, cells anchored their filopodia on the micro-roughened surface and spread into the depths of the niches; hence, the cell area appeared smaller than for cells on all other surfaces, as previously seen for primary human osteoblasts (HOB) [38].

For the gene expression of RNA markers, the downregulation of ALP after 3 d, stable COL and upregulated OCN in MG-63 cells, as observed, can be considered typical reactions in osteoblast maturation. COL and ALP are early differentiation markers in the osteoblast lineage; hence, mRNA expression of these markers is increased in preosteoblasts and declines during osteoblast maturation [42]. mRNA expression of COL was significantly increased for cells on Zmt when compared to PPAAm-coated specimens after 24 h. OCN is first expressed at very low levels and, later in osteoblast maturation, transcription is enhanced [43]. In the present study, after 3 d, this was noticeable for ZA0 but further progressed for Zmt and Z14. In general, gene expression on Zmt and Z14 was comparable. Another study compared the gene expression of the same markers of human primary osteoblasts on machined as well as on sandblasted, etched and heat-treated zirconia [38]. ALP and COL mRNA expression of primary human osteoblasts on both zirconia surfaces was downregulated after 3 d. OCN was further upregulated for cells on the machined surface than on the sandblasted, etched and heat-treated surface after 3 d [38]. These gene expression results support the finding of the present study that smoother zirconia surfaces are favorable for initial cell behavior. Contrary to previous findings on titanium, calcium phosphate and Y-TZP, a coating of zirconia with PPAAm in the presence of serum in the culture medium does not improve cell behavior further. Five-year aging of PPAAm-coated zirconia resulted in the oxidation of the layer, but did not affect cell behavior differently than for freshly coated zirconia.

5. Conclusions

Within the limitations of this study, it can be concluded that zirconia coated with PPAAm does not additionally improve osteoblast behavior in cell culture experiments due to the high biocompatibility of zirconia.

Author Contributions: Conceptualization, N.R. and J.F.; methodology, N.R., J.B.N. and K.F.; formal analysis, N.R. and J.F. investigation, N.R., C.B. and K.F.; writing—original draft preparation, N.R. and K.F.; writing—review and editing N.R., K.F., C.B., J.B.N. and J.F. All authors have read and agreed to the published version of the manuscript.

Funding: The zirconia discs were kindly provided by Vita Zahnfabrik, Bad Säckingen.

Acknowledgments: The authors would like to thank Fredy Schmidli, Martina Grüning, Petra Müller and Petra Seidel for the laboratory support.

Conflicts of Interest: The authors declare no conflict of interest. The funders had no role in the design of the study; in the collection, analyses, or interpretation of data; in the writing of the manuscript, or in the decision to publish the results.

References

1. Mombelli, A.; Hashim, D.; Cionca, N. What is the impact of titanium particles and biocorrosion on implant survival and complications? A critical review. *Clin. Oral Implant. Res.* **2018**, *29*, 37–53. [CrossRef] [PubMed]
2. Cadosch, D.; Gautschi, O.P.; Chan, E.; Filgueira, L.; Simmen, H.-P. Titanium induced production of chemokines CCL17/TARC and CCL22/MDC in human osteoclasts and osteoblasts. *J. Biomed. Mater. Res. Part A* **2009**, *9999*, 475–483. [CrossRef] [PubMed]
3. Cadosch, D.; Al-Mushaiqri, M.S.; Gautschi, O.P.; Meagher, J.; Simmen, H.-P.; Filgueira, L. Biocorrosion and uptake of titanium by human osteoclasts. *J. Biomed. Mater. Res. Part A* **2010**, *95*, 1004–1010. [CrossRef]
4. Bormann, K.H.; Gellrich, N.-C.; Kniha, H.; Schild, S.; Weingart, D.; Gahlert, M. A prospective clinical study to evaluate the performance of zirconium dioxide dental implants in single-tooth edentulous area: 3-year follow-up. *BMC Oral Health* **2018**, *18*, 181. [CrossRef] [PubMed]
5. Balmer, M.; Spies, B.C.; Kohal, R.-J.; Hämmerle, C.H.; Vach, K.; Jung, R.E. Zirconia implants restored with single crowns or fixed dental prostheses: 5-year results of a prospective cohort investigation. *Clin. Oral Implant. Res.* **2020**, *31*, 452–462. [CrossRef] [PubMed]
6. Adánez, M.H.; Nishihara, H.; Att, W. A systematic review and meta-analysis on the clinical outcome of zirconia implant–restoration complex. *J. Prosthodont. Res.* **2018**, *62*, 397–406. [CrossRef] [PubMed]
7. Binon, P.P. Implants and components: Entering the new millennium. *Int. J. Oral Maxillofac. Implant.* **2000**, *15*, 76–94.
8. Wennerberg, A.; Albrektsson, T. Effects of titanium surface topography on bone integration: A systematic review. *Clin. Oral Implant. Res.* **2009**, *20*, 172–184. [CrossRef]
9. Fischer, J.; Schott, A.; Martin, S. Surface micro-structuring of zirconia dental implants. *Clin. Oral Implant. Res.* **2015**, *27*, 162–166. [CrossRef]
10. Pieralli, S.; Kohal, R.-J.; Hernandez, E.L.; Doerken, S.; Spies, B.C. Osseointegration of zirconia dental implants in animal investigations: A systematic review and meta-analysis. *Dent. Mater.* **2017**, *34*, 171–182. [CrossRef]
11. Nishihara, H.; Adanez, M.H.; Att, W. Current status of zirconia implants in dentistry: Preclinical tests. *J. Prosthodont. Res.* **2019**, *63*, 1–14. [CrossRef] [PubMed]
12. Kniha, K.; Schlegel, K.; Kniha, H.; Modabber, A.; Hölzle, F. Evaluation of peri-implant bone levels and soft tissue dimensions around zirconia implants—A three-year follow-up study. *Int. J. Oral Maxillofac. Surg.* **2018**, *47*, 492–498. [CrossRef] [PubMed]

13. Morra, M. Biomolecular modification of implant surfaces. *Expert Rev. Med Devices* **2007**, *4*, 361–372. [CrossRef] [PubMed]

14. Narayanan, R.; Seshadri, S.K.; Kwon, T.Y.; Kim, K.H. Calcium phosphate-based coatings on titanium and its alloys. *J. Biomed. Mater. Res. Part B Appl. Biomater.* **2008**, *85*, 279–299. [CrossRef] [PubMed]

15. Faucheux, N.; Tzoneva, R.; Nagel, M.D.; Groth, T. The dependence of fibrillar adhesions in human fibroblasts on substratum chemistry. *Biomaterials* **2006**, *27*, 234–245. [CrossRef]

16. Aziz, G.; De Geyter, N.; Morent, R. Incorporation of Primary Amines via Plasma Technology on Biomaterials. In *Advances in Bioengineering*; Serra, P.A., Ed.; InTech: London, UK, 2015. [CrossRef]

17. Liu, X.; Feng, Q.; Bachhuka, A.; Vasilev, K. Surface Modification by Allylamine Plasma Polymerization Promotes Osteogenic Differentiation of Human Adipose-Derived Stem Cells. *ACS Appl. Mater. Interfaces* **2014**, *6*, 9733–9741. [CrossRef]

18. Gallino, E.; Massey, S.; Tatoulian, M.; Mantovani, D. Plasma polymerized allylamine films deposited on 316L stainless steel for cardiovascular stent coatings. *Surf. Coatings Technol.* **2010**, *205*, 2461–2468. [CrossRef]

19. Crespin, M.; Moreau, N.; Masereel, B.; Feron, O.; Gallez, B.; Vander Borght, T.; Michiels, C.; Lucas, S. Surface properties and cell adhesion onto allylamine-plasma and amine-plasma coated glass coverslips. *J. Mater. Sci. Mater. Electron.* **2011**, *22*, 671–682. [CrossRef]

20. Testrich, H.; Rebl, H.; Finke, B.; Hempel, F.; Nebe, B.; Meichsner, J. Aging effects of plasma polymerized ethylenediamine (PPEDA) thin films on cell-adhesive implant coatings. *Mater. Sci. Eng. C* **2013**, *33*, 3875–3880. [CrossRef]

21. Manakhov, A.; Landová, M.; Medalová, J.; Michlíček, M.; Polčák, J.; Nečas, D.; Zajíčková, L. Cyclopropylamine plasma polymers for increased cell adhesion and growth. *Plasma Process. Polym.* **2016**, *14*, 1600123. [CrossRef]

22. Chan, K.V.; Asadian, M.; Onyshchenko, I.; Declercq, H.; Morent, R.; De Geyter, N. Biocompatibility of Cyclopropylamine-Based Plasma Polymers Deposited at Sub-Atmospheric Pressure on Poly (ε-caprolactone) Nanofiber Meshes. *Nanomaterials* **2019**, *9*, 1215. [CrossRef] [PubMed]

23. Buddhadasa, M.; Lerouge, S.; Girard-Lauriault, P.-L. Plasma polymer films to regulate fibrinogen adsorption: Effect of pressure and competition with human serum albumin. *Plasma Process. Polym.* **2018**, *15*, 1800040. [CrossRef]

24. Schweikl, H.; Müller, R.; Englert, C.; Hiller, K.-A.; Kujat, R.; Nerlich, M.; Schmalz, G.; Müller, R. Proliferation of osteoblasts and fibroblasts on model surfaces of varying roughness and surface chemistry. *J. Mater. Sci. Mater. Electron.* **2007**, *18*, 1895–1905. [CrossRef] [PubMed]

25. Nebe, J.B.; Finke, B.; Lüthen, F.; Bergemann, C.; Schröder, K.; Rychly, J.; Liefeith, K.; Ohl, A. Improved initial osteoblast functions on amino-functionalized titanium surfaces. *Biomol. Eng.* **2007**, *24*, 447–454. [CrossRef]

26. Finke, B.; Luethen, F.; Schroeder, K.; Mueller, P.D.; Bergemann, C.; Frant, M.; Ohl, A.; Nebe, J.B. The effect of positively charged plasma polymerization on initial osteoblastic focal adhesion on titanium surfaces. *Biomaterials* **2007**, *28*, 4521–4534. [CrossRef]

27. Rebl, H.; Finke, B.; Lange, R.; Weltmann, K.-D.; Nebe, J.B. Impact of plasma chemistry versus titanium surface topography on osteoblast orientation. *Acta Biomater.* **2012**, *8*, 3840–3851. [CrossRef]

28. Staehlke, S.; Rebl, H.; Finke, B.; Mueller, P.; Gruening, M.; Nebe, J.B. Enhanced calcium ion mobilization in osteoblasts on amino group containing plasma polymer nanolayer. *Cell Biosci.* **2018**, *8*, 22. [CrossRef]

29. Hoene, A.; Walschus, U.; Patrzyk, M.; Finke, B.; Lucke, S.; Nebe, B.; Schroeder, K.; Ohl, A.; Schlosser, M. In vivo investigation of the inflammatory response against allylamine plasma polymer coated titanium implants in a rat model. *Acta Biomater.* **2010**, *6*, 676–683. [CrossRef]

30. Finke, B.; Rebl, H.; Hempel, F.; Schäfer, J.; Liefeith, K.; Weltmann, K.-D.; Nebe, J.B. Aging of Plasma-Polymerized Allylamine Nanofilms and the Maintenance of Their Cell Adhesion Capacity. *Langmuir* **2014**, *30*, 13914–13924. [CrossRef]

31. Rebl, H.; Finke, B.; Schmidt, J.; Mohamad, H.S.; Ihrke, R.; Helm, C.A.; Nebe, J.B. Accelerated cell-surface interlocking on plasma polymer-modified porous ceramics. *Mater. Sci. Eng. C* **2016**, *69*, 1116–1124. [CrossRef]

32. Nebe, J.B.; Rebl, H.; Schlosser, M.; Staehlke, S.; Gruening, M.; Weltmann, K.-D.; Walschus, U.; Finke, B. Plasma Polymerized Allylamine—The Unique Cell-Attractive Nanolayer for Dental Implant Materials. *Polymers* **2019**, *11*, 1004. [CrossRef]

33. Moerke, C.; Mueller, P.; Nebe, J.B. Sensing of micropillars by osteoblasts involves complex intracellular signaling. *J. Mater. Sci. Mater. Med.* **2017**, *28*, 171. [CrossRef]

34. Rohr, N.; Zeller, B.; Matthisson, L.; Fischer, J. Surface structuring of zirconia to increase fibroblast viability. *Dent. Mater.* **2020**, *36*, 779–786. [CrossRef]

35. Staehlke, S.; Rebl, H.; Nebe, B. Phenotypic stability of the human MG-63 osteoblastic cell line at different passages: MG-63 cells: Receptors, cell cycle, signaling. *Cell Boil. Int.* **2018**, *43*, 22–32. [CrossRef]

36. Luo, F.; Hong, G.; Matsui, H.; Endo, K.; Wan, Q.; Sasaki, K. Initial osteoblast adhesion and subsequent differentiation on zirconia surfaces are regulated by integrins and heparin-sensitive molecule. *Int. J. Nanomed.* **2018**, *13*, 7657–7667. [CrossRef]

37. Rohr, N.; Nebe, J.B.; Schmidli, F.; Müller, P.; Weber, M.; Fischer, H.; Fischer, J. Influence of bioactive glass-coating of zirconia implant surfaces on human osteoblast behavior in vitro. *Dent. Mater.* **2019**, *35*, 862–870. [CrossRef]

38. Bergemann, C.; Duske, K.; Nebe, J.B.; Schöne, A.; Bulnheim, U.; Seitz, H.; Fischer, J. Microstructured zirconia surfaces modulate osteogenic marker genes in human primary osteoblasts. *J. Mater. Sci. Mater. Med.* **2015**, *26*, 26. [CrossRef]

39. Hempel, U.; Hefti, T.; Dieter, P.; Schlottig, F. Response of human bone marrow stromal cells, MG-63, and SaOS-2 to titanium-based dental implant surfaces with different topography and surface energy. *Clin. Oral Implant. Res.* **2011**, *24*, 174–182. [CrossRef]

40. Lüthen, F.; Lange, R.; Becker, P.; Rychly, J.; Beck, U.; Nebe, J.B. The influence of surface roughness of titanium on β1- and β3-integrin adhesion and the organization of fibronectin in human osteoblastic cells. *Biomaterials* **2005**, *26*, 2423–2440. [CrossRef]

41. Hidalgo-Bastida, L.A.; Cartmell, S. Mesenchymal Stem Cells, Osteoblasts and Extracellular Matrix Proteins: Enhancing Cell Adhesion and Differentiation for Bone Tissue Engineering. *Tissue Eng. Part B Rev.* **2010**, *16*, 405–412. [CrossRef]

42. Schünemann, F.H.; Galárraga-Vinueza, M.E.; Magini, R.; Fredel, M.; Silva, F.; De Souza, J.C.M.; Zhang, Y.; Henriques, B. Zirconia surface modifications for implant dentistry. *Mater. Sci. Eng. C* **2019**, *98*, 1294–1305. [CrossRef]

43. Billiard, J.; Moran, R.A.; Whitley, M.Z.; Chatterjee-Kishore, M.; Gillis, K.; Brown, E.L.; Komm, B.; Bodine, P. Transcriptional profiling of human osteoblast differentiation. *J. Cell. Biochem.* **2003**, *89*, 389–400. [CrossRef]

Journal of
Clinical Medicine

Article

Adapting the Pore Size of Individual, 3D-Printed CPC Scaffolds in Maxillofacial Surgery

David Muallah [1], Philipp Sembdner [2], Stefan Holtzhausen [2], Heike Meissner [3], André Hutsky [4], Daniel Ellmann [4], Antje Assmann [5], Matthias C. Schulz [6], Günter Lauer [1] and Lysann M. Kroschwald [1,7,*]

[1] Department of Oral and Maxillofacial Surgery, Faculty of Medicine "Carl Gustav Carus", Technische Universität Dresden, Fetscherstraße 74, 01307 Dresden, Germany; David.Muallah@uniklinikum-dresden.de (D.M.); Guenter.Lauer@uniklinikum-dresden.de (G.L.)

[2] Department of Mechanical Engineering, Institute of Machine Elements and Machine Design, Technische Universität Dresden, 01062 Dresden, Germany; Philipp.Sembdner@tu-dresden.de (P.S.); stefan.holtzhausen@tu-dresden.de (S.H.)

[3] Department of Prosthetic Dentistry, University Hospital "Carl Gustav Carus", Technische Universität Dresden, Fetscherstraße 74, 01307 Dresden, Germany; Heike.Meissner@uniklinikum-dresden.de

[4] Organical CAD/CAM, Ruwersteig 43, 12681 Berlin, Germany; Andre.Hutsky@organical-cadcam.com (A.H.); Daniel.Ellmann@organical-cadcam.com (D.E.)

[5] Zahntechnik Schönberg, Altseidnitz 19, 01277 Dresden, Germany; Kontakt@Zahntechnik-Schoenberg.de

[6] Department of Oral and Maxillofacial Surgery, University Hospital Tübingen, Eberhard Karls Universität Tübingen, Osianderstraße 2-8, 72076 Tübingen, Germany; Matthias.Schulz@med.uni-tuebingen.de

[7] Centre for Translational Bone, Joint and Soft Tissue Research, University Hospital "Carl Gustav Carus", Technische Universität Dresden, Fetscherstraße 74, 01307 Dresden, Germany

* Correspondence: Lysann.Kroschwald@ukdd.de

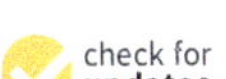

Citation: Muallah, D.; Sembdner, P.; Holtzhausen, S.; Meissner, H.; Hutsky, A.; Ellmann, D.; Assmann, A.; Schulz, M.C.; Lauer, G.; Kroschwald, L.M. Adapting the Pore Size of Individual, 3D-Printed CPC Scaffolds in Maxillofacial Surgery. *J. Clin. Med.* **2021**, *10*, 2654. https://doi.org/10.3390/jcm10122654

Academic Editor: Tim Joda

Received: 25 May 2021
Accepted: 15 June 2021
Published: 16 June 2021

Publisher's Note: MDPI stays neutral with regard to jurisdictional claims in published maps and institutional affiliations.

Abstract: Three dimensional (3D) printing allows additive manufacturing of patient specific scaffolds with varying pore size and geometry. Such porous scaffolds, made of 3D-printable bone-like calcium phosphate cement (CPC), are suitable for bone augmentation due to their benefit for osteogenesis. Their pores allow blood-, bone- and stem cells to migrate, colonize and finally integrate into the adjacent tissue. Furthermore, the pore size affects the scaffold's stability. Since scaffolds in maxillofacial surgery have to withstand high forces within the jaw, adequate mechanical properties are of high clinical importance. Although many studies have investigated CPC for bone augmentation, the ideal porosity for specific indications has not been defined yet. We investigated 3D printed CPC cubes with increasing pore sizes and different printing orientations regarding cell migration and mechanical properties in comparison to commercially available bone substitutes. Furthermore, by investigating clinical cases, the scaffolds' designs were adapted to resemble the in vivo conditions as accurately as possible. Our findings suggest that the pore size of CPC scaffolds for bone augmentation in maxillofacial surgery necessarily needs to be adapted to the surgical site. Scaffolds for sites that are not exposed to high forces, such as the sinus floor, should be printed with a pore size of 750 μm to benefit from enhanced cell infiltration. In contrast, for areas exposed to high pressures, such as the lateral mandible, scaffolds should be manufactured with a pore size of 490 μm to guarantee adequate cell migration and in order to withstand the high forces during the chewing process.

Keywords: calcium phosphate cement; pore size; augmentation; additive manufacturing

1. Introduction

Maxillofacial surgeons are often challenged by complex bone defects caused by trauma, tumors, inflammation or long lasting edentulism. In many cases, the reconstruction of bony structures is necessary for the rehabilitation of the shape, function and aesthetics of the orofacial system. For this purpose, different materials, such as ready-made bone substitution materials (e.g., granulate, membranes or cones) or autologous bone, are used. An eligible bone substitute with outstanding properties is the autologous bone graft [1]. However, this requires harvesting from other anatomical sites, which is associated with donor site morbidity and limited capacity [2–5]. For this reason, the necessity to develop new synthetic bone substitute materials is increasing.

Three dimensional (3D) printing is an emerging technology in the medical field that offers new opportunities for tissue engineering and the reconstruction of bone [6–8]. Based on three-dimensional imaging, patient specific scaffolds can be manufactured additively. Different materials can be used for this purpose; one of them is calcium phosphate cement (CPC). CPC is a hydroxyapatite forming, synthetic bone substitution material, which mimics the inorganic part of human bone. Due to its material properties, such as its pasty consistency, biocompatibility and biodegradability, it has gained the attention, not only of scientists, but also of many clinicians [9–11]. CPC is osteoconductive, which means it is capable of guiding the growth and proliferation of osteoblasts on its surface. While the synthesis of other bio ceramics involves high temperature sintering, CPC can set and harden at room or body temperature at a nearly neutral pH. Its clinical benefit has been proven in several trials [12–15].

For successful bone reconstruction, the bone substitution material has to be permanently integrated into the defect site. This can be achieved by way of two mechanisms. The bone substitution material can be resorbed and replaced by the organism's host bone [16]. Materials such as bovine collagen, or autologous as well as allogenic bone grafts, become integrated in this way [16,17]. The duration of this mechanism depends on the material's biodegradability. For other materials, such as BioOss® or CPC, the human body needs years to perform complete remodeling [18,19]. The functional integration of these materials is realized in a different way. Through integrated macro pores, blood-, bone- and stem cells infiltrate those scaffolds directly after implantation. This leads to the incorporation of the scaffold after a few months. Therefore, the pores in these bone substitution materials play a major role in a successful outcome. Bigger pore sizes, which may increase the infiltration of cells, come along with a smaller scaffold surface. The printing orientation also affects the surface area of the scaffolds. By printing the strands at a 45° angle to the scaffold's edges, the surface can additionally be increased. A smaller surface could lead to decreased osteoconductivity of the scaffold. Additionally, wider pores and a smaller surface mean less stability. In maxillofacial surgery especially, the stability of bone scaffolds is crucial since high pressures emerge during the chewing process.

Using 3D printing, the size of the pores can be adapted to a specific purpose. In regions such as the maxillary sinus, the scaffold's stability might play a secondary role, whereas on the alveolar ridges, high stability is absolutely essential due to the high forces that emerge during the chewing process [20,21]. Considering these aspects, a defined pore size of CPC scaffolds for maxillofacial surgery could be of substantial clinical relevance. The optimal compromise between porosity, surface and stability needs to be determined. Many studies have investigated and approved CPC as a promising bone substitution material, but the ideal porosity of CPC scaffolds for specific indications has not yet been described [15,22,23]. This study aims to find the above mentioned optimal compromise between porosity, surface area and scaffold stability. We therefore investigated CPC scaffolds with six different porosities (0.1 mm, 0.23 mm, 0.36 mm, 0.49 mm, 0.62 mm and 0.75 mm) and two printing orientations (90° and 45°) in vitro, and compared the results with commercially available bone grafts. The scaffolds were colonized with human mesenchymal stem cells (hMSC) and were investigated regarding the depth of cell infiltration. Furthermore, the influence of strand arrangement and pore size on the scaffolds' stability was studied. Moreover, in order to replicate the various in vivo conditions as accurately as possible, individual cone-beam computed tomography (CBCT) based CPC scaffolds presenting different parts of the maxillofacial region were printed and analyzed regarding their stability. In this regard, we hypothesized that an increasing pore size significantly influences not only migration but also the graft´s stability. Our results show the importance of the external and internal structure, especially for individual scaffolds in maxillofacial surgery.

2. Materials and Methods

2.1. Virtual Scaffold Planning

Cubic shaped and individual scaffolds were digitally planned. For individual scaffolds, geometries were designed based on patients' cone beam data, in the manner of backward planning. First, the prosthetic restoration was set into the ideal position, determining the position needed for the dental implant. Based on the dental planning, boundary conditions were defined using cone-beam data (CBCT) for the individual scaffold to be designed. These boundary conditions represent geometric elements, such as planes or curves, that limit the dimension of the scaffold from a medical point of view and define the principal location of design features to be integrated (holes, cavities, etc.). Subsequently, the CPC scaffold was designed around the dental implant according to all clinical and geometric specifications. The use of patient data was approved by the local ethical review board (IRB00001473; file reference: EK1450420019).

2.2. Scaffold Fabrication

The scaffolds were fabricated from plottable CPC paste (INNOTERE Paste-CPC), manufactured by INNOTERE GmbH (Radebeul, Germany), by using a 3D plotting device (KOSY4, Elektronik and Mechanik GmbH, Thalheim, Germany), and were sterilized with γ-irradiation (25 kGy). For colonization studies, cubic-shaped scaffolds ($10 \times 10 \times 10$ mm) were plotted utilizing a 310 µm needle with a plotting speed of 8 mm/s and an air pressure of approx. 4 bar. The inner geometry of the cubic-shaped scaffolds was adjusted as follows: 3 layers with a strand-to-strand distance of 0.3 mm and a further 36 layers with a strand-to-strand distance as follows: 0.43 mm (Scaffold A), 0.56 mm (Scaffold B), 0.69 mm (Scaffold C), 0.82 mm (Scaffold D), 0.95 mm (Scaffold E) or 1.08 mm (Scaffold F). Layer-to-layer orientation was 90° or 45° in relation to the scaffolds' edges. To test the stability, individualized scaffolds were plotted using the same technique. The inner geometry of the individualized scaffolds was adjusted as follows: the drilling axis was aligned parallel to the direction of fabrication (Z-axis). The filling pattern was then set at $-45°/45°$. After plotting, scaffolds were incubated for 72 h in a water-saturated atmosphere (humidity 95%, temperature 37 °C), followed by three intensive washing steps in acetone to remove residual oil from the CPC paste. Afterwards, the scaffolds were dried under a fume cupboard. Bio-Oss®Blocks (Geistlich Biomaterials Vertriebsgesellschaft mbH, Baden-Baden, Germany) were used as a control group. The blocks were bisected into $10 \times 10 \times 10$ mm cubes for Zwick testing and colonization.

2.3. Scaffold Characterization

The shape and macro porosity of the printed scaffolds were initially studied by stereo microscopic investigation using a Leica M205C equipped with a DFC295 camera (Leica, Wetzlar, Germay). Scanning electron microscopy (SEM) was performed to assess the microporosity and colonization of the scaffolds. For this purpose, the samples were coated with gold using a Cressington Sputter Coater 108 auto (Crawley, UK). The following process sputtering parameters were applied: $p = 0.1$ mbar, $I = 30$ mA and a target-sample surface distance of 55–60 mm. Surface morphology and cell colonization were imaged with a Philips XL 30 ESEM scanning electron microscope (Philips Electron optics GmbH, Kassel, Germany) utilizing an SE detector. For the image acquisitions, depending on the material and imaging type (overview or detail), the voltage varied from 10 kV to 20 kV and the working distance varied from 4.5 to 20 mm. The mechanical characterization was performed via a uniaxial compressive test with a speed of 100 NM in the vertical direction by using a Zwick universal testing machine (Z010 equipped with a 10 kN load cell; Zwick, Ulm, Germany). Compressive modulus and compressive strength were calculated from the obtained data ($n = 5$) and representative curves are shown.

2.4. Colonization

After 24 h of re-equilibration to culture conditions in DMEM, scaffolds were seeded with hMSC at a density of 1×106 cells per scaffold, in order to study colonization. Therefore, cells were expanded in DMEM containing 15% fetal calf serum (FCS), 1% L-glutamate and 1% Pen/Strep (all from GIBCO, Germany).

For seeding, the immersed scaffolds were placed into 5 mL tubes and a 5000 µL cell suspension containing 1×106 cells. Scaffold colonization was performed by way of a rotation method over a period of 6 h, as previously described by Korn et al. [24]. Tubes were rotated every 30 min by 540°, while being stored at 37 °C and 5% CO_2. Finally, the scaffolds were placed in 24-well plates, covered with culture medium and incubated for up to 12 weeks. The medium was replaced twice a week. Live/Dead staining was performed by using a Live/Dead Cell Staining Kit II (Promocell, Germany) according to the manufacturer´s instructions. For fluorescence microscopic analyses, colonized scaffolds were fixed using 4% formaldehyde. Actin cytoskeletons and cell nuclei were stained with AlexaFlour 488® phalloidin (Invitrogen, Waltham, MA, USA) and DAPI (Sigma Aldrich, Taufkirchen, Germany). All microscopic investigations were performed with a Keyence BZ9000E (Keyence, Neu-Isenburg, Germany). For the determination of cell number and LDH activity, frozen samples were thawed, followed by cell lysis with PBS containing 1% Triton X-100. During cell lysis, each sample was sonicated for 1 min at 80 W. One aliquot of the cell suspension was used to determine LDH activity via Cytotox96 kit (Promega, Madison, WI, USA) according to the manufacturer´s instructions. The LDH activity was correlated with the cell number using a calibration curve. Total DNA was quantified for the calculation of cell number using a calibration curve of cells. Therefore, DNA was quantified via Quantifluor assay (Promega, Madison, WI, USA) according to the manufacturer´s instructions. All measurements were performed by using a spectrofluorometer (Infinite M200pro; Tecan Trading AG, Männedorf, Switzerland).

2.5. Statistics

For statistical analyses, GraphPad Prism 6.0 software (San Diego, CA, USA) was used. All experiments were performed at defined time points using replicates as indicated in the figure captions. The results were expressed as mean $\pm$ standard deviation (SD). One-way analysis of variance (ANOVA) with Bonferroni adjustment of p-values was performed to analyze statistical significance. Therefore, $p \leq 0.05$ was considered to indicate statistical significance.

3. Results

3.1. Scaffold Fabrication and Mechanical Testing

Cube shaped scaffolds with six different strand distances, resulting in six different macro porosities, possessed a well-defined porous structure (Figure 1).

Comparing the CPC scaffolds, significant differences in the mechanical properties were observed (Figure 2). Data are shown in Table 1.

The results showed that the energy absorption of the different scaffold types is greatly reduced with increasing pore size (Table 1). This intense decrease is also shown in the compressive strength data (Figure 2c). Nevertheless, all investigated scaffolds, independently from pore size, showed a higher energy absorption and strength in comparison to the control group (BioOss®).

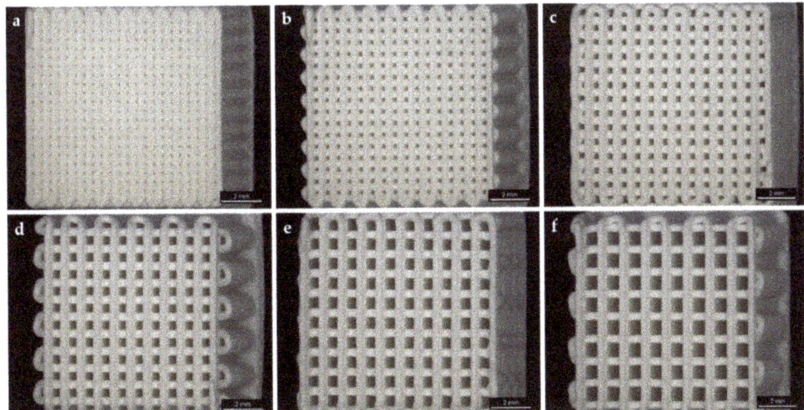

Figure 1. Three dimensional (3D) plotted scaffolds with different pore sizes. Strand-to-strand-distance μm/pore size μm: (**a**) 430/100; (**b**) 560/230; (**c**) 690/360; (**d**) 820/490; (**e**) 950/620; (**f**) 1080/750. Scale bars: 2 mm.

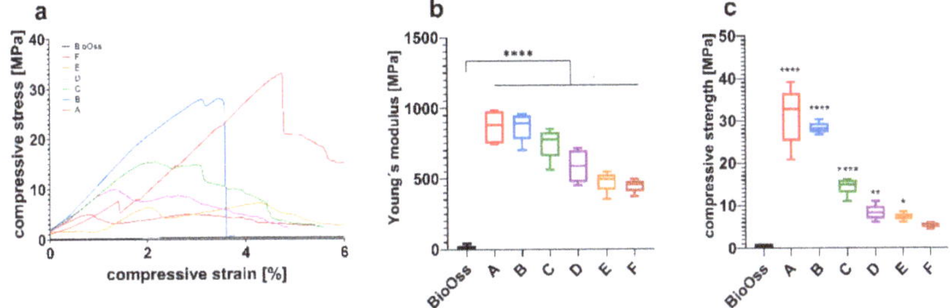

Figure 2. Mechanical properties of CPC scaffolds A–F (see also Table 1) with different pore sizes in comparison to BioOss®. (**a**) Representative compressive stress–strain curves. (**b**) Young's modulus and (**c**) compressive strength determined from the curves (* $p \leq 0.05$, ** $p \leq 0.01$, **** $p \leq 0.0001$, mean ± standard deviation, n = 5).

Table 1. Mechanical properties of 3D printed scaffolds A–F with different strand-to-strand distances and pore sizes and control.

Scaffold	Strand-to-Strand-Distance (μm)	Pore Size (μm)	Young's Modulus # (MPa)	Compressive Strength # (MPa)
A	430	100	870 ± 117	31.3 ± 6.8
B	560	230	870 ± 101	28.3 ± 1.3
C	690	360	749 ± 110	14.5 ± 2.0
D	820	490	586 ± 118	8.3 ± 1.8
E	950	620	477 ± 73	7.4 ± 0.9
F	1080	750	444 ± 44	5.2 ± 0.6
Control			7 ± 4	0.5 ± 0.007

mean ± standard deviation (SD), n = 5.

3.2. Colonization of Scaffolds

In order to enable colonization within porous scaffolds, it is important to equilibrate the scaffolds for 24 h in a cell culture medium before seeding. Following the equilibration, the scaffolds were seeded with hMSCs and incubated for up to 12 weeks in order to study colonization. Determination of DNA and LDH was performed to evaluate the proliferation of cells cultivated on the scaffolds (Figure 3). After four weeks, cells completely covered the CPC strands of the topmost layer, but the cell number of the CPC scaffolds was significantly lower in comparison to the control (BioOss®). After 12 weeks, the cell number of scaffolds A, B and C was significantly lower in comparison to the control, while scaffolds D, E and F showed a colonization comparable to that of BioOss®.

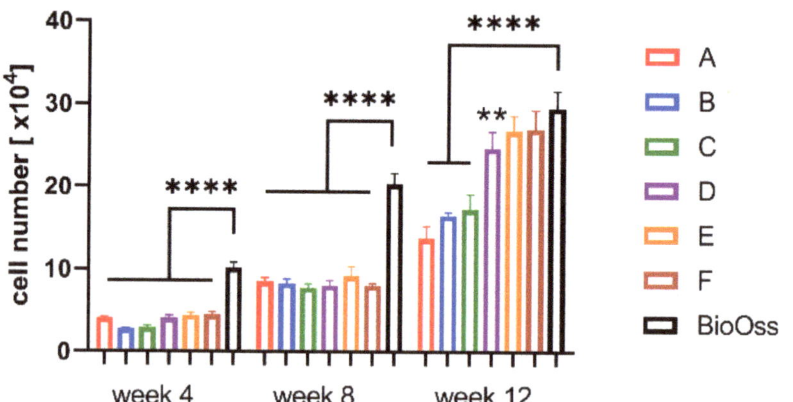

Figure 3. Proliferation of MSCs on/in the scaffolds depending on the various porosities after 4, 8 and 12 weeks in comparison to BioOss® (strand-to-strand-distance μm/pore size μm: A: 430/100; B: 560/230; C: 690/360; D: 820/490; E: 950/620; F: 1080/750; ** $p \leq 0.01$, **** $p \leq 0.0001$, mean ± standard deviation, n = 5).

Live/Dead staining was carried out to assess the viability of the cells. Through the culture period, the density of living cells (stained green) increased. Furthermore, microscopically, a widespread colonization of scaffolds was observed earlier in those with a higher pore size in comparison to those with a smaller pore size. The cells covered the superficial cement strands and also those in subjacent layers. Scaffold D, with a strand-to strand-distance of 820 μm and a pore size of 490 μm, exhibited a colonization similar to that of the control (BioOss®) (Figure 4). In all cases, no dead cells (stained red) were detected.

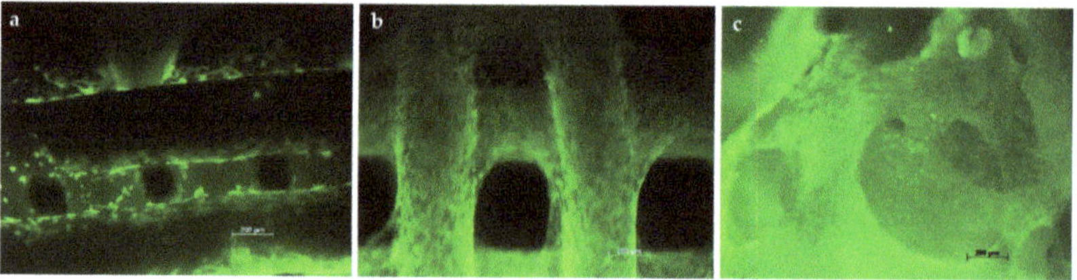

Figure 4. Colonization of cubic scaffolds with MSCs: Scaffold B (**a**), Scaffold D (**b**) and BioOss® (**c**) after 28 days. (Live/Dead-staining).

Microscopic SEM evaluation of colonized scaffolds after 28 days revealed that, similar to the Life/Dead staining, cells completely covered the scaffold strands. For Scaffold D especially, cell clusters bridging the interspaces between strands were observed (Figure 5).

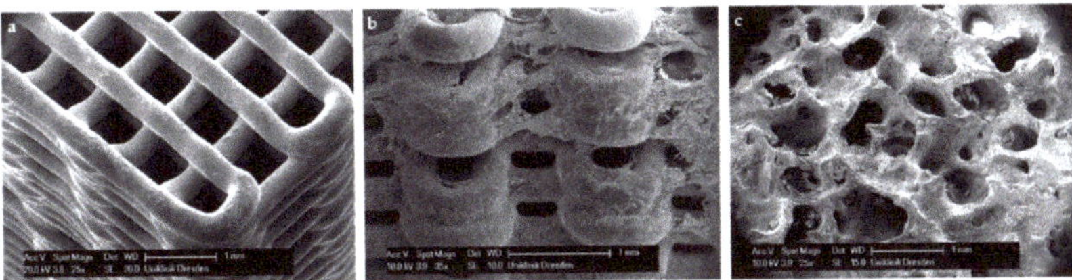

Figure 5. SEM imaging of CPC scaffold D non-colonized (**a**) and after 28 days (**b**) in comparison to BioOss® (**c**).

3.3. Scaffold Design for Intraoral Applications

In addition to the above mentioned scaffolds with strands laying 90° in relation to the scaffold's edges, we printed scaffolds with strands laying 45° related to the scaffold's edges to enhance the surface area (Figure 6).

Figure 6. Three dimensional (3D) plotted scaffold with defined pore size (0.49 mm) and strand distance (0.82 mm): (**a**) strand orientation 90° related to edge, (**b**) strand orientation 45° related to edge, (**c**) BioOss®; scale bars: 1 mm.

For this purpose, we used a pore size of 0.49 mm and a strand-to-strand-distance of 0.82 mm. Since the chewing process causes the application of forces from different directions, we investigated both strand orientations applying uniaxial strength from above and laterally. Young's modulus (Figure 7b) was estimated from the initial slope of the stress–strain curves (Figure 7a) in the elastic region. Compressive strength (Figure 7c) was evaluated from the stress–strain curves (Figure 7a). Data are presented in Table 2.

The results have shown that the energy absorption of the different scaffold types varies not only depending on strand orientation but also on the direction that the strength is applied from. Scaffolds with a 90° strand orientation seem to be more stable compared to 45° scaffolds (Figure 7c).

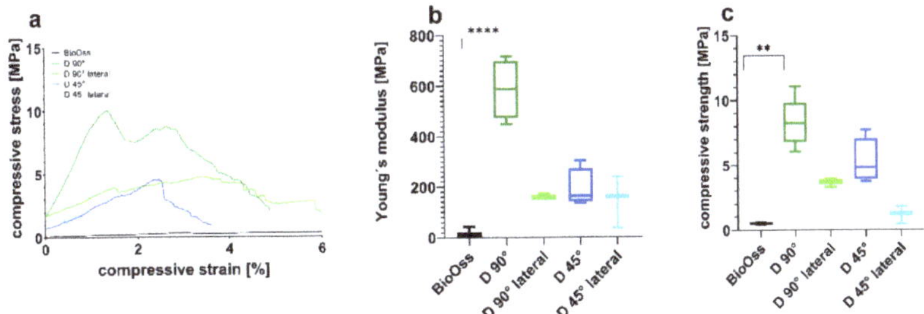

Figure 7. Mechanical properties of CPC scaffolds with different strand orientations in comparison with BioOss®. (**a**) Representative compressive stress–strain curves. (**b**) Young's modulus and (**c**) compressive strength determined from the curves (** $p \leq 0.01$, **** $p \leq 0.0001$, mean ± standard deviation, n = 5).

Table 2. Mechanical properties of scaffolds with different printing directions in comparison to the control.

Scaffold	Young's Modulus # (MPa)	Compressive Strength Resistance # (MPa)
90°	586 ± 118	8.3 ± 1.8
90° lateral	159 ± 11	3.6 ± 0.3
45°	191 ± 77	5.3 ± 1.7
45° lateral	145 ± 101	1.2 ± 0.7
Control	7 ± 4	0.5 ± 0.007

mean ± standard deviation (SD).

It also seems as if the compressive strength resistance of both strand orientations is significantly lower when the strength is applied laterally. Nevertheless, all investigated scaffolds have shown a higher energy absorption in comparison with the control group (BioOss®).

3.4. Preliminary Investigations of Individual Scaffolds for Clinical Cases

For the manufacturing of patient-specific scaffolds in order to reconstruct individual bone defects, a high-resolution model of the defect site is necessary. Based on patients' computed tomography data, irregular shaped macroporous scaffolds were designed in close collaboration with maxillofacial surgeons via digital backward planning. For this study, we analyzed different clinical cases—sinus floor elevation and onlay osteoplasty—in different regions, shapes and sizes (Figure 8).

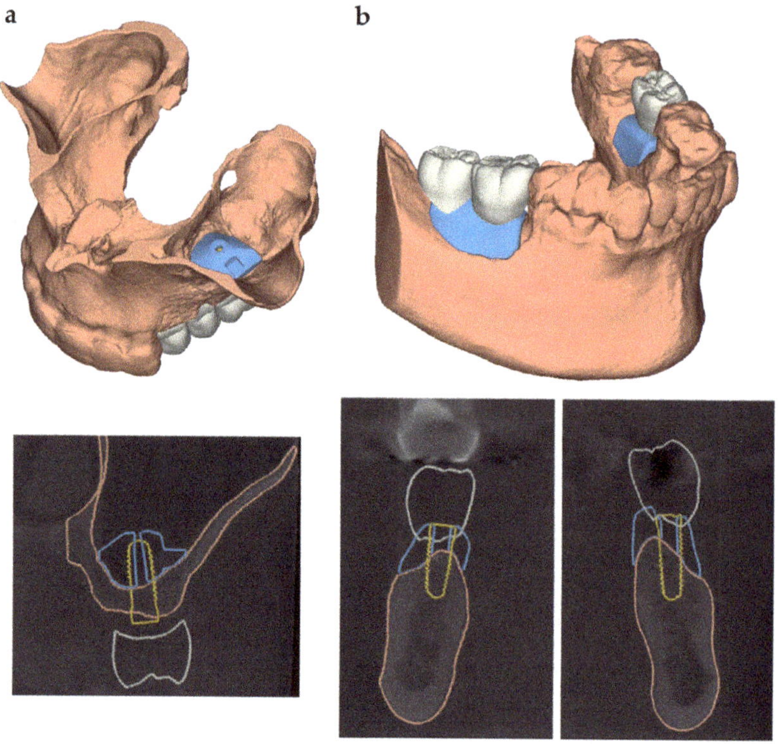

Figure 8. Digital backward planning based on CBCT data. (**a**) Sinus floor elevation in region 26, (**b**) bilateral onlay graft in the posterior mandible (white: prosthetic restoration; yellow: dental implant; orange: bone contour; blue: planned CPC scaffold.

According to the anatomical requirements of the defect site, the digital planned scaffolds were printed by INNOTERE GmbH (Radebeul, Germany). Since blood and bone cells are expected to migrate into the scaffold from the prepared adjacent bone, the bone-facing part of the scaffolds was printed with the approved pore size of 0.49 mm and a strand distance of 0.82 mm. To avoid fibroblasts or keratinocytes infiltrating the scaffold from the soft tissue facing side, this part was printed with a strand distance of 0.3 mm (Figure 9) to achieve a similar effect to that of using the membrane technique.

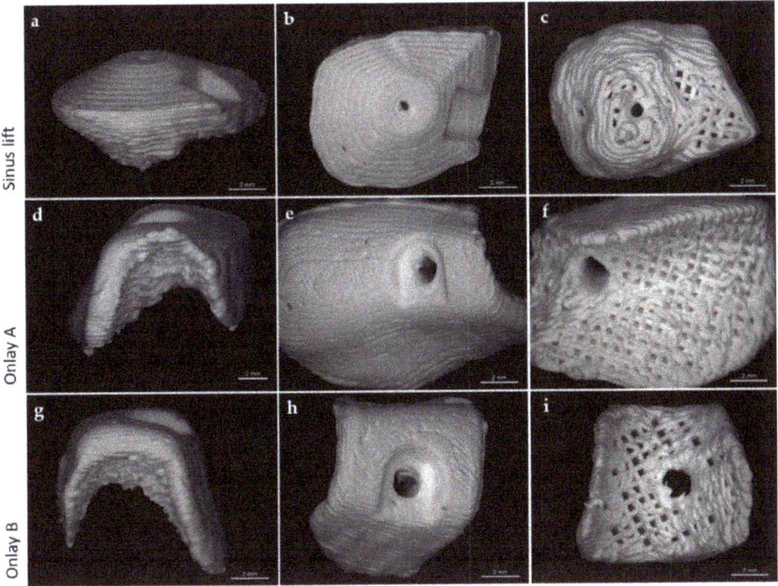

Figure 9. Individual scaffold for sinus floor elevation (**a–c**) and two onlay grafts in the lower jaw (region 36, 46: (**d–f**); region 47: (**g–i**)).

The mechanical properties of the individual scaffolds were tested by conducting uniaxial compression tests. Data are shown in Table 3.

Table 3. Mechanical properties of patient individual scaffolds and control group.

Scaffold	Young's Modulus # (MPa)	Compressive Strength Resistance # (MPa)
Sinus lift	135 ± 21	1.7 ± 0.3
Onlay A	239 ± 45	1.2 ± 0.3
Onlay B	127 ± 22	1.0 ± 0.2
Control	7 ± 4	0.5 ± 0.007

\# mean ± standard deviation (SD).

Young's modulus (Figure 10b) was estimated from the initial slope of stress–strain curves (Figure 10a) in the elastic region. Compressive strength (Figure 10c) was evaluated from stress–strain curves (Figure 10a). The results showed that the energy absorptions of the sinus lift scaffold and onlay B are comparable, while onlay A showed a higher energy absorption. Nevertheless, all investigated scaffolds, independently from their shape, showed a higher energy absorption and strength in comparison with the control group (BioOss®).

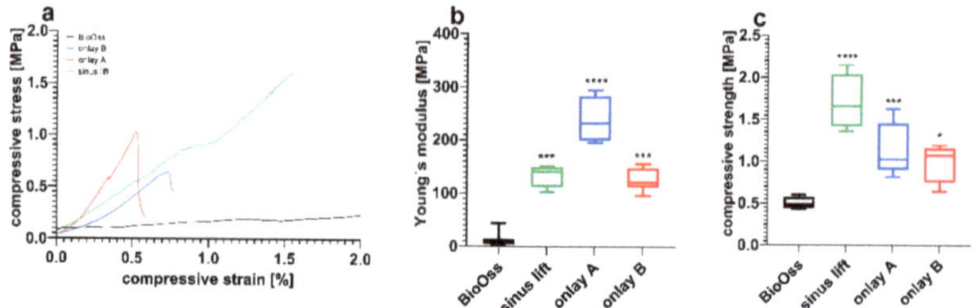

Figure 10. Mechanical properties of individual scaffolds in comparison with BioOss®. (**a**) Representative compressive stress–strain curves. (**b**) Young's modulus and (**c**) compressive strength determined from the curves (* $p \leq 0.05$, *** $p \leq 0.001$, **** $p \leq 0.0001$, mean ± standard deviation, n = 5).

4. Discussion

This study aimed to determine the optimal porosity of CPC scaffolds for bone augmentation in maxillofacial surgery according to specific indications. Regarding stability and cell infiltration, the data presented suggest that pore sizes of 750 µm allow for a significantly higher increase in cell colonization compared to smaller pores after 12 weeks (see Figure 3). Furthermore, the stability of the CPC cubes increases up to a pore size of 100 µm with an observed compressive strength of 31.3 ± 6.8 MPa and a Young's Modulus of 870 ± 117 MPa (see Figure 2). Nevertheless, individual CPC scaffolds, which are closer to clinical conditions, have shown a much lower compressive strength resistance depending on the respective site of destination. Thus, in certain cases, the porosity of individual scaffolds in maxillofacial surgery needs to be adapted with an acceptance of the concomitant decrease of cell infiltration.

Bone augmentation is performed when bony defects compromise the function and aesthetics of the orofacial system [25–28]. One of its main functions is to grind food as the first step of digestion. This is conducted by frequently repeated contraction of the chewing muscles. The chewing muscles belong to the strongest muscles in the human body. During the chewing process, forces beyond 200 N emerge depending on the region within the oral cavity [20,21]. The highest pressure can be measured in the lateral region of the jaws since this is the chewing center [20]. In other sites, such as the sinus floor, the anterior parts of the jaws or parts of the facial bone, the pressure is much lower [20]. Assuming a full dentition with an average chewing surface of approx. 6 cm², this corresponds to a pressure of approx. 0.4 MPa per tooth.

By Zwick universal testing, uniaxial compression can be applied to the test object. This makes it an appropriate testing procedure to resemble the in vivo situation, as teeth and the adjacent bone are stressed in a similar way. By testing standardized CPC cubes, 32 MPa was measured as the highest compressive strength withstood by scaffold A, which had a strand-to-strand distance of 430 µm and a pore size of 100 µm. This easily exceeds the essential requirements (0.4 MPa) for an in vivo application. Compared to the control group (BioOss®), which is commonly used for bone substitution, the applicable compressive strength of scaffold A was 60 times higher.

Nevertheless, the colonization experiments have shown that pore sizes of 100 µm (scaffold A) are too small to let cells quickly migrate into the scaffold. Due to the small pores, instead of infiltrating, cells instead colonized the scaffold's outer surface.

As depicted in Figure 3, in our case, a porosity of 750 µm (scaffold F) seems to be the best for cell infiltration. Nevertheless, cell numbers observed in the control group (BioOss®) were still superior from week 4 onwards. The reason for this could be the surface of BioOss®, which mimics the surface of natural bone better than CPC does (Figure 11). As seen in Figure 11, the CPC's surface is smooth, whereas the surfaces of BioOss® and natural bone have many micro irregularities. These irregularities lead to an enhanced attachment area for cells.

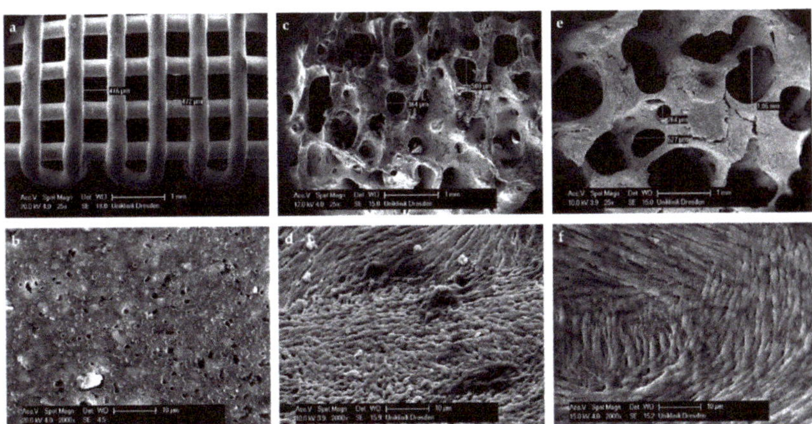

Figure 11. SEM imaging of CPC scaffold D (**a,b**), BioOss ® (**c,d**) and natural bone (**e,f**).

However, this advantage seems to decrease over time. Up to week 12, the difference in cell numbers between BioOss® and the CPC scaffolds with high porosity decreases continuously until they are nearly the same after week 12 (Figure 3). The influence of surface roughness on cell adhesion and function has been discussed in several studies [29,30]. The observed micro irregularities not only offer more surface for cell binding, but they also strengthen the adsorption of proteins and the extracellular matrix, which enhances the cells' adhesion and function. This effect was observed for different biomaterials and cells [31–33]. To improve the early cell adhesion on CPC scaffolds, the CPC could be enriched with nanoparticles such as bioactive glass, as shown by Richter et al. [34]. Thereby, the CPC's surface could be enriched with irregularities to better resemble natural bone.

In contrast, due to a decreased surface and increased strand-to-strand distance, a higher porosity goes along with a significant decrease of the compressive strength resistance and Young's Modulus. As depicted in Figure 2, scaffold F (pore size 750 µm) shows a compressive strength of 5.2 ± 0.6 MPa, which is much higher compared to that of the control group (0.5 ± 0.007 MPa). In Figure 5, it is shown that the porosity of scaffold D seems to be similar to that of BioOss®. Nevertheless, scaffold D is much more stable. This superior stability of the CPC scaffolds compared with BioOss® may be caused by the differences in their architecture. Microscopically, a natural spongious bone, similar to the architecture of BioOss®, can be observed. As shown in Figure 6, the spongious trabeculae are arranged irregularly in contrast to the strands of the CPC scaffolds. This regular arrangement of the CPC strands may be the reason for the higher compressive strength resistance. The pressure can be evenly deviated above the whole surface.

The Young's Modulus of the observed CPC cubes ranged from 444 ± 44 MPa to 870 ± 117 MPa. Human bone has a Young's Modulus of about 4.42 MPa as shown by Boughton et al. [35]. It is worth noting that Boughton et al. investigated cortical bone samples from femoral necks, the mechanical properties of which may differ from jaw and facial bone. Furthermore, the donors from which the bone was harvested had a mean age of 69 years. Due to the fact that age and chronic diseases have a significant impact on bone density, architecture and mechanical properties [36], these values may not be comparable to the jaw bones of patients undergoing maxillofacial surgery.

Nevertheless, the Young's Modulus of natural human bone seems to be much lower in comparison to the observed CPC scaffolds. The orofacial system is in permanent motion and underlies continuous dynamics. In such dynamic systems, differences of the Young's modulus can be crucial. They could lead to micro movements between scaffold and bone and thereby compromise the scaffold's integration. Here, BioOss® is much closer to natural bone, due to its natural origin and closely mimics natural bone tissue. In contrast to this, the CPC scaffolds consist of artificial tri-calcium phosphate and are manufactured by using amorphous paste. This may be why they are more brittle and less elastic. Besides the differences in the Young's Modulus, a high brittleness could compromise the intraoperative handling since the scaffolds have to be fixed with titanium screws. If the scaffolds are too brittle, they may break when the screw is inserted. This could probably be avoided by integrating a screw channel preliminarily and thereby decreasing the stress in the CPC while inserting the screw (as shown in Figure 9). Nevertheless, a pore size of 750 µm seems to be adequate for facilitating a high infiltration of cells and still meeting the mechanical requirements in the orofacial system. These findings were, however, observed in regularly shaped, cubic CPC scaffolds.

Knowing that this design might fail in simulating in situ settings with complex shaped defects, we additionally investigated clinical cases. Three cases were selected that displayed typical intraoral regions with mechanical requirements different to those of a bone scaffold (Figure 8): sinus floor elevation, onlay osteoplasty located posterior to the remaining teeth (onlay A) and onlay osteoplasty embraced by remaining teeth (onlay B).

Sinus floor elevation is a procedure that is used to create a sufficient base for dental implants in the posterior maxilla [37–39]. This is realized by inserting the bone substitution material through a bony window that has to be cut into the lateral wall of the maxillary sinus. During the healing period prior to implant insertion, it is not affected by pressure or movement. Due to these highly protected conditions during the healing period, a CPC scaffold for sinus floor elevation does not need to withstand a high compressive strength. Therefore, in such cases it could be advantageous to choose large pore sizes to gain the maximum cell infiltration. According to our findings, 750 µm would be the appropriate pore size in this case. Nevertheless, due to its pyramidal and compact geometry, the scaffold reaches high compressive strength resistance (1.7 ± 0.3 MPa) and therefore exceeds the essential requirements of the maxillary sinus. Considering this, even larger pore sizes could be assumed for such cases. The control group also seems to be a good choice for sinus floor elevation. As discussed above, the low compressive strength resistance of BioOss® can be neglected. According to Figure 3, BioOss® would even allow for a faster and larger increase of cell colonization on its surface compared to CPC scaffolds. This advantage of BioOss® could probably be compensated for by coating the CPC scaffolds with collagen as shown by Lee et al. [40]. Moreover, there are several advantages that favor CPC scaffolds. In contrast to BioOss®, CPC scaffolds can be individually designed based on a CBCT scan. Patient-specific geometries can be printed [24,41,42], thus they will fit perfectly to the defect site. The surgeon saves time during surgery since there is no need to prepare or adapt the scaffold intraoperatively. The planning of the augmentation is conducted before the surgery, which minimizes the risk of over- or under treatment. Additionally, CPC scaffolds can be printed with a graded porosity. Thus, the outer "soft tissue facing side" of the scaffold can be printed densely so that fibroblasts are not able to immigrate. Usually for this purpose additional membranes need to be placed to cover the defect site [43–45]. These membranes always come with the risk of early dehiscences and inflammation [46]. To summarize, for sinus floor elevation, CPC scaffolds with a pore size of 750 µm seem to be a sufficient tool.

Onlay osteoplasty in combination with dental implants is a standard procedure for the functional rehabilitation of highly atrophic jaws [27]. For onlay osteoplasty, the surgeon prepares a mucoperiosteal flap and fixes the bone substitution material directly to the defect site. In these cases, the bone scaffold is located submucosally. Hence, it is exposed to motions and forces directly after surgery. In this study, we simulated two clinical cases: region of teeth 46 and 47 (onlay A: Figure 9D–F) and region 36 (onlay B: Figure 10G–I). Both sites are under permanent pressure due to their location in the chewing center. Remarkably, there is an important difference between both cases. Onlay A covers an area of two teeth and there are no teeth posterior to the defect. Therefore, it needs to hold those forces that emerge in the chewing center on its own. In contrast, onlay B covers the area of one tooth and is embraced by teeth, anteriorly and posteriorly. The adjacent teeth may protect the scaffold from high compressive strength. Nevertheless, both scaffolds need to resist a higher compressive strength compared to that resisted by the sinus floor scaffold. Zwick testing of onlay A and onlay B revealed that the favored pore size of 750 µm is not stable enough to withstand the forces during the chewing process. The same effect was shown with a pore size of 620 µm (data not shown).

The pore size that was found to be strong enough to withstand the forces in the chewing center and also showed excellent colonization data was 490 µm with a compressive strength resistance of at least 1.0 ± 0.2 MPa. The individual scaffolds with pores of 490 µm have a Young's Modulus of 239 ± 45 MPa and 127 ± 22 MPa for onlay A and onlay B, respectively. As mentioned above, the Young's Modulus of human bone is approx. 4.42 MPa [35]. The mandible, especially, is known to be flexible and moved by different surrounding muscles. Unfortunately, the mechanical testing has shown that even very wide pore sizes, such as 750 µm, cannot affect the Young's Modulus to the extent that it would be comparable to human bone (Figure 2B). Nevertheless, CPC scaffolds with pore sizes of 490 µm seem to be a solid option for onlay osteoplasty in the lower lateral jaw. The superior compressive strength resistance compared to the control group especially makes CPC scaffolds an appropriate alternative to autologous bone, which is mostly used for onlay osteoplasty. Nevertheless, this study has certain limitations. The outer soft tissue facing layer was printed as densely as possible to prevent the migration of mucosal cells. Our experimental setting does not clarify whether our scaffold design fulfills this requirement properly. Furthermore, there are various other patient specific aspects that influence the integration of the scaffolds, such as certain comorbidities or lifestyle habits. Moreover, the degradation time of scaffolds is of high clinical relevance and it could be hypothesized that pore size also affects degradation time. To answer this question, an in vivo study would have to be conducted.

5. Conclusions

Our findings suggest that the pore size of CPC scaffolds for bone augmentation in maxillofacial surgery should be adapted for the planned site. CPC scaffolds for augmentation sites that are not exposed to high forces, such as the sinus floor, could be printed with a pore size of 750 µm to benefit from the enhanced cell infiltration. In contrast, CPC scaffolds for bone augmentation in areas exposed to high pressures, such as the lateral mandible, should be planned with a pore size of 490 µm. This pore size facilitates adequate cell infiltration and simultaneously meets the mechanical requirements in these highly stressed areas.

Author Contributions: Conceptualization, M.C.S.; methodology, L.M.K., D.M., P.S., S.H., H.M.; software, L.M.K., P.S., S.H.; validation D.M., L.M.K.; formal analysis L.M.K., P.S., S.H.; investigation, L.M.K., D.M., P.S., S.H., H.M.; resources, G.L.; data curation, D.M., L.M.K., P.S., S.H.; writing—original draft preparation, D.M., L.M.K.; writing—review and editing, D.M., P.S., S.H., A.H., M.C.S., G.L., L.M.K., H.M.; visualization, D.M., L.M.K.; supervision, L.M.K., G.L.; project administration, G.L.; funding acquisition, M.C.S., S.H., P.S., A.H., D.E., A.A. All authors have read and agreed to the published version of the manuscript.

Funding: This research was funded by the German Federal Ministry for Economic Affairs and Energy (BMWi), grant number ZF4379203MC8.

Institutional Review Board Statement: The use of patient data was approved by the local ethical review board (IRB00001473; file reference: EK1450420019).

Informed Consent Statement: Informed consent was obtained from all subjects involved in the study. Written informed consent has been obtained from the patient(s) to publish this paper.

Data Availability Statement: The data presented in this study are available on request from the corresponding author. The data are not publicly available due to privacy restrictions.

Acknowledgments: The authors thank Diana Jünger (Department of Oral and Maxillofacial Surgery, University Hospital "Carl Gustav Carus", Technische Universität Dresden) for their excellent technical assistance.

Conflicts of Interest: The authors declare no conflict of interest. The funders had no role in the design of the study; in the collection, analyses, or interpretation of data; in the writing of the manuscript, or in the decision to publish the results.

References

1. Shamsoddin, E.; Houshmand, B.; Golabgiran, M. Biomaterial selection for bone augmentation in implant dentistry: A systematic review. *J. Adv. Pharm. Technol. Res.* **2019**, *10*, 46–50. [CrossRef]
2. Starch-Jensen, T.; Deluiz, D.; Deb, S.; Bruun, N.H.; Tinoco, E.M.B. Harvesting of Autogenous Bone Graft from the Ascending Mandibular Ramus Compared with the Chin Region: A Systematic Review and Meta-Analysis Focusing on Complications and Donor Site Morbidity. *J. Oral Maxillofac. Res.* **2020**, *11*, e1. [CrossRef]
3. Scheerlinck, L.M.; Muradin, M.S.; van der Bilt, A.; Meijer, G.J.; Koole, R.; Van Cann, E.M. Donor site complications in bone grafting: Comparison of iliac crest, calvarial, and mandibular ramus bone. *Int. J. Oral Maxillofac. Implant.* **2013**, *28*, 222–227. [CrossRef] [PubMed]
4. Saha, A.; Shah, S.; Waknis, P.; Bhujbal, P.; Aher, S.; Vaswani, V. Comparison of minimally invasive versus conventional open harvesting technique for iliac bone graft in secondary alveolar bone grafting in cleft palate patients: A systematic review. *J. Korean Assoc. Oral Maxillofac. Surg.* **2019**, *45*, 241–253. [CrossRef] [PubMed]
5. Jakoi, A.M.; Iorio, J.A.; Cahill, P.J. Autologous bone graft harvesting: A review of grafts and surgical techniques. *Musculoskelet. Surg.* **2015**, *99*, 171–178. [CrossRef] [PubMed]
6. Mishra, A.; Srivastava, V. Biomaterials and 3D printing techniques used in the medical field. *J. Med. Eng. Technol.* **2021**, *45*, 290–302. [CrossRef] [PubMed]
7. Aimar, A.; Palermo, A.; Innocenti, B. The Role of 3D Printing in Medical Applications: A State of the Art. *J. Healthc. Eng.* **2019**, *2019*, 5340616. [CrossRef] [PubMed]
8. Parmar, H.; Khan, T.; Tucci, F.; Umer, R.; Carlone, P. Advanced robotics and additive manufacturing of composites: Towards a new era in Industry 4.0. *Mater. Manuf. Process.* **2021**, 1–35. [CrossRef]
9. Öztürkmen, Y.; Caniklioğlu, M.; Karamehmetoğlu, M.; Fiükür, E. Calcium phosphate cement augmentation in the treatment of depressed tibial plateau fractures with open reduction and internal fixation. *Acta. Orthop. Et. Traumatol. Turc.* **2010**, *44*, 262–269. [CrossRef]
10. Ji, C.; Ahn, J.G. Clinical experience of the brushite calcium phosphate cement for the repair and augmentation of surgically induced cranial defects following the pterional craniotomy. *J. Korean Neurosurg. Soc.* **2010**, *47*, 180–184. [CrossRef]
11. Xu, H.H.K.; Wang, P.; Wang, L.; Bao, C.; Chen, Q.; Weir, M.D.; Chow, L.C.; Zhao, L.; Zhou, X.; Reynolds, M.A. Calcium phosphate cements for bone engineering and their biological properties. *Bone Res.* **2017**, *5*, 11056. [CrossRef]
12. Reitmaier, S.; Kovtun, A.; Schuelke, J.; Kanter, B.; Lemm, M.; Hoess, A.; Heinemann, S.; Nies, B.; Ignatius, A. Strontium(II) and mechanical loading additively augment bone formation in calcium phosphate scaffolds. *J. Orthop. Res.* **2018**, *36*, 106–117. [CrossRef]
13. Cha, J.K.; Kim, C.; Pae, H.C.; Lee, J.S.; Jung, U.W.; Choi, S.H. Maxillary sinus augmentation using biphasic calcium phosphate: Dimensional stability results after 3–6 years. *J. Periodontal. Implant. Sci.* **2019**, *49*, 47–57. [CrossRef]
14. Wach, T.; Kozakiewicz, M. Fast-versus slow-resorbable calcium phosphate bone substitute materials-texture analysis after 12 months of observation. *Materials* **2020**, *13*, 3854. [CrossRef] [PubMed]
15. Marongiu, G.; Verona, M.; Cardoni, G.; Capone, A. Synthetic bone substitutes and mechanical devices for the augmentation of osteoporotic proximal humeral fractures: A systematic review of clinical studies. *J. Funct. Biomater.* **2020**, *11*, 29. [CrossRef]
16. Rolvien, T.; Barbeck, M.; Wenisch, S.; Amling, M.; Krause, M. Cellular Mechanisms Responsible for Success and Failure of Bone Substitute Materials. *Int. J. Mol. Sci.* **2018**, *19*, 2893. [CrossRef]
17. Pepelassi, E.; Perrea, D.; Dontas, I.; Ulm, C.; Vrotsos, I.; Tangl, S. Porous Titanium Granules in comparison with Autogenous Bone Graft in Femoral Osseous Defects: A Histomorphometric Study of Bone Regeneration and Osseointegration in Rabbits. *Biomed. Res. Int.* **2019**, *2019*, 8105351. [CrossRef]
18. Duda, M.; Pajak, J. The issue of bioresorption of the Bio-Oss xenogeneic bone substitute in bone defects. *Ann. Univ. Mariae Curie-Skłodowska. Sect. D Med.* **2004**, *59*, 269–277.
19. Schlegel, A.K.; Donath, K. BIO-OSS®-A resorbable bone substitute? *J. Long-Term Eff. Med. Implant.* **1998**, *8*, 201–209.
20. Ledogar, J.A.; Dechow, P.C.; Wang, Q.; Gharpure, P.H.; Gordon, A.D.; Baab, K.L.; Smith, A.L.; Weber, G.W.; Grosse, I.R.; Ross, C.F.; et al. Human feeding biomechanics: Performance, variation, and functional constraints. *PeerJ* **2016**, *26*, e2242. [CrossRef]
21. Righetti, M.A.; Taube, O.L.S.; Palinkas, M.; Gonçalves, L.M.N.; Esposto, D.S.; de Mello, E.C.; Regalo, I.H.; Regalo, S.C.H.; Siéssere, S. Osteoarthrosis: Analyze of the Molar Bite Force, Thickness and Masticatory Efficiency. *Prague Med. Rep.* **2020**, *121*, 87–95. [CrossRef]
22. La Monaca, G.; Iezzi, G.; Cristalli, M.P.; Pranno, N.; Sfasciotti, G.L.; Vozza, I. Comparative Histological and Histomorphometric Results of Six Biomaterials Used in Two-Stage Maxillary Sinus Augmentation Model after 6-Month Healing. *Biomed. Res. Int.* **2018**, *2018*, 9430989. [CrossRef]

23. Yamada, M.; Egusa, H. Current bone substitutes for implant dentistry. *J. Prosthodont Res.* **2018**, *62*, 152–161. [CrossRef]
24. Korn, P.; Ahlfeld, T.; Lahmeyer, F.; Kilian, D.; Sembdner, P.; Stelzer, R.; Pradel, W.; Franke, A.; Rauner, M.; Range, U.; et al. 3D Printing of Bone Grafts for Cleft Alveolar Osteoplasty-In vivo Evaluation in a Preclinical Model. *Front. Bioeng. Biotechnol.* **2020**, *25*, 217. [CrossRef]
25. Sakkas, A.; Wilde, F.; Heufelder, M.; Winter, K.; Schramm, A. Autogenous bone grafts in oral implantology—is it still a "gold standard"? A consecutive review of 279 patients with 456 clinical procedures. *Int. J. Implant. Dent.* **2017**, *3*, 1–17. [CrossRef] [PubMed]
26. Toledano-Serrabona, J.; Sánchez-Garcés, M.Á.; Sánchez-Torres, A.; Gay-Escoda, C. Alveolar distraction osteogenesis for dental implant treatments of the vertical bone atrophy: A systematic review. *Med. Oral Patol. Oral Cir. Bucal.* **2019**, *24*, e70. [CrossRef] [PubMed]
27. Nguyen, T.T.H.; Eo, M.Y.; Kuk, T.S.; Myoung, H.; Kim, S.M. Rehabilitation of atrophic jaw using iliac onlay bone graft combined with dental implants. *Int. J. Implant. Dent.* **2019**, *5*, 11. [CrossRef] [PubMed]
28. Salmen, F.S.; Oliveira, M.R.; Gabrielli, M.A.C.; Piveta, A.C.G.; Pereira Filho, V.A.; Ganrielli, M.F.R. Bone grafting for alveolar ridge reconstruction. Review of 166 cases. *Rev. Do Colégio Bras. De Cir.* **2017**, *44*, 33–40. [CrossRef] [PubMed]
29. Stepanovska, J.; Matejka, R.; Rosina, J.; Bacakova, L.; Kolarova, H. Treatments for enhancing the biocompatibility of titanium implants. *Biomed. Pap.* **2020**, *164*, 23–33. [CrossRef] [PubMed]
30. Samavedi, S.; Whittington, A.R.; Goldstein, A.S. Calcium phosphate ceramics in bone tissue engineering: A review of properties and their influence on cell behavior. *Acta Biomater.* **2013**, *9*, 8037–8045. [CrossRef]
31. Hu, X.; Mei, S.; Wang, F.; Qian, J.; Xie, D.; Zhao, J.; Yang, L.; Wu, Z.; Wei, J. Implantable PEKK/tantalum microparticles composite with improved surface performances for regulating cell behaviors, promoting bone formation and osseointegration. *Bioact. Mater.* **2021**, *6*, 928–940. [CrossRef] [PubMed]
32. Deligianni, D.D.; Katsala, N.D.; Koutsoukos, P.G.; Missirlis, Y.F. Effect of surface roughness of hydroxyapatite on human bone marrow cell adhesion, proliferation, differentiation and detachment strength. *Biomaterials* **2000**, *22*, 87–96. [CrossRef]
33. Zhou, K.; Li, Y.; Zhang, L.; Jin, L.; Yuan, F.; Tan, J.; Yuan, G.; Pei, J. Nano-micrometer surface roughness gradients reveal topographical influences on differentiating responses of vascular cells on biodegradable magnesium. *Bioact. Mater.* **2021**, *6*, 262–272. [CrossRef]
34. Richter, R.F.; Ahlfeld, T.; Gelinsky, M.; Lode, A. Development and characterization of composites consisting of calcium phosphate cements and mesoporous bioactive glass for extrusion-based fabrication. *Materials* **2019**, *12*, 2022. [CrossRef]
35. Boughton, O.R.; Ma, S.; Zhao, S.; Arnold, M.; Lewis, A.; Hansen, U.; Cobb, J.P.; Giuliani, F.; Abel, R.L. Measuring bone stiffness using spherical indentation. *PLoS ONE* **2018**, *13*, e0200475. [CrossRef] [PubMed]
36. Oftadeh, R.; Perez-Viloria, M.; Villa-Camacho, J.C.; Vaziri, A.; Nazarian, A. Biomechanics and Mechanobiology of Trabecular Bone: A Review. *J. Biomech. Eng.* **2015**, *137*, 0108021–01080215. [CrossRef]
37. Maska, B.; Lin, G.-H.; Othman, A.; Behdin, S.; Travan, S.; Benavides, E.; Kapila, Y. Dental implants and grafting success remain high despite large variations in maxillary sinus mucosal thickening. *Int. J. Implant. Dent.* **2017**, *3*, 1–8. [CrossRef]
38. Starch-Jensen, T.; Jensen, J.D. Maxillary Sinus Floor Augmentation: A Review of Selected Treatment Modalities. *J. Oral Maxillofac. Res.* **2017**, *8*, e3. [CrossRef]
39. Thoma, D.S.; Cha, J.K.; Jung, U.W. Treatment concepts for the posterior maxilla and mandible: Short implants versus long implants in augmented bone. *J. Periodontal Implant. Sci.* **2017**, *47*, 2–12. [CrossRef]
40. Lee, M.H.; You, C.; Kim, K.H. Combined effect of a microporous layer and type I collagen coating on a biphasic calcium phosphate scaffold for bone tissue engineering. *Materials* **2015**, *8*, 1150–1161. [CrossRef]
41. Ahlfeld, T.; Köhler, T.; Czichy, C.; Lode, A.; Gelinsky, M. A Methylcellulose Hydrogel as Support for 3D Plotting of Complex Shaped Calcium Phosphate Scaffolds. *Gels* **2018**, *4*, 68. [CrossRef]
42. Ahlfeld, T.; Akkineni, A.R.; Förster, Y.; Köhler, T.; Knaack, S.; Gelinsky, M.; Lode, A. Design and Fabrication of Complex Scaffolds for Bone Defect Healing: Combined 3D Plotting of a Calcium Phosphate Cement and a Growth Factor-Loaded Hydrogel. *Ann. Biomed. Eng.* **2017**, *45*, 224–236. [CrossRef] [PubMed]
43. Lyu, C.; Shao, Z.; Zou, D.; Lu, J. Ridge Alterations following Socket Preservation Using a Collagen Membrane in Dogs. *BioMed Res. Int.* **2020**, *2020*, 1487681. [CrossRef] [PubMed]
44. Guarnieri, R.; Stefanelli, L.; De Angelis, F.; Mencio, F.; Pompa, G.; Di Carlo, S. Extraction Socket Preservation Using Porcine-Derived Collagen Membrane Alone or Associated with Porcine-Derived Bone. Clinical Results of Randomized Controlled Study. *J. Oral Maxillofac. Res.* **2017**, *8*, e5. [CrossRef] [PubMed]
45. Kolerman, R.; Qahaz, N.; Barnea, E.; Mijiritsky, E.; Chaushu, L.; Tal, H.; Nissan, J. Allograft and collagen membrane augmentation procedures preserve the bone level around implants after immediate placement and restoration. *Int. J. Environ. Res. Public Health* **2020**, *17*, 1133. [CrossRef]
46. Garcia, J.; Dodge, A.; Luepke, P.; Wang, H.L.; Kapila, Y.; Lin, G.H. Effect of membrane exposure on guided bone regeneration: A systematic review and meta-analysis. *Clin. Oral Implants Res.* **2018**, *29*, 328–338. [CrossRef]

MDPI

St. Alban-Anlage 66

4052 Basel

Switzerland

Tel. +41 61 683 77 34

Fax +41 61 302 89 18

www.mdpi.com

Journal of Clinical Medicine Editorial Office

E-mail: jcm@mdpi.com

www.mdpi.com/journal/jcm